D1061146

Paul-Gerhard Reinhard, Eric Suraud

**Introduction to Cluster Dynamics**

Paul-Gerhard Reinhard, Eric Suraud

# Introduction to Cluster Dynamics

WILEY-VCH Verlag GmbH & Co. KGaA

**Authors**

*Prof. Dr. Paul-Gerhard Reinhard*
Friedrich-Alexander-Universität Erlangen,
Germany
e-mail: reinhard@theorie2.physik.uni-erlangen.de

*Prof. Dr. Eric Suraud*
Université Paul Sabatier, Toulouse, France
e-mail: suraud@irsamc.ups-tlse.fr





**Library of Congress Card No.: applied for**
**British Library Cataloging-in-Publication Data:**
A catalogue record for this book is available from the British Library

**Bibliographic information published by Die Deutsche Bibliothek**
Die Deutsche Bibliothek lists this publication in the Deutsche Nationalbibliografie; detailed bibliographic data is available in the Internet at <http://dnb.ddb.de>.

**Cover Picture**
The cover picture shows four snapshots from a collision of an $Ar^{8+}$ ion on the cluster $Na_9^+$. The ions ($Na^+$ and $Ar^{8+}$) are indicated by red dots. The electronic density is drawn in colour scale with dark yellow representing the highest value. The time steps are arranged counterclockwise. The sequence starts at the upper right panel showing the unperturbed ground state of $Na_9^+$. The approaching $Ar^{8+}$ and its first polarizing effects on the electron cloud are seen in the next time slot (upper left panel). Short after the encounter (lower left panel), one notes that the $Ar^{8+}$ takes away a sizeable amount of cluster electrons. The thus highly ionized cluster undergoes a Coulomb fragmentation. The last snapshot (lower right) demonstrates the global expansion of the cluster.



Printed in the Federal Republic of Germany
Printed on acid-free paper

**Printing** Strauss Offsetdruck GmbH, Mörlenbach
**Bookbinding** Großbuchbinderei J. Schäffer GmbH & Co. KG, Grünstadt

**ISBN** 3-527-40345-0

*To our families*

# Preface

Cluster physics, and even more so cluster dynamics, is a young and fast developing field of science. It addresses fundamental questions in physics and chemistry, but leads also to new technological developments. Because it is such a recent field and also because it has links to many areas in physics and chemistry, cluster science is not yet routinely taught in universities, at least not as a well identified field. Thus there are only few textbooks presenting the topic as such and even less books with a sufficient treatment of dynamics. The aim of this book is to try to fill this gap whereby we concentrate on dynamical aspects. The material presented here stems from lectures given by the authors at undergraduate and graduate levels, both in France and in Germany. The presentation is thus to a large extent tutorial, particularly in the first chapters.

Cluster science was recognized as an independent field of physics and chemistry only in the last decades of the twentieth century. This recognition is, to a large extent, connected to the experimental capability of producing free clusters without constraining environments such as substrates or matrices. The discovery of fullerenes which are large carbon cages with $C_{60}$ as the most famous example was an important step in this direction, also in view of the technological applications of carbon compounds. Somewhat surprisingly, it turns out that clusters had been known and used for centuries, for example by craftworkers (exploiting their remarkable optical properties) or in photography (exploiting photosensitive AgBr clusters embedded in a film). In spite of these important applications, it was only at the beginning of the twentieth century that the notion of a cluster as an independent object was suggested by G. Mie in the context of the optical response. Decades later, many investigations were performed on deposited or embedded clusters and clusters finally became a central object of investigations in the late twentieth century. As already mentioned, cluster physics opens numerous technological applications, in particular in relation to carbon compounds (for example with the so called carbon nanotubes). But the impact of cluster physics is not only focused on applications. Clusters constitute special objects with specific properties which require specific investigations, both from experimental and theoretical sides. But they also have an extremely important and unique property: they are scalable. One can vary the size of a cluster from the dimer close to the bulk. Clusters thus provide a useful laboratory for investigating the structure of matter and in particular how matter properties do evolve with size. They constitute, so to say, a bridge between microscopic and macroscopic worlds.

As cluster physics is a merger between various fields of physics and chemistry, it would be illusory to conceive an introductory book covering all the many facets. We shall focus on one particular aspect, namely cluster dynamics. Even then, there is far too much material for one book like this. We aim here at a tutorial introduction reporting basic experimental and theoretical tools without entering too deeply into the details covered by the various subfields. And we will illustrate our discussions with results directly taken from the most recent researches in the field, in order to keep contact with these rapid developments. Such a choice implies a certain degree of superficiality and incompleteness. We try to avoid technicalities in order to focus on physical insights. Some more involved, but necessary, details will be outlined in a few appendices. Even with such simplifying assumptions, one is bound to make choices, which de facto eliminate or less fully cover some aspects. Our bias here goes into three (physically motivated) directions. First, we shall devote a large fraction of our discussions and examples to the particular case of metal clusters. The reason is that they have played an important role in cluster studies, that many complementing experimental results are available for these systems, and that the metallic bonding gives these systems systematic behaviors, which are not so easily found in other types of clusters. Metal clusters thus constitute ideal examples of illustration of physical concepts, particularly in the context of a textbook. The second bias concerns our focus on dynamical questions. Many different theoretical methods, with origins in various fields of physics and chemistry, have been developed to describe clusters. In the case of dynamics, the task becomes so involved that only few of these methods are really applicable to practical problems, in particular when one wants to exploit the above mentioned scalability of clusters. This gives naturally more weight to time-dependent density functional theory as the presently most efficient theoretical method for describing truly dynamical processes in clusters. Of course, we also discuss examples obtained with other techniques when applicable. The third bias concerns an experimental tool. Laser excitation as a doorway to dynamical processes will acquire a large weight in the examples. This, again, is natural as the fast progress of laser technology stimulates research in cluster dynamics very much.

The book is primarily written for undergraduate students in their last-year course. It should also be useful for graduate students and, because we have included most recent developments, the book might also be used by researchers. In order to merge both aspects, namely tutorial introduction and account of the latest developments, we have decided to start at an elementary level and to finish with up to date developments. The first four chapters thus require only elementary notions of quantum physics and basic knowledge of solid state and molecular physics. The last chapter is devoted to discussions of actual research with bias on dynamical problems. This chapter probably requires a deeper knowledge of the field, but should in principle be accessible to a reader having gone through the first four chapters. Finally the text is complemented by several appendices providing more details on several key methods.

The first chapter, *About clusters*, is meant to be propaedeutic. It aims at introducing clusters in relation to atoms, molecules and condensed matter. It provides a first overview of cluster structure, its chemical classification, and a few simple dynamical features. Metal clusters are discussed separately at the end of the chapter, addressing especially the dominant optical response. It is assumed that the reader has here a basic knowledge of atomic physics and chemistry. Crucial input from these fields is supplied in an appendix.

The second chapter, *From clusters to numbers: experimental aspects*, discusses the various experimental methods in cluster physics, focusing the discussions on cluster production and analyzing tools (by means of various spectroscopies). The basic setups are introduced accompanied by a brief discussion of their practical handling. The typical results from each area are presented and discussed. At the end, the reader should have a good oversight of structural and dynamical properties of the various sorts of clusters. A lot of material does exist for $C_{60}$ and for metal clusters, in which dynamical effects are especially well documented. These will dominate in the presentation. Results from other species will also be listed where available.

The third chapter, *The cluster many-body problem: a theoretical perspective*, gives a brief overview of the basic theoretical concepts for the computation of cluster structure and dynamics. A detailed discussion of the various levels of approximation for ions and electrons, and of their validity, is presented. This chapter also gives many results as examples, complementing (and, when possible, in direct relation with) the experimental findings from the previous chapter, emphasizing again observables in direct relation to cluster dynamics.

Chapter 4, *Gross properties and trends*, summarizes the results of experiments and theory from a general point of view. Typical time-, length- and frequency-scales are worked out. Trends with system size, material density, and strength of excitation are established. Electronic and ionic shell effects are addressed. Because of the numerous available systematics a strong bias on metal clusters exists in this chapter. The close relation between metal clusters and nuclei is also discussed.

The last chapter, *New frontiers in cluster dynamics*, deals with the most recent developments, in particular with dynamics in the regime of high excitations. It covers the transient regime where the dynamics can still be sorted in terms of multi-photon processes. It continues to the most violent excitation processes in strong laser fields or in ionic collisions which lead to high ionization states and to fast cluster fragmentation by Coulomb explosion. This corresponds to a very active field of research nowadays. The chapter has thus an open end.

The five chapters are complemented by eight appendices containing various practical and theoretical data. The titles of the appendices are to a large extent self explanatory: Appendix A *Conventions of notations, symbols, units, acronyms*, Appendix B *Gross properties of atoms and solids*, Appendix C *Some details on basic techniques from molecular physics and quantum chemistry*, Appendix D *More on pseudo-potentials*, Appendix E *More on density functional theory*, Appendix F *Fermi gas model and semiclassics*, Appendix G *Linearized TDLDA and related approaches* and Appendix H *Numerical considerations*.

An enterprise such as a book is necessarily the result of numerous interactions with many colleagues. They all have played an important role in the process, many of them on our way of learning cluster physics and some of them specifically giving advice for writing this book. We would thus like to acknowledge the help of all these colleagues and tell them how much they helped us and brought to us, both in terms of science and personal contacts. We would in particular like to mention here: J. Alonso, M. Belkacem, G. F. Bertsch, S. Bjornholm, M. Brack, A. Bulgac, F. Calvayrac, M. Chabot, J. P. Connerade, G. Gerber, B. Gervais, E. Giglio, E. K. U. Gross, C. Guet, H. Haberland, J. M. L'Hermite, P. H. Hervieux, B. Huber, B. von Issendorf, U. Kreibig, S. Kümmel, P. Labastie, M. Manninen, K.-H. Meiwes-Broer, V. Nesterenko, G. Pastor, R. Poteau, R. Schmidt, L. Serra, L. Schweikhard, F. Stienkemeier, J. Tiggesbäumker, C. Toepffer, C. Ullrich, D. Vernhet, K. Wohrer, and G. Zwicknagel. This book emerges from a long-standing collaboration between the authors. This would not have been possible without the help of funding from the French-German exchange program PROCOPE, the Institut Universitaire de France, and the Alexander-von-Humboldt foundation. We are thankful to these institutions to have supported us in our common efforts.

*Paul-Gerhard Reinhard* and *Eric Suraud*

September 2003

# Contents

# 1 About clusters

Although we usually do not realize it, clusters belong to our everyday life. They have been exploited practically in many situations without people being aware of the underlying details. The tailoring of fine dispersed pieces of material (clusters !) inside bulk has, for example, been turned to an art by craft-workers for centuries. Already the Romans knew, empirically, how to play with the size of dispersed particles in a glass to produce various shining colors. Depending on the size of the gold inclusions, a glass could thus exhibit red as well as yellow reflections, for an example see [KV93]. In a different domain, photography also represents a typical application of cluster physics. Depending on the size of the AgBr clusters deposited on the film, they will more or less quickly and finely respond to light and thus perpetuate the properties of the produced image. Early photographers of the nineteenth century quickly realized and controlled such physical behaviors. Still, even at that time, clusters were not considered as objects of scientific studies, even not recognized as specific objects. The work of Mie at the turn of the twentieth century probably constitutes one of the earliest speculations on the existence and specificity of metal clusters — or rather, because the word "cluster" did not exist in this context at that time, of "small particles" [Mie08]. The question raised by Mie concerned the response of small metal particles to light, and how this response might depend on the size of the considered particle. Let us quote Mie: "Because gold atoms surely differ in their optical properties from small gold spheres", it would "probably be very interesting to study the absorption of solutions with the smallest submicroscopical particles; thus, in a way, one could investigate by optical means how gold particles are composed of atoms". A non negligible part of today's investigations on clusters relies on, or is very close to, Mie's intuition. As we shall see throughout this book, light represents a particularly useful means for the investigation of both structure and dynamics of clusters. And the response of clusters to light is indeed extremely sensitive to their properties. The beautiful achievements of Roman and later glass makers precisely reflect such properties. And photography, as well, is an art dealing with light.

The inspiring intuition of Mie did not suffice to promote cluster science to the status of a well recognized field of physics. Indeed the times of the early twentieth century were busy with the identification and analysis of "elementary" constituents of matter and only simple molecules seemed in reach of understanding in the mid century, or alternatively the other extreme, namely the problem of bulk material, of course with other techniques than the ones used in atomic physics. Modern cluster physics, as we know it today, only appeared in the last quarter of the twentieth century with the possibility of producing free clusters and tracking such small particles with several, still developing, techniques. This allowed the initiation of studies on clusters as such, which was an important step for the field. Indeed, embedded

*Introduction to cluster dynamics.* Paul-Gerhard Reinhard, Eric Suraud

ISBN: 3-527-40345-0

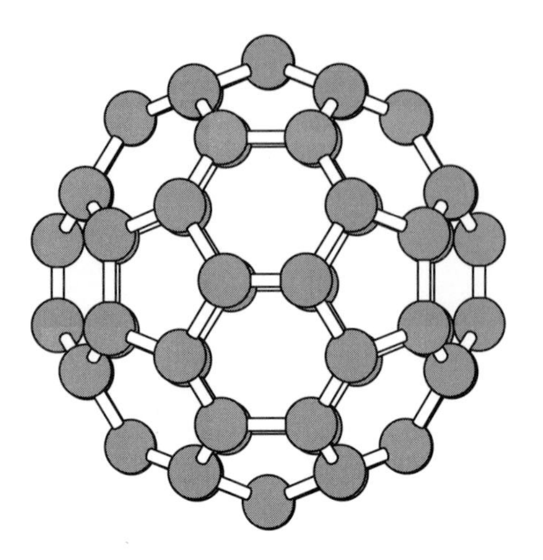

**Figure 1.1:** Ionic structure of the famous $C_{60}$ cluster. The 60 atoms are arranged as 12 pentagons and 20 hexagons yielding a truncated icosahedron. That shape consists in 12 vertices bound together by 20 equilateral (and equal) triangles. At each vertex of the icosahedron, 5 triangles meet. Truncating these vertices by a plane, leads to the pentagonal faces. The total number of vertices becomes 60. For this $C_{60}$, 120 symmetry operations can be identified. And this high degree of symmetry has long been used by artists, the actual name of "Buckminster fullerenes" going back to the architect Richard Buckminster Fuller, renowned for his geodesic domes based on pentagons and hexagons.

clusters (as the fine gold particles in Roman glasses), or clusters deposited on surfaces, were experimentally accessible since long and of course the subject of numerous studies. But these rarely concentrated on the clusters, independently from the matrix or substrate. The case was more or less appended to surface or material science and did not constitute a true independent field. The capability of producing free clusters from dedicated sources, allowed the true starting of cluster science on a systematic basis. One of the startup events was the identification of $C_{60}$ clusters, the famous fullerenes [YPC+87, Kro87], with their remarkable geometry shown in Figure 1.1. At about the same time, free metal clusters had been produced and investigated, see e.g. [KCdH+84]. The many original results obtained from then on for metal clusters, carbon clusters and, increasingly, other materials, established cluster physics as an independent, although cross-disciplinary, field among the well defined branches of physics and chemistry. Of course the production and analysis of free clusters gave new impetus to activities on deposited or embedded clusters as well. At the same time, amazing developments in the nanoscopic analysis of surfaces opened new views and much refined analysis of supported clusters. An example is given in Figure 1.2 showing in detail Ag nano-clusters sitting on a HOPG surface. As we will see, the combination of these quickly developing methods of nano-analysis with nano-particles, called clusters, constitutes a powerful tool for fundamental and applied physics.

The physics and chemistry of clusters, with its many facets covering free as well as embedded or deposited clusters, addresses an impressive set of problems, ranging from fundamental to applied ones. In all that, clusters are a species of their own asking for specific understanding of their properties. Although molecular and solid state physics, as well as chemistry or nuclear physics, do add helpful aspects, clusters belong to none of these fields and thus require devoted methods, both at the experimental and theoretical levels. A specific feature is, e.g., that cluster size can be varied systematically between atoms and bulk: they are, so to say, "scalable" objects. Clusters thus play an essential role from a fundamental physical point of view. They do represent an exceptional opportunity for testing the many-body problem, which is a generic quantum mechanical task and lies at the heart of the understanding of most complex systems. Bear in mind that solid state physics deals with virtually infinite, although

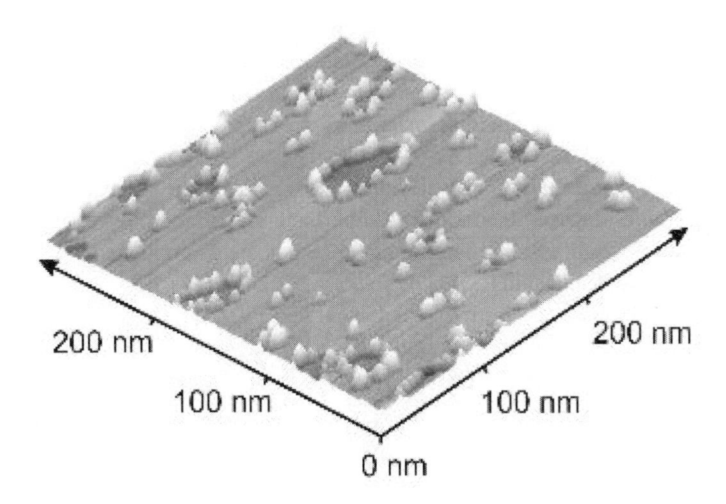


**Figure 1.2:** Topography of silver nanoparticles deposited on highly oriented pyrolytic graphite (HOPG), recorded with an *in situ* scanning tunneling microscope (STM). From [LMP$^+$00].

symmetric, samples while in molecular or nuclear physics the systems never contain more than a few hundreds of constituents. Clusters bridge the gap. But the interest in them is not purely fundamental. As outlined above, clusters play a role in many practical situations like photography or artwork. More precisely, they have applications in many fields of science as, e.g., astrophysics, chemistry or material science. For example, clusters seem to play an important role in the formation mechanisms and the properties of cosmic dust [CTB89]. Carbon clusters are also expected to be present in the interstellar medium. The most striking property of clusters for applications in chemistry is their size. Indeed, because clusters may be quite small, but not too small, they can exhibit a large and tunable surface to volume ratio. They may thus provide ideal catalysts and play a crucial role in reaction kinetics [SAH$^+$99]. A typical example of application here is photography. In material science, the discovery of fullerenes and carbon nanotubes opened the way to the design of new materials [Kai01]. This breakthrough renewed chemistry and physics of carbon to such a level that this field is almost becoming independent from the mother field of cluster physics, probably in part because of its many industrial applications.

Cluster physics with its many achievements now belongs to one of the most active fields in physics, and offers, through related domains like the physics of nanotubes or fullerenes, one of the fastest developing areas in applied as well as in fundamental science. It is close to impossible to cover in a single book all the topics related to cluster physics. We shall thus focus here on one important aspect of the field: cluster dynamics. As in atomic and molecular physics, detailed studies on the dynamics of quantum many-body systems were boosted by the rapid progress of laser technology and the possibility of studying electronic motion at the femtosecond (fs) level. Clusters add to these studies the variability of sizes, as discussed above. Of course, the field of cluster science is so recent that we shall devote a large part of the book to the principle methods of cluster physics, experimental ones (in Chapter 2) as well as theoretical ones (in Chapter 3). Before that, we are going discuss in this chapter here the nature of clusters in relation to more established objects such as atoms and molecules, on the one side, and bulk, on the other side. Not surprisingly, we shall see that size turns out to be a key quantity, influencing many cluster properties. And we shall see that clusters are more than just big molecules or "small" pieces of bulk. They are indeed objects of their own, and cluster physics thus has to combine expertise from various fields of physics and chemistry into an excitingly new area of research.

## 1.1 Atoms, molecules and solids

Before considering clusters made of atoms, it is useful to briefly summarize what we need to know on the more "traditional" systems such as atoms, molecules and solids. Clusters range between these extremes and we shall see that understanding binding mechanisms between atoms or inside bulk provides the necessary keys to understand binding of clusters. Starting point is the atoms, then we discuss their combination in terms of molecules and in the infinite limit in terms of solids. We can then address clusters. For all species, we give here a brief overview with bias on the electronic structure, mostly at a qualitative level. For more thorough discussions, we refer the reader to standard textbooks of atomic, molecular, chemical physics or solid-state physics as cited at the relevant places.

### 1.1.1 Atoms

#### 1.1.1.1 Qualitative aspects

Atoms consist of a central nucleus and a neutralizing electronic cloud. The atomic number $Z$ labels the charge of the nucleus which is the number of protons inside the nucleus. The other constituents are the neutrons which, however, play a negligible role for the electronic problem, at least at the level we are interested in. For our purpose, we can safely reduce the effect of the nucleus to that of a point charge $Ze$ (thus neglecting hyper-fine structure as effects of finite size and magnetic coupling [YH96]). Electrons are then supposed to feel only the Coulomb field of the (point-) nucleus. At the level of the fine structure, there are the relativistic effects on the electrons as spin-orbit splitting and Breit interaction [Wei78]. These are, in fact, crucial for heavy elements or when going for quantitative details of bonding. Nonetheless, we neglect fine structure in the discussions of this book to keep the presentation simple.

What remains is the non-relativistic many-electron problem in the central field of the nucleus. It is well known that the electrons arrange themselves in shells around the nucleus. The quantitative understanding of the arrangement of these shells is a non-trivial problem, except for the case of the hydrogen atom where the problem reduces to one single electron in a central field. The case of many-electron atoms quickly becomes complex because of the two-body coulomb interactions between electrons. It is usually treated in a mean-field approximation where each electron is supposed to feel the net effect of all other electrons as one common central mean field. This allows one to sort electrons in shells denoted by their orbital angu-

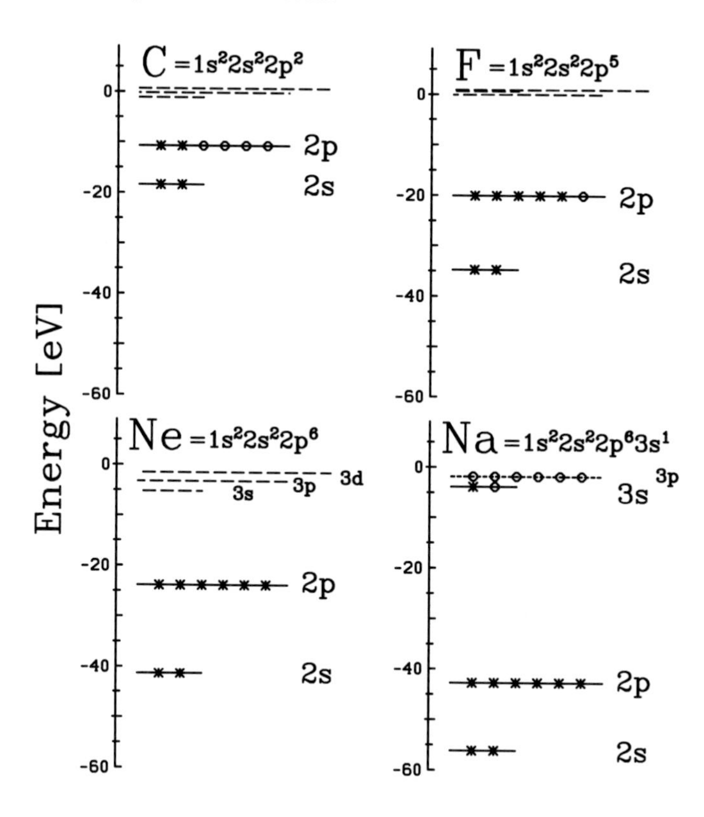


**Figure 1.3:** Level schemes for the C, F, Ne, and Na atoms. Open circles and/or dashed lines indicate empty states, crosses occupied states. The same energy scale is used for all 4 atoms for better comparison. The spectroscopic notation is given for each atom.

lar momentum and sorted according to increasing single-electron energy. Without entering details which can be found elsewhere (see [Wei78] and Appendix B), we want to remind the basic "Aufbau" principle of electronic shells. Let's imagine that we fill the atom successively, electron by electron. The first electron will feel only the nuclear attraction as a pure Coulomb field. But already the second electron experiences both the nuclear attraction and the repulsion due to the first electron. Both electrons will stabilize to form the most deeply lying electronic level, the $1s$ shell, filled with these two electrons and distinguished by spin up or down. The third electron has to fight the conflicting influence of the attracting nucleus and the two repelling $1s$ electrons. It will feel a screened nuclear charge $(Z-2)|e|$ and form, together with the next electron to come, the $2s$ electronic shell. Because the nuclear charge is screened, the $2s$ shell is much less bound than the $1s$ shell of the pure nuclear field. Mind also that the Pauli principle does hinder these $2s$ electrons to approach the nucleus in the area occupied by the $1s$ electrons. The $2s$ shell will thus be pushed outside the $1s$ one. Carrying on, the building principle works in a similar way. Electrons do gather in shells, characterized by a principal ($n = 1, 2, 3\ldots$) and an orbital quantum number ($l = 0 \equiv s$, $l = 1 \equiv p$, $l = 2 \equiv d$, $l = 4 \equiv f\ldots$) where each $l$ shell contains $2(2l+1)$ electrons. In light atoms (lighter than Ca (Z=20)) the sequence of successively occupied shells is $1s, 2s, 2p, 3s, 3p, 4s, 3d\ldots$. An example, for the electronic structures of Carbon (C), Fluorine (F), Neon (Ne), and Sodium (Na), is given in Figure 1.3. The single-electron energies have been obtained by density-functional methods as discussed in Chapter 3 which suffices for the purpose of the schematic discussion here. The lowest level, the $1s$ shell, lies for all examples so deep that it does not fit into the given scale. We track here mainly the evolution of the $2s$, $2p$ and $3s$ shells in the vicinity of the rare gas Ne. The nuclear charge, and thus the electron number, increases from C to Na as $Z(\mathrm{C}) = 6$, $Z(\mathrm{F}) = 9$, $Z(\mathrm{Ne}) = 10$, and $Z(\mathrm{Na}) = 11$. The filling of the shells proceeds accordingly as indicated in the figure. Next to the element symbol we show also the spectroscopic notation for the shell filling. It is self-explanatory. The example also allows one to understand a widely used naming in molecular physics: one speaks of the highest occupied molecular orbital (HOMO) and lowest unoccupied molecular orbital (LUMO). The case is clearest for the rare gas Ne here. The HOMO is the $2p$ shell and the LUMO the $3s$. The energy difference, which often dominates excitation properties, is called the HOMO–LUMO gap.

The comparison of the various spectra in Figure 1.3 is enlightening. Indeed, although the atoms are quite close to each other in terms of size, they have very different properties. Compare, e.g., the cases of F and Na in which the HOMO in Na is much more loosely bound than in F. As we shall see below, this feature is at the origin of the different chemical reactivity of these two atoms. The energy difference between occupied and empty states also varies significantly from one element to the other. In all cases the empty states are only bound by a few eV at most, while the binding energy of the HOMO may vary between about 5 and 20 eV. Such differences in the spectra imply differences in the capability of electrons to be removed from an occupied to an empty state. In other words, not only binding properties of an assembly of atoms, but for example their response to a laser will significantly vary from one atom to the other. We shall see examples of such differences at various places below. Furthermore, we shall see that these four atoms exemplify the basic bonding types. And they have been used extensively in cluster studies (except maybe for the fact that the preferred rare gas for building clusters is Ar rather than Ne).

Going to heavier and heavier atoms the filling of electrons in shells becomes more and more involved. Nevertheless there exist the so called Hund's rules as reasonably robust guidelines for an estimate of the sequence of occupied levels, and of their degree of occupancy. The first rule states that electrons successively occupy the levels $1s$, $2s$, $2p$, $3s$, $3p$, $4s$, $3d$, $4p$, $5s$, $4d$, $5p$, $6s$ .... The rule is strictly correct as long as no $d$ states are involved. As soon as $d$ states enter the game, the actual electronic structures do not exactly match the filling of the shells monotonously but tend to oscillate between $d$ and $s$ while stepping up the elements. This first rule does not yet predict the electronic structure fully. An open question concerns degenerate levels. For example, in carbon, the occupied levels are indeed $1s$, $2s$, and $2p$ with the corresponding electronic structure $1s^2 2s^2 2p^2$. But the $2p$ state is $2{\times}3$-fold degenerate ($2p_x, 2p_y, 2p_z$ times spin) and the above electronic structure allows 2 out of 6, i.e. 15, equivalent choices. The second Hund's rule states that electrons occupy all degenerate subshells once *before* occupying any level twice. This nevertheless does not yet tell with which spin configuration this "democratic" filling is performed. This last open point is addressed by the third rule which states that the ground state configuration of an atom is the one with maximum net spin (namely aligned spins). There is a subtle detail to be observed, though, when applying the third rule. One has to count electron spins up to half filling and switch to hole spin for the upper half towards the next rare gas. All in all, Hund's rules, although with some exceptions, provide a useful guideline for predicting the electronic structure and the level filling in an atom. They are especially pertinent in small atoms.

The heavier the atom, the more shells are filled with electrons. However the simple systematic filling of a given sequence of levels is no more ensured in these heavy atoms and from one element to the next one may observe fluctuations in the filling. This is typically the case in the vicinity of the noble metals such as Ag, Au or Pt. For example, the last occupied shells of Ag (having $Z = 47$) are successively $4d$ with 10 electrons and $5s$ with one electron. On the basis of light atoms one would have expected a full $s$ shell ($5s^2$) more bound than an open $d$ shell ($4d^9$). This change is caused by small rearrangements of the mean field which influence the relative energy of $4d$ and $5s$ shells. The bigger surprise rather comes when looking at the whole sequence: Mo ($Z = 42$, $4d^5 5s$), Tc ($Z = 43$, $4d^5 5s^2$), Ru ($Z = 44$, $4d^7 5s$), Rh ($Z = 45$, $4d^8 5s$), Pd ($Z = 46$, $4d^{10}$), Ag ($Z = 47$, $4d^{10} 5s$). The $5s$ shell is alternatively filled or emptied depending on minimal energetic changes. Not surprisingly, in these cases where the two shells, $4d$ and $5s$, are energetically so close, the $4d$ and $5s$ electrons are likely to simultaneously play a role in low energy phenomena. The simplified discussion which we use here thus becomes insufficient for heavier elements. One needs to think of the spin-orbit splitting which can easily span the few eV energy differences discussed in the filling of transitional elements. The spin-orbit force is also crucial for the magnetic properties which also show their most dramatic consequences for transitional elements, particularly Fe, Ca, and Ni.

#### 1.1.1.2 The atomic many-body problem

It is obvious from the above qualitative discussion that there is no trivial model to understand the details of the filling of electron shells. The problem is complex and of many-body nature. The atomic nucleus provides an attractive potential. The quantum mechanics of that is well known as it constitutes a central field problem for *one* electron. The difficulty comes from the electron–electron interactions for *several* electrons simultaneously. Qualitatively, and to some extent quantitatively, this many body nature of the problem manifests itself in terms of screening. In other words, one electron essentially feels the nuclear field screened by the electrons belonging to the more bound shells. But the balance is more delicate. First, as already mentioned, one has to account for the Pauli principle which repels outer electrons from the central part of the atom. This exclusion effect still preserves the sphericity of the problem. More delicate is the fact that electrons belonging to the same shell also repel each other via the Coulomb interaction. And such an effect has to be accounted for on a one to one basis. Stated so crudely, the problem might appear as awfully complicated. It is indeed, if one looks for spectroscopic details. Global features, however, are more forgiving. The key mechanism of simplification lies here in the presence of the external field provided by the nucleus. Indeed, in spite of the electron–electron corrections, this central attracting field somewhat shapes the gross structure of the electronic spectrum and basically sets a hierarchy in the way one accounts for electron–electron effects. To an acceptable approximation, the simple screening picture can thus be recast into a self consistent mean field approach in which each electron feels an average potential field created by the nucleus and the electron–electron interactions (including both intra and inter-shell interactions). This is the spirit of the so called "central field" approximation. Let us briefly explain what lies behind this concept.

We start from the many body electron Hamiltonian for an atom with $N$ electrons (positions $\mathbf{r}_i$) bound to a nucleus of charge $Z$

$$H = \sum_i^N \left(-\frac{\hbar^2}{2m_e}\nabla_i^2 - \frac{Ze^2}{r_i}\right) + \sum_{i<j}^N \frac{e^2}{|\mathbf{r}_i - \mathbf{r}_j|} \quad (1.1)$$

where $m_e$ is the electronic mass (note that we are using the Gaussian system of units, see Appendix A.1). The central field approximation of Hartree and Slater [Har27, Sla51] is based on an independent particle picture. One admits that electrons do move in an effective potential representing the nuclear attraction together with the repulsive interactions with the remaining $N-1$ other electrons. As discussed above, the major effects of these $N-1$ electrons is to screen the nuclear charge. Thus the inter-electron repulsive term contains a large spherically symmetric component which we shall write as $\sum_i U_{\mathrm{H}}(r_i)$. This means that we can approximate the effective potential field felt by one electron by a spherically symmetric potential

$$U_{\mathrm{total}}(r) = -Ze^2/r + U_{\mathrm{H}}(r) \quad . \quad (1.2)$$

The actual form of $U_{\mathrm{H}}(r)$ remains to be specified. The asymptotic behaviors can be easily fixed by requiring recovery of the nuclear field at short distance and of the net charge felt by the last electron at large distance, namely

$$U_{\mathrm{total}}(r) \underset{r \to 0}{\longrightarrow} \frac{-Ze^2}{r} \quad , \quad U_{\mathrm{total}}(r) \underset{r \to \infty}{\longrightarrow} \frac{-(Z-N+1)e^2}{r} \quad . \quad (1.3)$$

The determination of the effective potential at intermediate range is of course more involved as it requires one to account for all the electrons. Many methods have been developed to attack this problem. Let us cite the most famous Thomas-Fermi and Hartree-Fock methods, which both provide a well defined way to compute a central field, actually a mean-field potential common to all the electrons constituting the system (for a detailed discussion and references see Chapter 3). The central field approximation turns the initial three dimensional (3D) problem into an effective spherical 1D problem, which can fairly easily be solved with standard techniques of basic quantum mechanics. The $N$-electron wavefunction is factorized into $N$ one-electron wavefunctions, which in turn can be factorized into a radial and an angular part $\varphi_{nlm}(\mathbf{r}) = R_{nl}(r)Y_{lm}(\theta,\phi)$ (with obvious notations). The resolution of the radial Schrödinger equation provides the $R_{nl}(r)$ and the corresponding principal ($n = 1,2,3\ldots$) and orbital ($l = 0,1,2\ldots$) quantum numbers. There is one subtle feedback loop in that treatment: the effective central field is composed from the Coulomb force of actual electron densities; these, in turn, depend on the central field. The problem is then resolved by iterating the solution with subsequent update of the central field until the feedback loop has converged. After all one disposes of the sequence of one-electron energies in levels $1s, 2s, 2p, 3s, 3p, 4s\ldots$, as already discussed above. Of course the details of the level filling may vary from one atom to the next, an effect which is related to the actual form of the central field $U_{\mathrm{total}}(r)$ and its self-consistent arrangement. Such mean field models are widely used in atomic physics and provide a sound starting basis for more elaborate approaches, taking better into account correlations between electrons. The density functional theory in which these correlations are expressed in terms of the local electronic density (see Section 3.3.3) offers here a robust starting point for such investigations. A first example of applications of this method was given in Figure 1.3 showing the spectra of the four simple atoms C, F, Ne, and Na.

A comment is in order here concerning the meaning of single electron energies, and in particular its relation to experimental observables. When an atom is excited, for example by a laser, an electron may be promoted from one state of initial energy $\varepsilon_1$ to another of energy $\varepsilon_2$ (the latter being possibly unbound). Experimentally, one only can access the transition energy $\varepsilon_2 - \varepsilon_1$, for example through a photon. This provides the well known series of transition lines observed in various atoms. But one has to keep in mind that this electronic transition is a complex dynamical process which affects not only the "promoted" electron but the whole atom. Indeed, the energy levels of an atom with an empty electronic slot (called a *hole*, $h$) at level $\varepsilon_1$ and an (originally empty but now) occupied level $\varepsilon_2$ (called a *particle*, $p$) are different from the energy levels of the original atom ($0ph$). All the electrons in the excited ($1ph$) atom have to rearrange themselves to account for the electronic modification, and the energy levels do the same. This means in particular that, if the excited atom remains bound, the final energy of the "promoted" electron will not exactly be $\varepsilon_2$. As a consequence observing the transition energy gives a direct access neither to $\varepsilon_1$ nor to $\varepsilon_2$ of the original atom but includes corrections from a complex (specific) rearrangement process. An ideally instantaneous "promotion" of the electron (if possible) might leave the rest of the atom frozen and thus indeed give information on the actual energies of occupied levels in the original atom. In the other extreme case of an infinitely slow electronic excitation all other electrons do rearrange continuously and one will rather measure the energies of the final excited atom. The actual experimental situation usually lies in between these two extremes. We shall discuss similar examples in the case of clusters, see e.g. Sections 2.3.2 and 2.3.8. It is however to be noted that rearrangement effects are much larger in clusters than they are in atoms, because of possible ionic effects.

## 1.1.2 Molecules

### 1.1.2.1 The Mendeleev classification of elements

The successive filling of electronic shells provides the microscopic basis for the understanding of chemical properties of elements. Indeed, the early Mendeleev classification, although initially founded on mass rather than on electron number, was precisely trying to explain the regularity observed in the chemical reactivity of apparently different atoms. Since then we have learned that these chemical similarities reflect the behavior of the tiny fraction of least bound electrons in atoms, the so called valence electrons. And as long as valence electrons belong to shells with the same angular momenta ($s, p, d \ldots$ shells) the corresponding atoms behave similarly at the chemical level. This is, after all, not so surprising. The value of the angular momentum of the shell fixes the shape of the electronic wavefunctions and hence the capability of a given electron of a given atom to interact, and possibly bind, with electrons of another atom. The degree of filling also fixes the capability of an atom to accept or to release valence electrons. The atoms with completed shells (shell closure) are especially stable. The so called rare gases (He ($Z = 2$), Ne ($Z = 10$), Ar ($Z = 18$), Kr ($Z = 36$), Xe ($Z = 54$), and Rn ($Z = 86$)) form the family of the most stable atoms. They have the highest values of ionization potential. This explains why these species are particularly inert from the chemical point of view. In the "vicinity" of these rare gases lie two categories of atoms either with one electron more or one electron less than the corresponding rare gas. In turn, these two families of atoms are extremely active from the chemical point of view because the atoms belonging to these families have a strong tendency either to grab or to release an electron, in order to attain an electronic configuration similar to that of the neighbor rare gas. The alkaline atoms (Li ($Z = 3$), Na ($Z = 11$), K ($Z = 19$), Rb ($Z = 37$), Cs ($Z = 55$), Fr ($Z = 87$)) have one valence electron belonging to a $s$ shell as compared to the neighbor rare gas. This $s$ electron is weakly bound, which implies a small ionization potential. Correlatively, it has a strong tendency to leave its parent atom, as we shall see below. On the contrary, the halogen atoms (F ($Z = 9$), Cl ($Z = 17$), Br ($Z = 35$), I ($Z = 53$), At ($Z = 85$)) exhibit a valence $p$ shell missing one electron for closure. They thus have a strong tendency to find this electron in their surrounding. Obviously alkaline and halogen atoms are likely to "complement" each other. An alkaline atom will tend to "give" its peripheral $s$ electron to a halogen atom to complete its $p$ shell. This exchange of electrons forms the basis of the most robust chemical bond, the ionic bond in which electron exchange ensures a robust electrostatic binding between two charged species, a positively charged alkaline and a negatively charged halogen. We shall come back to this below. But before exploring more systematically binding possibilities between atoms, one has first to come back to the Mendeleev classification for a few more remarks. Up to now we have discussed only 3 columns of atoms (rare gas, alkalines, halogens) while the Mendeleev classification contains up to 18 columns. In fact, halogens and alkalines, as "ends" of each row do represent the extremes cases of tendencies which develop along each row of the classification. Starting from an alkaline atom, successively adding electrons (namely going in the direction of the next heavier halogen) will progressively attenuate the tendency of the atom to release electrons and increase its tendency to grab them. There is thus a continuous path between electron donors and electron acceptors. In the middle of the path lie atoms with an "equal" tendency to grasp or to release electrons. The typical representatives of this class are carbon (C, $Z = 6$) silicon (Si, $Z = 14$) and germanium (Ge, $Z = 32$). We shall see that they tend to bind between each other in a specific way giving rise to a very rich palette of molecular structures, particularly for carbon.

#### 1.1.2.2 A pedestrian view of bonding

The first step along the path from atoms to molecules (and to clusters) is represented by dimer molecules and we shall here focus on this simple case. As already mentioned, depending on their electronic structures two atoms may, or may not, bind together, with different type and strength of binding. Chemical binding is by nature a low energy phenomenon, involving at most a few eV energy. For this reason only valence electrons significantly play a role, and the Mendeleev classification thus provides a gross map of which atom may possibly bind with which other and how. Of course, binding between two atoms is a matter of electrons and energy. The key question is thus whether the valence electron(s) of a given atom may find it energetically interesting to "leave" their parent atom to "bind" to another atom, or whether there are "intermediate" solutions for these electrons, like partial attachment to both atoms. According to such a simple picture, one may envision four classes of binding reflecting the possibility of sharing the set of valence electrons between the two partner atoms, as illustrated in Figure 1.4. As in any scheme, these bonding types are idealized extremes. Real molecules are usually dominated by one of these bondings with admixtures of the complementing types.

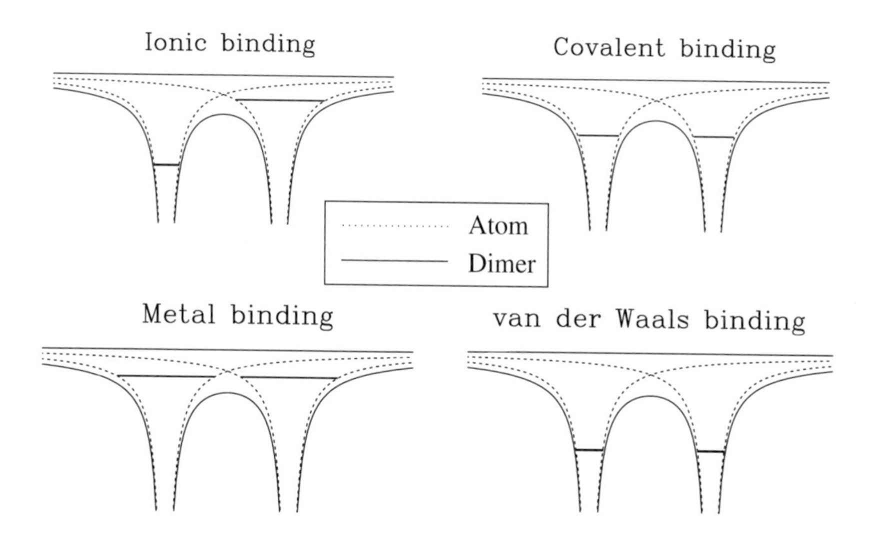


**Figure 1.4:** Schematic view of potential and valence levels in dimers with the four basic types of binding as indicated. The molecular potential is drawn with solid lines and the atomic one with dashed lines. The valence levels are indicated by horizontal straight lines.

The first possibility is ionic binding. One loosely bound electron is transferred from its parent atom to a host atom in which it finds an empty slot on a more deeply bound level. The electronic energy level in the parent atom is less bound than the host level and the transfer becomes easier if the possible potential barrier between the two atoms is smaller. In other words, the smaller the ionization potential of the parent atom, the easier the transfer. Because the difference in the electronic levels of the donor and the acceptor is large the exchange is indeed effective. The electron localizes predominantly on the host atom. The exchange of the electron thus charges positively the donor atom and negatively the acceptor atom. This gives rise to what is known as an ionic bond, in which a positive charge is concentrated on one atom and a negative charge on the other atom, making the system almost a +/- ionic dimer. Typical examples of ionic binding are realized between alkaline (electron donor) and halogen (electron acceptor) atoms, as explained above. Everyday kitchen salt (NaCl) crystals are bound with such a mechanism. The large energy gain associated with the electron exchange explains the robustness of the ionic bond.

When the energy difference between the valence shells of the two atoms is not so large, electron transfer becomes unfavorable and electrons tend to remain close to their parent atoms. This does not exclude binding. But then binding stems from charge sharing rather than charge transfer. The valence electrons tend to form a common electronic cloud establishing the binding of the two atoms together. The properties of this common valence cloud again depend on the nature of the partner atoms and may give rise to two different types of binding. The binding energy of the atomic valence electrons determines the behavior. If the valence electrons are weakly bound the electronic wavefunctions can easily spread outside the parent atoms. As soon as the proximity of the partner atoms has sufficiently decreased the barrier between them (see Figure 1.4), the electronic wavefunctions become delocalized all around the two atoms and one then speaks of a metallic bond (as a precursor of the behavior of such systems in the bulk, discussed below). Alkalines are the typical elements which establish metallic bonds between each other. They are called simple metals because of the well separated valence electron (see Na in Figure 1.3). We shall often consider the generic example of $Na_n$ clusters as typical simple metallic systems. Metallic binding is also observed in more complicated metals such as, e.g., Cu, Ag, Au, or Pt. The valence binding mechanism in these systems is qualitatively similar to the simple case of alkalines, but there is then a $d$ shell energetically close to the valence $s$ shell (see the example of Cu in Figure 3.2). This causes strong polarization effects with impact on binding and other dynamical properties (see e.g. the discussion of optical response in Section 4.3.4.1).

The metallic binding does not exhaust all possibilities of sharing a common valence cloud. If the electrons are initially more strongly bound to their parent atom, electron delocalization cannot fully develop and electrons gather in the region of smallest potential energy, between the two partner atoms. One then speaks of a covalent bond. The typical covalent bond is realized between atoms like C or Si in which atomic valence electrons are fairly well bound (see also Figure 1.3). However, one has to keep in mind that covalent and metallic bonds are idealized situations. There is a smooth transition between these extremes. It is usually hard to disentangle these two types in small molecules. The distinction becomes better defined in bulk material where interstitial density and conductivity add useful criteria [AM76].

When valence electrons are too deeply bound in the atom, neither charge transfer nor charge sharing are possible between the two atoms. Because the HOMO electrons are too deeply bound, the electrons cannot escape the attraction of their parent atom. One may then wonder how bonding is possible at all. Although electrons remain localized around their parent atom, the electronic cloud of each atom is influenced by the partner atom. This virtual polarization of the electronic clouds results in mutual dipole-dipole interactions between the two atoms, which establishes a binding of the system. As is obvious the resulting bond is much weaker than the previous ionic, metallic or covalent bonds. And yet, it suffices for the binding of rare gas molecules up to possibly large compounds, as we shall see below. This type of loose binding is known as van der Waals (or molecular) binding, in reference to the interactions between two neutral atoms or molecules. The typical cases of van der Waals bonding thus concern rare gas atoms mostly. It should be noted that the details of the binding mechanism are somewhat involved. The polarization of the electronic cloud has to be understood as a dynamical correlation effect, with constantly fluctuating dipoles, rather than as a mere static polarization. To properly account for this effect requires a quantum mechanical treatment which we shall not discuss here.

All in all, simple energetic considerations for the HOMO, as exemplified in Figure 1.3, allow one to identify four major types of bonding between atoms, as schematically represented in Figure 1.4. These energetic aspects reflect themselves in the degree of localization of the electrons binding the two atoms: the more bound the electrons, the more localized their wavefunctions. Of course, just like the behavior of atoms as electron donors or acceptors, there is no clear separation between the various binding mechanisms, but rather a continuous path between them (see Figure 1.8 for an example in the bulk). Still, the four classes identified above can be seen as robust guidelines for understanding the binding of most molecular systems. Finally, it should be noted that one sometimes introduces a few other classes of binding. The term “molecular binding” for example refers to the binding between molecules, while the term "van der Waals" usually concerns atoms, although in both cases the basic binding mechanism is the same. Similarly one sometimes speaks of "hydrogen bonding" to label covalent binding involving hydrogen atoms. Because its particularly large ionization potential, hydrogen usually establishes especially robust bonds (as compared to standard covalent bonds) with other species, hence the specific label. Again, such fine distinctions do not basically alter our global classification scheme. We shall thus ignore such details in the following.

#### 1.1.2.3 Deeper into microscopic mechanisms

Before proceeding to the bulk (in Section 1.1.3) and to clusters (in Section 1.2) we would like to discuss in some more detail the binding mechanism for simple dimers, in order to exhibit the dominant mechanisms in a semi-quantitative fashion. We keep the discussion at a minimal level of formalism. More detailed elaborations of that sort can be found in many standard textbooks on molecular physics or chemistry (see Appendix A.4).

We consider the case of a dimer formed from two atoms $A$ and $B$, each having *one* $s$ valence electron (as is the case for H or Na). The atoms are placed at a distance $R$ from each other. We denote by $\varphi_A$ and $\varphi_B$ the respective atomic single-electron wavefunctions following the Schrödinger equations

$$-\frac{\hbar^2}{2m}\Delta\varphi_X + V_X\varphi_X = \varepsilon_X\varphi_X \quad , \quad X \in \{A, B\} \quad , \tag{1.4}$$

where $V_X$ labels the potential felt by the valence electron in atom $X$ and $\varepsilon_X$ is the corresponding eigenvalue of energy. In order to simplify the problem further, we restrict ourselves to the case of an ionized dimer, in which only *one* active electron is left (as in $H_2^+$ or $Na_2^+$). The goal is then to solve the Schrödinger equation for the dimer electron

$$-\frac{\hbar^2}{2m}\Delta\varphi_{AB} + V_{AB}\varphi_{AB} = \varepsilon_{AB}\varphi_{AB} \quad , \tag{1.5}$$

where $\varepsilon_{AB}$ labels the energy of the electron level binding the dimer. The total potential $V_{AB}$ is, in principle, a self consistent quantity as it depends on the wavefunction $\varphi_{AB}$ itself, because the formation of the bond induces an electronic rearrangement and thus a modification of the net atomic fields. For the sake of simplicity and because these rearrangement effects can be assumed reasonably weak, at least for an exploratory approach, we approximate $V_{AB}$ by

$$V_{AB} \simeq V_A + V_B + \frac{e^2}{R} \quad . \tag{1.6}$$

For the solution we assume that we can represent the molecular electronic wave function $\varphi_{AB}$ in terms of the given atomic wave functions. This means that we make the ansatz of Linear Combination of Atomic Orbitals (LCAO)

$$\varphi_{AB} = c_A\varphi_A + c_B\varphi_B \quad . \tag{1.7}$$

It remains to determine the expansion coefficients $c_A$ and $c_B$.

Variation of the total energy with respect to these coefficients yields in a straightforward manner the secular equation

$$\begin{pmatrix} \tilde{\varepsilon}_A + \frac{e^2}{R} & (\bar{\varepsilon} + \frac{e^2}{R})S + h \\ (\bar{\varepsilon} + \frac{e^2}{R})S + h & \tilde{\varepsilon}_B + \frac{e^2}{R} \end{pmatrix} \begin{pmatrix} c_A \\ c_B \end{pmatrix} = \varepsilon_{AB} \begin{pmatrix} 1 & S \\ S & 1 \end{pmatrix} \begin{pmatrix} c_A \\ c_B \end{pmatrix} \quad (1.8)$$

where

$$S = \int d\mathbf{r}\varphi_A^*\varphi_B \quad , \quad h = \int d\mathbf{r}\varphi_A^*\bar{V}\varphi_B \quad ,$$

$$\tilde{\varepsilon}_A = \varepsilon_A + \int d\mathbf{r}\varphi_A^* V_B \varphi_A \quad , \quad \tilde{\varepsilon}_B = \varepsilon_B + \int d\mathbf{r}\varphi_B^* V_A \varphi_B \quad ,$$

$$\bar{\varepsilon} = \frac{\tilde{\varepsilon}_A + \tilde{\varepsilon}_B}{2} \quad , \quad \bar{V} = \frac{V_A + V_B}{2} \quad .$$

The key quantities for the mixing are the overlap ($S$) and bond ($h$) integrals. They provide measures of how much the two electronic wavefunctions "communicate", either "directly" (overlap integral) or via the interaction (bond integral). In the limit of a sufficiently small overlap integral $S$ the two solutions $\varepsilon_{AB}^{+,-}$ of the secular equation Eq. (1.8) take the simple forms ($\Delta\varepsilon = \varepsilon_A - \varepsilon_B$)

$$\varepsilon_{AB}^{\pm} = \bar{\varepsilon} + |h|S \mp \sqrt{h^2 + \Delta\varepsilon^2/4} \quad (1.9)$$

where the most bound level $\varepsilon_{AB}^{+}$ is known as the bonding state while the least bound one $\varepsilon_{AB}^{-}$ is the antibonding state. Both the bond and the overlap integrals enter the expressions of $\varepsilon_{AB}^{\pm}$. Because by construction the bond integral is less than 1, the bonding state is always more bound than the most bound atomic level $\varepsilon_{AB} \leq \min(\varepsilon_A, \varepsilon_B)$ (note that the cross-over elements are negative, e.g. $\int d\mathbf{r}\varphi_A^* V_B \varphi_A < 0$), which is what establishes binding. The net gain of binding then depends on $S$ and $h$, which points out the importance of the capability of electrons to explore the partner atom ($S$ and $h$). Note finally that the energy separation between bonding and antibonding states increases with the bond integral. These behaviors are illustrated for the case of homonuclear dimers in the schematic Figure 1.5. Note that here $\varepsilon_A = \varepsilon_B$ and thus $\Delta\varepsilon = 0$. The lower part shows how the matrix elements $S$ and $|h|$ decrease quickly as a function of molecular distance $R$. Similarly behaves the convergence $\varepsilon_{AB}^{\pm} \longrightarrow \bar{\varepsilon} = \varepsilon_A = \varepsilon_B$ for large $R$.

The energetic information provided by Eq. (1.9) can be nicely complemented by looking at the spatial extension of the electronic cloud. The interesting quantity is here the electronic density $\rho_{AB}^{+}(\mathbf{r}) = |\varphi_{AB}^{+}(\mathbf{r})|^2$ of the bonding state. It reads

$$\rho_{AB}^{+}(\mathbf{r}) = (1 + \alpha_i)\rho_A(\mathbf{r}) + (1 - \alpha_i)\rho_B(\mathbf{r}) + \alpha_c \rho_{\text{bond}}(\mathbf{r}) \quad , \quad (1.10a)$$

$$\rho_{\text{bond}} = 2\varphi_A^*\varphi_B - S(\rho_A + \rho_B) \quad , \quad (1.10b)$$

$$\rho_X = \varphi_X^*\varphi_X \quad \text{for} \quad X \in \{A, B\} \quad , \quad (1.10c)$$

$$\alpha_c = \left[1 + \frac{\Delta\varepsilon}{2|h|}\right]^{-1/2} \quad , \quad \alpha_i = \frac{\Delta\varepsilon}{2|h|}\alpha_c \quad . \quad (1.10d)$$

The relations simplify in the homonuclear case, where $\Delta\varepsilon = 0$, to $\alpha_c = 1$, $\alpha_i = 0$ and finally to $\rho_{AB}^{-} = \rho_A + \rho_B + \rho_{\text{bond}}$. It shows that the bond density makes all the difference.

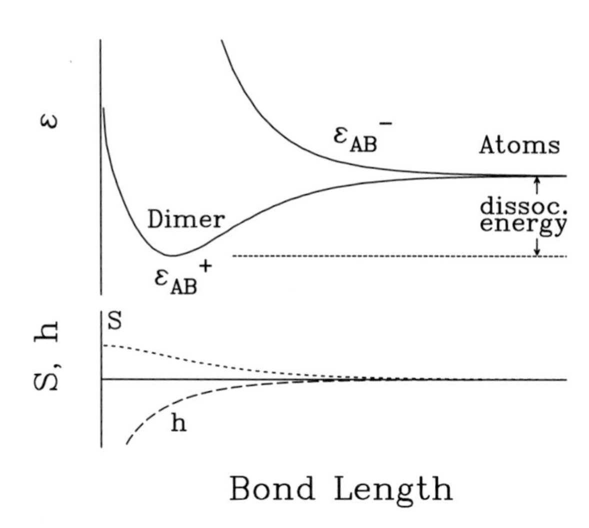


**Figure 1.5:** Schematic plot of the overlap ($S$) and bond ($h$) integrals (bottom part) and of the energies of the bonding ($\varepsilon_{AB}^{+}$) and antibonding ($\varepsilon_{AB}^{-}$) levels as a function of distance $R$ between the two atoms. Scales are arbitrary, but note that the overlap integral tends towards 1, at short $R$. Asymptotically (large $R$) both $S$ and $h$ integrals vanish and $\varepsilon_{AB}^{\pm} \rightarrow \bar{\varepsilon}$, namely the dimer becomes a collection of two independent atoms. The dip in ($\varepsilon_{AB}^{+}$) at finite distance corresponds to the bond length, namely the equilibrium distance $d_{\mathrm{b}}$ between the two bound atoms forming a dimer. The difference between $\varepsilon_{AB}^{\pm}(d_{\mathrm{b}})$ and $\varepsilon_A$ (or $\varepsilon_B$) represents the dissociation energy of the dimer.

Let us briefly discuss the case of a heteronuclear system ($A \neq B$) which is already prepared in the above equations. The difference $\Delta\varepsilon \simeq \varepsilon_A - \varepsilon_B$ shows up at the level of the bonding and antibonding energies. Note that the energy of the bonding state, and thus the nature of binding also depend on the energy difference between the two original atomic orbitals. It also sensitively enters the total electronic density $\rho_{AB}(\mathbf{r})$ given in Eq. (1.10). The now non-trivial parameter $\alpha_i$ provides a measure of the ionic charge renormalization, hence of the amount of electron density transferred from one atom to the other. Obviously the most covalent bond corresponds to $\alpha_i = 0$, as is the case in a homonuclear system. Otherwise, if $\alpha_i > 0$ ($\varepsilon_A \leq \varepsilon_B$), the electron density gathers around $A$ as in a more or less ionic bond. All in all, the two parameters $\alpha_i, \alpha_c$ thus provide a measure of the degree of ionicity of the bond, allowing one to span in a continuous way the various situations between pure covalent ($\alpha_i = 0$, $\alpha_c = 1$) and pure ionic ($\alpha_i = 1$, $\alpha_c = 0$) binding.

#### 1.1.2.4 More realistic examples

For sake of simplicity, we have restricted our discussion, up to now, to the case of ionized dimers, with only one active valence electron. Passing to neutral dimers is qualitatively easy to grasp, but computationally more involved. As soon as two electrons, one from each atom, bind together, one has to take into account the Coulomb repulsion between the two electrons. It affects both the energies $\varepsilon_{AB}^{\pm}$ and the repartition of the electronic density $\rho_{AB}(\mathbf{r})$. Nonetheless, the above discussions remain qualitatively valid, in particular concerning the roles of the parameters governing the basic properties of the bond (bond and overlap integrals, energy difference $\Delta\varepsilon$). To obtain quantitative results in practice, an explicit and detailed calculation of electronic structure has to be performed, starting from a microscopic description of each atom, usually including more than only valence electrons. And it will turn out that the details of electronic spectra (both occupied and empty states) are actually to be accounted for. We will address these theoretical methods in Chapter 3.

| type | v. d. Waals | metallic | covalent | ionic |
|---|---|---|---|---|
| example | $Ar_2$ | $Na_2$ | $C_2$ | NaF |
| $D$ [eV] | 0.010 | 0.74 | 6.2 | 5.0 |
| $d_b$ [$a_0$] | 7.1 | 5.8 | 2.3 | 3.5 |

**Table 1.1:** Dissociation energy $D$ and bond length $d_b$ for selected diatomic molecules, each one representing one bond type as indicated. Data from [AM76, Bei95, Kra96, GLC$^+$01].

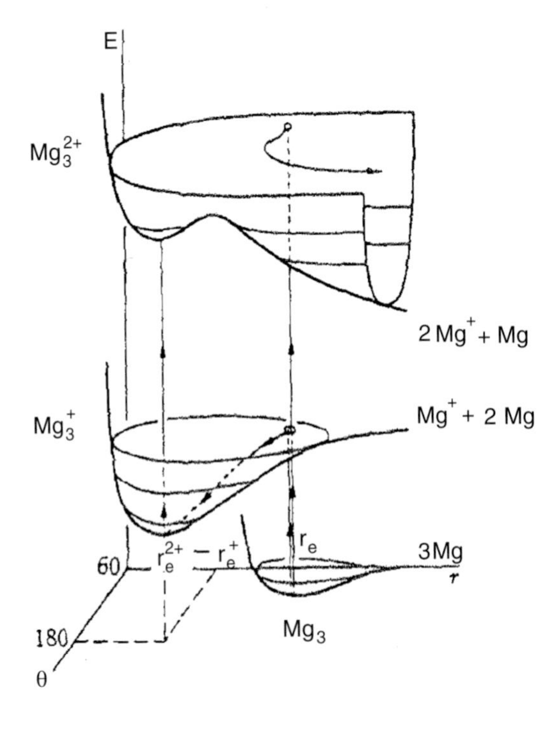


**Figure 1.6:** Born-Oppenheimer surface for the dissociation of $Mg_3$ possibly with electron emission. The considered dissociation channels are here $Mg_3 \longrightarrow Mg^+ + 2Mg + e^-$ and $Mg_3 \longrightarrow 2Mg^+ + Mg + 2e^-$. The potential energy surfaces of $Mg_3$, $Mg_3^+$ and $Mg_3^{2+}$ are thus drawn together with the corresponding dissociation paths. Note the various dissociation paths for the doubly charged species as a function of ionization mechanism (see text for details). After [DDM86].

Here we continue the discussion with a few realistic data for simple molecules. The energy curves look qualitatively all very similar to the schematic view in Figure 1.5. What changes is the depth and position of the potential well. It can be quantified in terms of the dissociation energy $D$ as the difference between the asymptotic energy and the bonding minimum (see Figure 1.5) and of the bond distance $d_b$ which is the distance where the minimum resides. Table 1.1 gives a few examples for the four typical bond types discussed above. The van der Waals binding is the weakest by far. Covalent and ionic binding are the strongest and the metallic example is slightly weaker than these.

A more complex case is demonstrated in Figure 1.6. It is the Born-Oppenheimer molecular energy surface (Appendix C.1) for the dissociation of $Mg_3$. Also are represented the potential energy surfaces for $Mg_3^+$ and $Mg_3^{2+}$.

The surfaces are plotted as a function of a typical distance and angle. The $Mg_3$ cluster structure can obviously be represented by two such parameters and the same holds for both $Mg_3^+$ and $Mg_3^{2+}$. One ought to realize, though, that the example of this figure corresponds to a particularly simple case, because of the small size of the system. A complex molecule (or cluster) has many ionic degrees of freedom. The molecular energies thus correspond to huge multi-dimensional Born-Oppenheimer surfaces from which one can inspect at most well taken cuts. Such large systems are treated in practice by more direct dynamical methods as will be outlined in Chapter 3.

Several dissociation channels are sketched in Figure 1.6, depending on the degree of ionization (singly or doubly charged $Mg_3$). As can be seen from the figure the corresponding potential energy surfaces do differ in shape. For example one sees that the more charged the cluster, the more compact the ground state configuration, corresponding to the minimum of the potential well. The figure also demonstrates the various possible dissociation paths as a function of the excitation process. When the cluster is only singly charged it still corresponds, not surprisingly, to a well stable species. When it is doubly charged the picture becomes more involved. Indeed, if electron removal is sequential, the system, having lost one electron, may have time to relax towards the corresponding equilibrium $Mg_3^+$ state, before a second ionization promotes it, again, towards an equilibrium (although very weakly bound, and thus probably metastable) state, this time for $Mg_3^{2+}$. On the contrary, in the case of a simultaneous double ionization the system is immediately promoted towards an unstable configuration which directly dissociates. We find here an illustration, in the case of a small cluster, of the importance of the details of the dynamics in the ionization process. We shall come back to this question at many places (see for example Sections 2.3.2 and 2.3.8).

### 1.1.3 The point of view of solid state physics

Complementing the discussion on atom and molecule, we now turn to bulk matter. Bulk might have two different meanings in the case of clusters. Clusters are often in a liquid state, because of their high formation temperature (Section 2.1.3.1). When extrapolating, bulk then refers to a macroscopic liquid. But the ground state (low temperature state) of a cluster stays in a frozen geometry, and bulk then refers to an infinite solid, more precisely a crystal. In the following discussion we shall only consider this zero temperature situation and bulk will thus mean a piece of solid material.

From the viewing angle of solid state physics, a key parameter of classification is the conductivity of a given piece of material. One will thus separate solids into two classes: conductors and insulators. The different bonding types will be associated to these two classes. We shall first, as in the case of atoms/molecules, consider the problem at a qualitative level, before addressing the question from a more formal point of view.

#### 1.1.3.1 Bond types in solids

The four classes of binding (metal, covalent, molecular and ionic) in the bulk are illustrated in Figure 1.7 for a set of typical examples. We will discuss them now step by step.

Metals constitute the class of conducting solids. In that case the valence electrons of each atom are fully delocalized, so that they form a nearly free uniform gas of electrons. Each electron thus loses any direct contact with the atom it was originally bound to. The typical (and simplest) examples of metallic bonding are again found in alkalines such as Na or K. The case of alkali metals is in some sense too clean. Indeed, one can find additional aspects

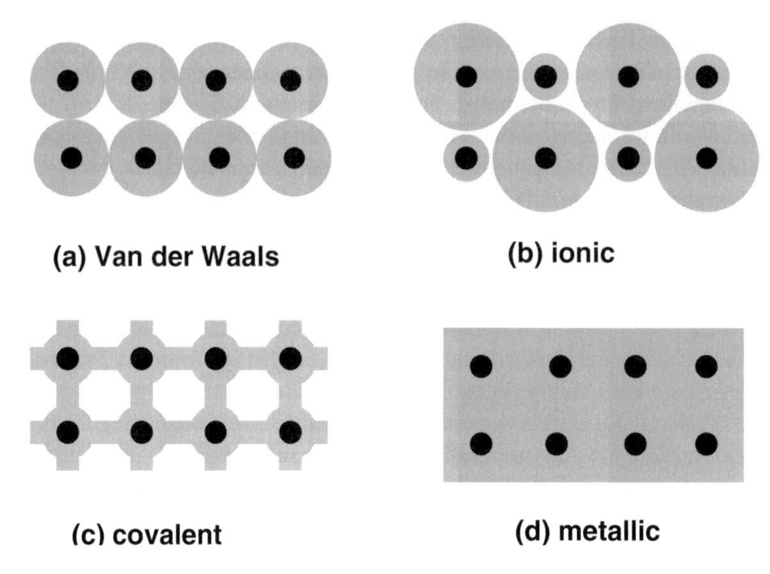


**Figure 1.7:** Schematic representation of electronic charge distributions in examples of solids. The small (filled) circles represent the ionic cores, with the corresponding "effective" charges seen by the electrons constituting the binding. The shaded areas represent regions in which one finds an appreciable amount of the electron density. Note that the electronic density is not uniform. The four cases correspond to the basic four types of binding: a) van-der-Waals (molecular) binding in an Ar crystal, b) ionic binding in KCl, c) covalent binding in C and d) metallic binding in K. After [AM76].

of van der Waals or covalent bonding in many metals, particularly in noble metals where loosely bound $d$ electrons participate in the bonding. A typical example is Ag, in which the least bound electron belongs to a $5s$ shell, while the loosely bound $4d$ orbitals are strongly distorted by the presence of neighbor atoms. In such a case the separation between valence and core electrons is much less clear than in alkalines. And yet, such materials do also exhibit a clearly metallic behavior.

Materials with a poor conductivity belong to the category of insulators. But contrarily to the class of metals, bonding in insulators covers different forms, essentially reflecting the degree of localization of electrons around their parent atom. Let us successively consider these various bindings as a function of the degree of localization of electrons, while keeping in mind that the largest degree of delocalization is attained in metals.

In covalent crystals, one faces a set of semi-localized electrons, which gather along the lines joining atoms together. The typical example of such a binding is the case of diamond. Just as in simple molecules, we find a covalent nature of binding in carbon based materials.

The last two classes of insulators leave electrons fully localized on atomic sites. In van-der-Waals crystals such as solid noble gas (e.g., Ne, Ar, Kr, or Xe), very few electrons gather between the sites. Electrons basically remain bound to the original atom they were attached to. Just as in simple molecules, polarization effects are responsible for the binding of the system. The last class of compounds are ionic crystals which associate metallic (electron donor) and non-metallic (electron acceptor) atoms in a regular manner, so that ionic bonds (see Section 1.1.2.2) can be reconstituted. In that case electrons are highly localized on atomic sites, once they have been transferred from the metal to the non-metal atoms. The typical example of such systems are alkaline halogen compounds such as NaCl, NaF, or LiF.

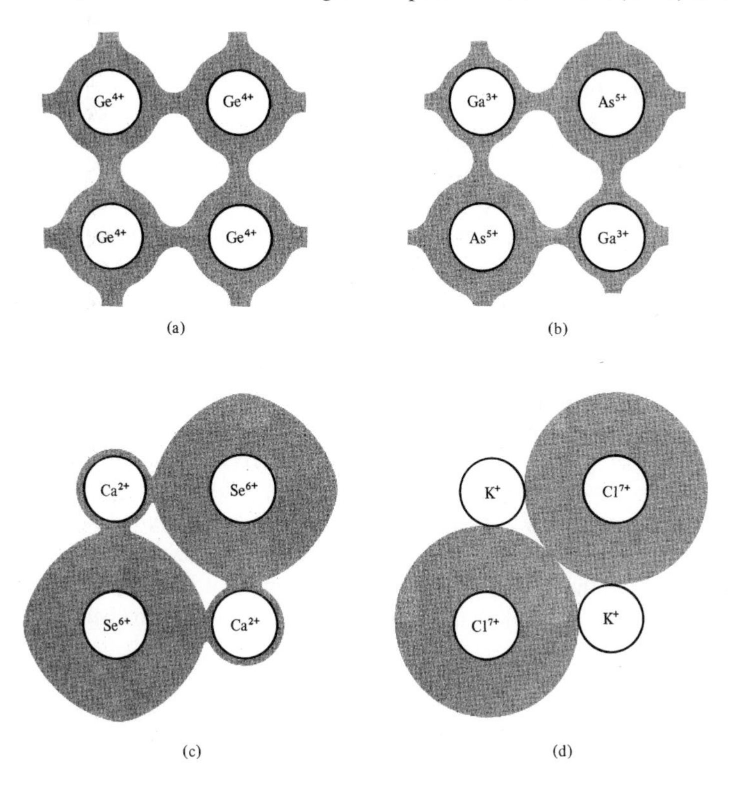


**Figure 1.8:** Schematic representation of the path between a perfectly covalent and a perfectly ionic crystal. a) Perfectly covalent Ge crystal: in this case 4 electrons are identically distributed about the $Ge^{4+}$ cores; the electronic density is large along the interstitial directions; b) Covalent GaAs crystal: again density along the interstitial directions, but the electrons gather more around $As^{5+}$ than around $Ga^{3+}$, which gives a slightly ionic character to the system. c) Even more ionic CaSe crystal: the trend of GaAs is even enhanced in that the $Ca^{2+}$ cores are almost denuded of electrons, to the benefit of the $Se^{6+}$ cores; still, the crystal keeps a slight covalent character. d) Perfectly covalent ionic KCl crystal: electrons have been fully stripped from the K and transferred to the $Cl^{7+}$ cores. The electronic structures of the atoms considered here are: Germanium (Ge, (Ar)$3d^{10}4S^{2}4p^{2}$), Gallium (Ga, (Ar)$3d^{10}4S^{2}4p$), Arsenic (As, (Ar)$3d^{10}4S^{2}4p^{3}$), Selenium (Se, (Ar)$3d^{10}4S^{2}4p^{4}$), Calcium (Ca, (Ar)$4s^{2}$), Chlorine (Cl, (Ne)$3s^{2}3p^{5}$) and Potassium (K, (Ar)$4s^{2}$). From [AM76].

It is clear from the above brief discussion that the solid state physics classification of bonding, not surprisingly, matches the classification of molecular physics. The analogies go even further. It was already discussed for molecules in Section 1.1.2 that the four types of bonding are idealizations and that reality often resides in between. Figure 1.8 complements that for crystals showing the path from covalent to ionic binding. Starting from the covalent Germanium crystal one observes, by associating successively atoms on both (complementing) sides of Ge, a continuous transition towards the ionic KCl crystal. As we shall see below, clusters provide here an even more interesting option. They allow one to span a continuous path from atom to bulk. And one finds situations where the nature of bonding changes with system size.

#### 1.1.3.2 Band theory in solids

A general theory of bonding in solids can be found in many specialized textbooks [AM76, Pet95]. We give here a brief summary of level structure in solids, and discuss how such a level structure is related to the nature of bonding. We start with a simple model crystal in 1D. The ionic lattice introduces an attractive potential $U(x)$ which alters the otherwise free motion of electrons and turns out to be responsible for the level structure. Taking a mesh size $a$, lattice periodicity implies a specific form for electron orbitals

$$U(x+na) = U(x) \quad \Longrightarrow \quad \varphi_k(x+na) = \exp(ikna)\varphi_k(x) \tag{1.11}$$

where $n$ means any integer and $k$ is the so-called lattice momentum. As is clear from this form of the wavefunction (Bloch ansatz), it is sufficient to consider values of $k$ in the first Brillouin zone defined by $-\pi/a \leq k \leq \pi/a$. Other values only repeat phase factors $\exp(ikna)$ from the first zone.

The problem thus amounts to solving the Schrödinger equation for $\varphi_k$ in an elementary cell $0 \leq x \leq a$ with the above defined periodic boundary conditions at $x = 0$ and $x = a$ (Bloch condition). For the sake of simplicity we assume that $U(x)$ is furthermore a square well potential of depth $U_0$ and width $a - b$, so that $b$ also represents the barrier width between two neighboring wells, i.e.

$$U(x) = \begin{cases} -U_0 & \text{for} \quad \left|\xi - \frac{a}{2}\right| \leq \frac{a-b}{2} \\ 0 & \text{for} \quad \left|\xi - \frac{a}{2}\right| > \frac{a-b}{2} \end{cases} \quad , \quad \xi = \mod(x, a) \quad . \tag{1.12}$$

Inside the well, as is well known, the wavefunction takes an oscillatory form $\varphi_k(x)$ =- $A\exp(iKx)$ $+B\exp(-iKx)$ where $K$ is directly associated to the level energy $E$ as $K = \sqrt{2mE/\hbar^2}$. Outside the well (or inside the barrier) the wavefunction takes an exponential form $\varphi_k(x) = A\exp(\kappa x) + B\exp(-\kappa x)$ with ($\kappa = \sqrt{2m(U_0 - E)/\hbar^2}$). The coefficients $A, B, C, D$ are found by matching $\varphi_k$ and $d\varphi_k/dx$ at boundaries $x = b$ and $x = a$ and imposing the Bloch periodic condition Eq. (1.11) on $\varphi_k$. This provides an implicit equation linking $K, \kappa, k, b$ and $a$. For a qualitative discussion, we consider here the simple case of vanishingly narrow potential barrier $b \to 0$ and increasing barrier height $U_0 \to \infty$ (within imposing constant $U_0 b$). The implicit equation then reduces to the simple form

$$\cos Ka + \mu\frac{\sin Ka}{Ka} = \cos ka \qquad \text{with} \qquad \mu = \frac{ma}{\hbar^2}U_0 b \tag{1.13}$$

The above equation links the electron energy $E = \hbar^2 K^2/2m$ to the so called Bloch vector $k$. Bloch-like solutions $\varphi_k$ therefore exist only for values of $K$, and hence $E$, which are solutions of Eq. (1.13) for a given $k$. Whatever $k$ may be, the r.h.s. of Eq. (1.13) is bound by $\pm 1$. The allowed values of $K$, and hence of electronic energies $E$, thus obey the constraint

$$|\cos Ka + \mu\frac{\sin Ka}{Ka}| \leq 1 \quad . \tag{1.14}$$

Except for the trivial case $\mu = 0$ which corresponds to the exact free electron gas (no ions, no potential), condition Eq. (1.14) provides the so called band structure of the electronic level scheme. Only certain bands of values for $K$ (and $E$) are possible. The width of the bands depends on the effective strength $\mu$.

**Table 1.2:** Lowest three bands (= allowed intervals) of $K$ and $E$ values in the simple 1D model for two values of $\mu$. The low $\mu = 3$ (upper part) provides good conditions for a conductor. The large $\mu = 40$ (lower part) is an example for an insulator. The $Ka$ are dimensionless. The translation into energies $E$ used $a = 6\,\mathrm{a_0}$ for $\mu = 3$ and $a = 8\,\mathrm{a_0}$ for $\mu = 40$. The width was in both cases $b = a/2$.

| | $\mu = 3 \longleftrightarrow$ conductor | | |
|---|---|---|---|
| $Ka$ | [2.03,3.13] | [4.38,6.28] | [7.13,9.38] |
| $E$ [eV] | [1.56,3.71] | [7.27,14.96] | [19.30,33.40] |
| | $\mu = 40 \longleftrightarrow$ insulator | | |
| $Ka$ | [3.03,3.13] | [6.03,6.28] | [9.03,9.38] |
| $E$ [eV] | [1.94,2.87] | [7.71,8.37] | [17.31,18.68] |

Two typical examples, for a small one and a large value of $\mu$, are given in Table 1.2. To translate the bands and gaps into energies, we have specified the lattice parameter $a$ in each case.

A graphical view of the potentials, lattices, and bands for these cases is given in Figure 1.9. There appear forbidden energy ranges which correspond to gaps between the allowed energy bands. A gap structure, as exemplified in the above simple model, is a generic feature of level structure in solids. The details of band structures and gap sizes turn out to vary with the type of bonding in the considered sample. It can be discussed qualitatively within this simple model by a mere variation of the key strength parameter $\mu$. The free electron gas limit is recovered for $\mu = 0$, in which case no gap appears and the electronic energy is of purely kinetic nature and can take any value. As soon as an interaction is introduced in the problem, forbidden energy regions appear, which means that, in spite of the fact that the system is infinite and the $k$ index continuous in the first Brillouin zone, only bands of well defined energies are accessible and gaps reside in between.

Small values of $\mu$ correspond to the case of a nearly free electron gas and represent metallic systems. In that case, the energy gaps are small. And the valence band, namely the band with the HOMO is only partially occupied, see left part of Figure 1.9. In a metal, the Fermi energy $\varepsilon_\mathrm{F}$ lies in the middle of an allowed band. Any vanishingly small excitation of the electrons in this band thus allows them to be easily promoted from below the Fermi energy to one of the closely lying empty states. This is the reason why electrons can so easily move in a metal with virtually no energy consumption. This also explains the electric conductivity and the metallic character of such a system.

Larger values of $\mu$ correspond to systems in which the electrons remain more localized (large value of the potential depth or the barrier width) and can thus exhibit successively covalent and ionic bonding (see also Figure 1.8). The behavior is schematically illustrated

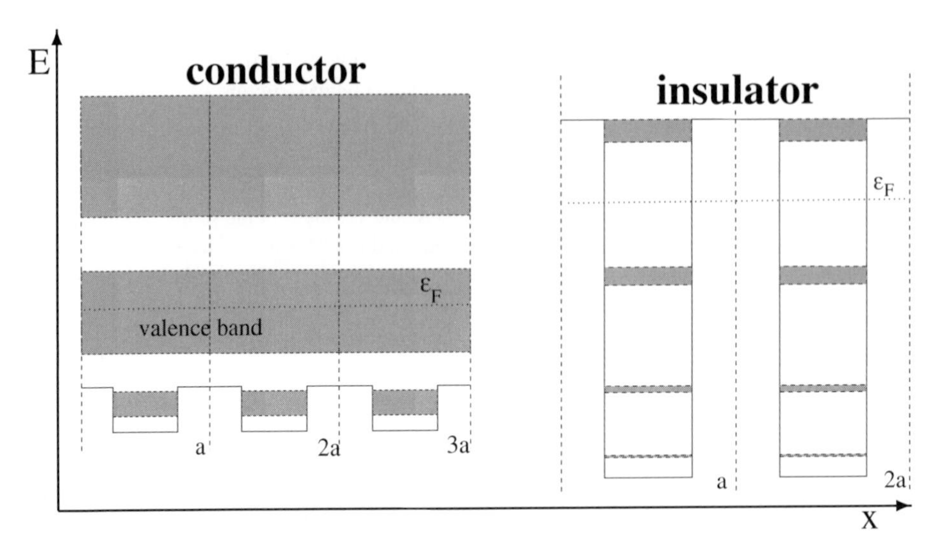


**Figure 1.9:** Schematic illustration of the band structure in a metal (left) and an insulator (right) according to the simple model discussed in the text (1D crystal with square potential wells). The allowed bands correspond to the shaded areas. Occupied levels are below the Fermi energy $\varepsilon_F$. Note that $\varepsilon_F$ lies in the middle of an allowed band in the case of a metal, while it falls into a gap between two bands for an insulator.

in the right part of Figure 1.9 for a typical insulator (as obtained from the above schematic model). The Fermi energy $\varepsilon_F$ now resides within a band gap. Promoting electrons from the highest occupied to the lowest unoccupied band now requires a sizable amount of energy. Conductivity in such systems is thus quite small, which explains the insulating character of the material.

Finally, the simple schematic model allows one to recover the case of molecular bonding in the limit of $\mu \to \infty$ which also corresponds to the case of isolated atoms (at least at the level of this schematic picture).

The metallic or insulating character of a material can thus be read off from the position of its Fermi energy with respect to bands and gaps. In a metal, $\varepsilon_F$ lies in the midst of the valence band which is only partially filled. In an insulator, it lies in between occupied bands, in other words in the middle of a gap. Of course, these two cases are extremes and there exists a bunch of intermediate situations in which the Fermi energy lies closer or farther from the edge of the bands. And this proximity to the band edge correlates with the more or less metallic nature of the system. A relevant measure thus turns out to be the degree of filling of the band in which the Fermi energy lies.

## 1.2 Clusters between atom and bulk

Now that we have discussed a few basic properties of atoms and bonding in simple molecules and in bulk, we are coming to clusters as objects which establish a link between all these cases. We try to work out their specificity and to explain why they are more than just large molecules or small pieces of bulk.

## 1.2.1 Clusters as scalable finite objects

Clusters are, by definition, aggregates of atoms or molecules with regular and arbitrarily scalable repetition of a basic building block. Their size is intermediate between atoms and bulk. One could thus loosely characterize them with a formula of the form $X_n$ ($3 \lesssim n \lesssim 10^{5-7}$). The actual upper limit in size is hard to fix. As we shall see below, the definition of bulk may vary from one observable to another, namely a given physical property may reach its bulk value for a size different from another observable.

### 1.2.1.1 Clusters are more than molecules

Molecules usually have a well defined composition and structure. Think, e.g., of the well known molecules such as $C_6H_6$. Such systems have only a small number of isomers, a few units at most, even if they are themselves not small molecules. This is different for clusters, which often possess a large number of energetically close isomers and in which the number of isomers grows huge with increasing cluster size. For example, in such a small cluster as $Ar_{13}$, one has found hundreds of isomers, the actual number slightly depending on the detail of the interatomic potential used in the calculations, while less than 10 isomers were identified in $Ar_8$ [DJB87]. Another example are metal clusters which are also swamped by isomers, because of the softness of the binding.

When facing such a huge number of isomers, it is obvious that it is very hard to simply assess which is the most stable structure. This holds the more so as clusters are formed at finite temperature where it may be a delicate task to precisely tune the actual temperature of the formed clusters (see Section 2.1). Such a finite temperature allows a given cluster to explore a huge variety of isomers (=shapes), by simple thermal activation. And this process is all the more important as the number of accessible isomers is large. Stated in another, more technical, way, the potential energy surface (Born-Oppenheimer surface as discussed in Section 3.4.2) is very flat and this makes it very hard to figure out the actual ground-state structure of the system. Standard quantum chemistry techniques, well adapted to molecules with few isomers, are here often at a loss. Note finally that the difficulty in identifying the actual ground state structure constitutes an example of a more general feature, namely that many cluster properties do strongly depend on cluster size, see e.g. the example of shapes in Section 4.1.3.

### 1.2.1.2 Clusters are more than finite pieces of bulk

There are several features which distinguish a small piece of bulk from bulk [Ber98]. The basic difference between bulk and clusters can be seen from the level structure. Finite systems are basically characterized by discrete levels, at least in the low energy part of their spectrum. As is well known (see also Section 1.1.3.2), single-electron spectra in solids come along in bands each containing a continuum of levels. Passing from a discrete to a continuous set of levels is, in fact, also a continuous process. We have seen in Section 1.1.2.3 that the binding

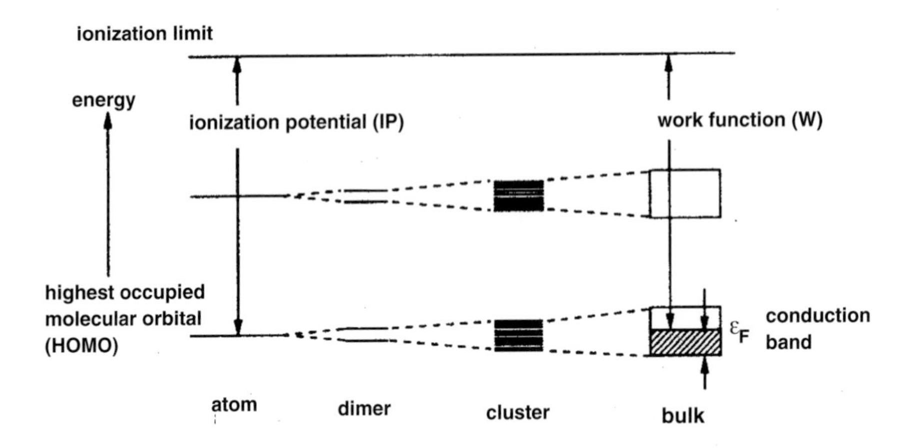


**Figure 1.10:** Schematic evolution of single-electron spectra in sodium clusters. After [KV93].

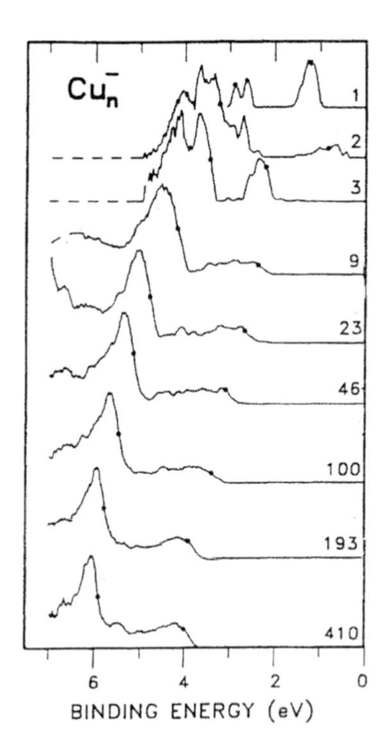


**Figure 1.11:** Evolution of single-electron spectra with system size for a series of $Cu_n^-$ cluster anions with increasing size. The spectra are deduced from photo-electron spectroscopy. The upper atomic level is the $4s$ state and the lower one the $3d$, see also Figure 3.2. One can track down how the corresponding bands develop out of the atomic states. The points in the spectra for larger clusters indicate the upper band edge. After [CTCS90].

of two atoms, each bringing one electronic level, gives rise to two separate electronic levels (bonding and antibonding states). When more atoms are brought together, this splitting process continues and produces more and more levels which turn out to be closer and closer to each other, the larger the number of involved atoms. There is a thus a continuous path between a fully discrete molecular level scheme and a continuous (although with occasional gaps) level

**Table 1.3:** Schematic classification of clusters according to the number $N$ of atoms. As a complement the diameter $d$ for Na clusters is given (second row), together with an estimate of the ratio of surface to volume atoms $f$ (surface fraction, in third row). After [KV93].

| Observable | Very small clusters | Small clusters | Large clusters |
|---|---|---|---|
| Number of atoms $N$ | $2 \leq N \leq 20$ | $20 \leq N \leq 500$ | $500 \leq N \leq 10^7$ |
| Diameter $d$ | $d \leq 1$ nm | 1 nm$\leq d \leq 3$ nm | 3 nm $\leq d \leq 100$ nm |
| Surface fraction $f$ | undefined | $0.9 \gtrsim f \gtrsim 0.5$ | $f \leq 0.5$ |

scheme in bulk. Such a path is illustrated in Figures 1.10 and 1.11. Figure 1.10 provides a schematic picture of this complexification starting from a simple dimer system (here $Na_2$) and going towards larger and larger clusters. The second figure (Figure 1.11) shows the evolution of experimental single electron spectra in Cu cluster anions. The figure nicely shows the evolution from atomic discrete levels to band-like pattern for heavier clusters. The levels had been recorded experimentally by photo-electron spectroscopy [CTCS90] (for photo-electron spectroscopy see also Figure 5.1 and Sections 2.2.3 and 2.3.8 as well as 5.2.3).

Another aspect where clusters are distinguished from bulk (and from molecules) are finite size effects in terms of surface to volume ratio. These features are much simpler to quantify than details of level structure. Most clusters are large enough such that one can speak about a surface zone. On the other hand, they are sufficiently small such that a sizeable fraction of the constituents lie on the surface of the system, while this is not true in bulk. Let us take as example the $Ar_{55}$ cluster. It has 32 atoms on its surface. Now consider a piece of bulk material of 1 mm$^3$ volume. Taking an order of magnitude for the bond of 1 Åtells us that there are about $(10^{-3}/10^{-10})^3 \sim 10^{18}$ atoms in the sample, out of which about $(10^{18})^{2/3} \sim 10^{12}$ lie on the surface of the system. The ratio of surface atoms to total is thus $32/55 \sim 0.6$ in $Ar_{55}$ as compared to about $10^{-6}$ in the bulk sample. Even taking a smaller sample of micrometer size the ratio would only increase to about $10^{-4}$. There is thus a huge difference in the fraction of surface atoms when passing from clusters to bulk. And it turns out that this ratio constitutes a simple and efficient way to classify clusters. For an example see Figure 4.13.

#### 1.2.1.3 Small and large clusters

As is obvious from the above discussion, clusters interpolate between atoms/molecules and bulk. But they are more than just trivial emanations of these two extremes cases. Of course, very small clusters look like molecules, though, and can thus be studied with chemistry methods, while very large clusters can be attacked with techniques stemming from solid state physics. In between, more specific techniques have to be developed for accessing clusters, as we shall see in this book. Already here, we can note that size is obviously a key parameter in cluster physics. Following the argument developed just above, a classification in terms of size can be given by considering the fraction of surface atoms to volume. Results are summarized in Table 1.3. Of course, this classification is schematic, and there is no strict boundary between the various classes of clusters. But it serves to sort the sizes and to give meaning to the terms small or large clusters in the following.

The question of size effects will be recurrent throughout this book, first, because clusters interpolate between atom and bulk with largely variable size, and second, because it turns out that most cluster properties do indeed strongly depend on size. But as already stressed above, convergence towards the bulk value with increasing cluster size essentially depends on the nature of the observable and on the resolution with which one looks at it. The best explanation is to consider a few examples. The experimental resolution of photo-electron spectra puts the transition from well separated discrete electron levels to quasi-continuous bands at about $N \approx 100$. Electronic shell effects, as e.g. magic HOMO-LUMO gaps, shrink $\propto N^{-1/3}$. And yet, their importance has triggered large efforts to resolve these up to the range of $N \approx 3000$, see Section 4.2.1.1. Atomic shell effects have been resolved up to $N \approx 10000$, see Section 4.2.1.2. The peak frequency of optical response converges also towards its bulk value with a term linear in $N^{-1/3}$. This means that colors keep drifting with size up to very large clusters in the range $> 10000$, see Figure 1.17 in Section 1.3.3.2. The cluster radius reaches the wavelength of visible light typically around $N \approx 10^9$, while the treatment of photo-excitation in the limit of long wavelengths is to be questioned much earlier. Many other examples will be found in this book.

### 1.2.2 Varying cluster material

There are, of course, various types of clusters in terms of bonding, depending on the nature of the atoms entering the cluster. As we have seen above, simple molecules as well as bulk can be grouped into the same four classes of bonds: metallic, covalent, ionic, and van-der-Waals bonding. It is natural then to classify bonding in clusters according to these four classes. And this turns out to constitute an important and relevant means for the classification of clusters. It also reflects deeper physical processes, even if these processes may sometimes differ from the ones observed either in molecules or in the bulk. The case of metalicity and its definition in terms of electron delocalization is a typical example of such a much debated criterion.

The first figure of this book, namely Figure 1.1, shows the famous $C_{60}$ fullerene which provides a beautiful example of covalent bonding in clusters. In that case the cluster exhibits a well defined structure with electrons localized along the various links between the atoms. As expected, and as observed, other clusters made out of atoms of the carbon group exhibit covalent bonding as well, consider e.g., the case of the silicon clusters [Sug98].

Van-der-Waals binding prevails in rare-gas clusters, as e.g. for Ar. Figure 1.12 shows the geometry of $Ar_{561}$. This cluster corresponds to a closed atomic shell (see Section 4.2.1). The material keeps the electrons tightly bound to each mother atom. One can describe the system by effective atom–atom potentials (see Section 3.4.5.4) which are fairly simple to use. The structure in the figure has been computed with such an approach. Similar calculations have been extensively performed, because of the simplicity of the interatomic potential. There exist also many dynamical studies using molecular dynamics simulations of such rare-gas clusters, for example to study temperature effects [DJB87, JBB86].

Clusters with ionic bonding have been in the focus of many studies over a few years, in particular in view of potential applications in photography (e.g. AgBr clusters). The example shown in Figure 1.13 corresponds to a structurally simpler case, as it associates "ideal" partners (alkaline + halogen) in an almost stoichiometric manner. In such $Na_nF_{n-1}$ structures, one Na electron is left "free" without a hosting F atom. It has thus to find a proper location in

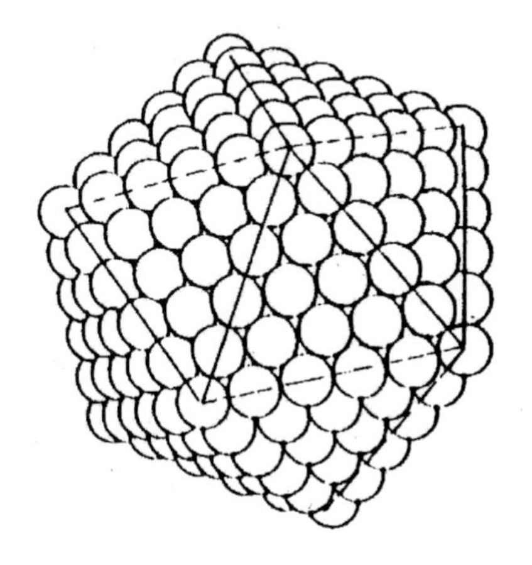

**Figure 1.12:** Geometrical configuration of the $Ar_{561}$ cluster computed by simulated annealing (Section 3.4.3.2 and Appendix H.3) with effective atom–atom potentials. After [Hab94].

the cluster. Systematic studies in $Na_nF_{n-1}$ clusters have shown that there basically exist four types of ionic structures corresponding to four different localizations of the excess electron provided by the excess Na atom. As illustrated in Figure 1.13, the regular NaF ionic structure clearly shines through, whatever the detail of the arrangement and the location of the excess electron. This regular ionic structure already appears in very small clusters such as $Na_5F_4$. In that particular case the excess electron remains localized around the "isolated" sodium atom, but there are other cases in which the electron may spread over a larger fraction of the structure such as for example in $Na_{14}F_{13}$.

The last class of clusters is made out of atoms which exhibit metallic bonding. The simplest example is here the case of alkaline atoms for which the metallic behavior is well assessed and simple to work out. Figure 1.14 shows the $Na_4$ cluster. It has a planar geometry which makes plotting and viewing simpler. Both electron density and ionic positions are explicitly represented. As expected in a metallic system, the electronic density extends more or less smoothly over the whole system. The small $Na_4$ cluster exhibits a clearly oblate shape. Correlatively, the electron cloud is strongly oblate, with a shape very similar to the ionic shape. The shape is determined here by electronic shell structure as discussed in Section 4.2.2.2. Larger metal clusters tend to favor spherical shapes due to the surface tension of the electron cloud. Crystalline shapes may emerge at very low temperature and/or for non-simple metals [Mar93]. In any case, the electrons have a large mobility and behave almost like an electron gas in a (spherical) container. Large metal clusters thus provide an ideal basis for the realization of the old Mie idea on the optical response of metallic spheres. This will be outlined below in Section 1.3.3. But before focusing on optical response of metal clusters let us briefly summarize our conclusions on the different types of binding.

A summary of the four classes of bonding and some of their consequences for cluster properties is presented in Table 1.4. It continues the energetic considerations of Section 1.1.2.4 on the dimer molecules. We recall the word of caution from Section 1.1 that such sorting schemes rely on idealizations and that reality falls often in between the categories. With that in mind, we can benefit from the table. We see that the least bound clusters are van-der-Waals (or molecular) clusters. They can be considered as a collection of weakly interacting atoms. The

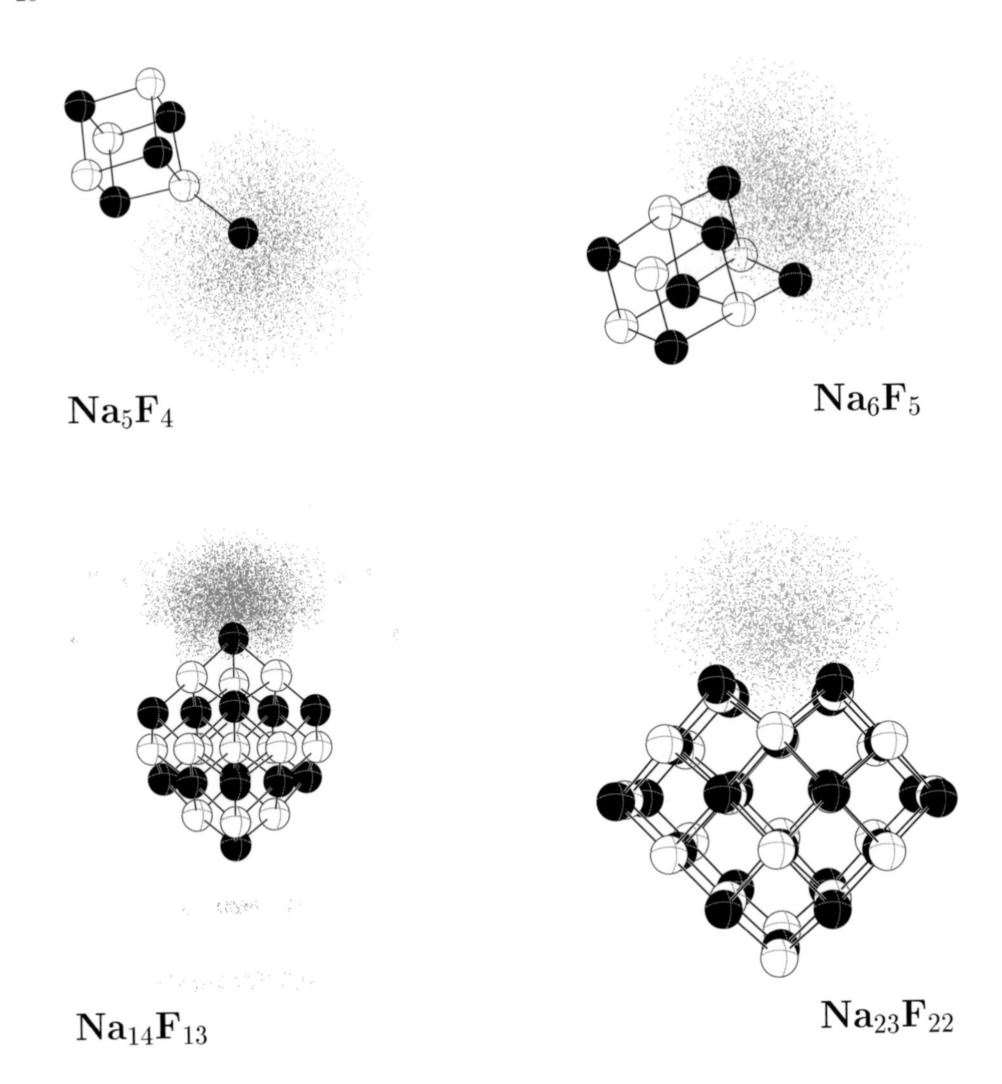


**Figure 1.13:** Typical structures of small $Na_nF_{n-1}$ clusters. Na atoms are denoted by black spheres, F atoms by white spheres. Dots represent the electronic cloud. Note that there is only one "free" electron in all these clusters. Note also the various localizations of the electron depending on the numbers of Na and F atoms. For details see text. After [DGGM+99].

strongest binding, on the contrary, usually appears in ionic clusters, although covalent clusters may bind almost as strongly. Metal clusters are generally a bit softer than covalent ones. They thus constitute an intermediate class between the almost unbound molecular clusters and the tightly bound ionic (or covalent) clusters.

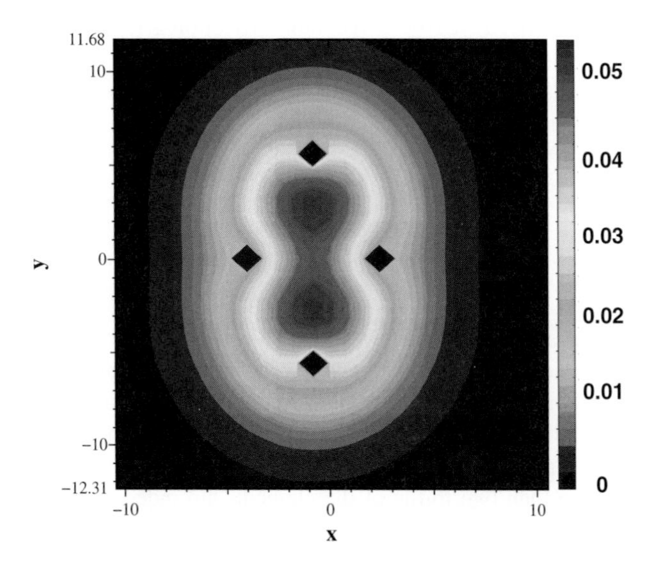


**Figure 1.14:** Equidensity plot of the electronic density of $Na_4$ computed with LDA (see Section 3.3.3). Plotted is the density in the $x$-$y$ plane integrated over all $z$. The ionic positions are marked by small diamonds. The $x$ and $y$ scales are in units of $a_0$.

**Table 1.4:** Classification of binding in clusters. For each one of the four types of binding, examples of clusters (second column), the nature (third column) and typical binding energies (last column) are given. After [Hab94].

| Type | Examples | Nature of binding | Binding energy |
|---|---|---|---|
| Ionic clusters | $(NaCl)_n$, $Na_nF_{n-1}$... | Ionic bonds | |
| | | Strong binding | $\sim 2-4$ eV |
| Covalent clusters | $C_{60}$, $S_n$ ... | Covalent bonding | |
| | | Strong binding | $\sim 1-4$ eV |
| Metal clusters | $Na_n$, $Al_n$, $Ag_n$ ... | Metallic bond | |
| | | Moderate to strong binding | $\sim 0.5-3$ eV |
| van der Waals | Rare gas clusters | Polarization effects | |
| | $Ar_n$, $Xe_n$, ... | Weak binding | $\lesssim 0.3$ eV |

Whatever cluster type, we see that binding energies lie in the eV range. Elementary bond lengths typically take values of a few $a_0$. Both these quantities thus fix the range of energies and distances characteristic of cluster physics. As complementing information, it is also interesting to evaluate a typical force in a cluster, in order to compare it to the intensity of an external field probing the system (laser or colliding projectile). Let us consider metal clusters, which exhibit a binding of intermediate strength. We take the example of a small alkaline cluster, $Na_9^+$, which we shall often use later on in applications. Its radius is of the order $R \sim 8a_0$

and we can estimate the typical electric field at surface as $E_0 \sim e^2/R^2 \sim 3.4\,\mathrm{eV/a_0}$. Such an electric field corresponds to a laser intensity $I \approx 10^{13-14}\mathrm{W/cm}^{-2}$ (Section 4.1.1.1). That means that non-destructive analysis of clusters by light has to be performed with laser intensities much below this value. In turn, with laser intensities of that order or larger, one will most probably destroy the cluster. For a more detailed view see Figure 3.3.

# 1.3 Metal clusters

## 1.3.1 Some specific properties

Metal clusters play a particularly pronounced role in cluster physics. They have attracted many investigators and the number of publications on that topic stays above average as compared to other materials. They will also appear quite often as examples throughout this book. This importance of metals is not the sole privilege of cluster physics. Following the introductory remarks of a famous book in solid state physics [AM76], one may recall that nature seems to have a "preference" for metals. More than two thirds of the elements are indeed metals. Furthermore, the understanding of metallic behavior turns out to constitute a key issue to understand the properties of solids, both conductors and insulators. Finally, as already pointed out before, the characteristic scalability of clusters finds particularly nice applications in the case of metal clusters. Electrons do form there a quasi free gas with generic properties, which lead to reasonably well understood behaviors with respect to size.

The features which make metals so different is the long mean free path of the valence electrons and the low melting point, ideally realized in alkalines and here at its best for Na. The nearly free propagation of the electrons throughout the cluster constitutes a generic test laboratory for the physics of an electron cloud, to be more precise, for a finite many-fermion system in the degenerate regime. The ionic background serves to deliver the finite bounds. Details of ionic structure can be easily wiped out, if necessary, by handling the cluster above the melting point. In other words, metal clusters are a near to perfect realization of a quantum system with self-adjusting bounds and shapes. One prominent consequence of the mobile electron cloud is a strong coupling to light with pronounced resonant behavior, the well known Mie surface plasmon resonance. We shall come back to this point at many places in this book. A first discussion in very simple terms is given in Section 1.3.3. The quantum mechanics of a confined electron cloud is dominated by shell effects. Metal clusters with their arbitrarily scalable size have provided here for the first time a laboratory to detect quantum mechanical shell effects up to system sizes of several thousands of fermions and with it shell beating and super-shells, see Section 4.2.1.1. The self-adjusting shape, as opposed to the fixed shapes of a quantum dot in a substrate, also delivers brilliant examples for the Jahn-Teller effect, as will be discussed in Section 4.2.2.2.

For pedagogical reasons we will thus use at many places throughout this book results obtained with metal clusters. This is motivated by two reasons. First, it does reflect the relative importance of metal clusters in experimental and theoretical investigations. Second, referring to the same test case at different places makes the links between the various approaches and applications more transparent. This makes the whole presentation more coherent and self contained. This can be seen, e.g., with respect to time scales. They will obviously play a

| | Li | Na | K | Rb | Cs |
|---|---|---|---|---|---|
| $r_s$ [a$_0$] | 3.3 | 4.0 | 4.9 | 5.2 | 5.6 |
| $\varepsilon_{\mathrm{F}}$ [eV] | 4.7 | 3.2 | 2.1 | 1.9 | 1.6 |
| $\frac{r_s}{v_{\mathrm{F}}}$ [fs] | 0.13 | 0.20 | 0.30 | 0.35 | 0.40 |

**Table 1.5:** Gross properties of alkaline systems: Wigner Seitz radius $r_s$, Fermi energy $\varepsilon_F$ and microscopic time scale $r_s/v_F$.

central role in the discussions of cluster dynamics. It is thus useful to have one case where one exemplifies in detail the various competing times. We will do this for alkaline metals.

### 1.3.2 On time scales

Before discussing the various time scales, we recall in table 1.5 the basic bulk parameters of alkaline metals. The different alkalines have the same order of magnitude within each observable. Changes stay within a factor of 2 to 3 and have monotonous trends with element number. The bulk parameters serve as "natural units" of length, energy and time. We will see that all electronic properties of alkaline clusters are about the same when expressed in terms of these natural units. The case is obvious for lengths where the Wigner-Seitz radius $r_s$ sets the scale. For energies, we chose the Fermi energy $\varepsilon_{\mathrm{F}}$ which is related to the basic material parameters in a simple Fermi gas picture as (see also Appendix F.1)

$$\varepsilon_{\mathrm{F}} = \frac{\hbar^2}{2m} k_F^2 = \frac{\hbar^2}{2m} (\frac{9\pi}{4})^{2/3} \frac{1}{r_s^2} \quad . \tag{1.15}$$

The Fermi momentum $k_F = m v_F/\hbar$ of the system is directly linked to the Wigner Seitz radius of the material through the density $\rho = 3/(4\pi r_s^3) = k_F^3/(3\pi^2)$. The unit for time is chosen as $r_s/v_F$ which is the time in which an electron at the Fermi surface travels through a distance $r_s$. The Fermi velocity can be deduced from the Fermi momentum and simply reads

$$v_F = \frac{\hbar}{m} k_F = \frac{\hbar}{m} (\frac{9\pi}{4})^{1/3} \frac{1}{r_s} \tag{1.16}$$

which fully defines our time unit in terms of $r_s$.

Various dynamical processes compete in metal clusters, both from the electronic and from the ionic side. A summary of typical time scales is given in Figure 1.15. Let us first concentrate on electrons. A key time scale is set by the all dominating Mie surface plasmon oscillations, in which the electron cloud collectively oscillates with respect to the ionic background (see Section 1.3.3). These collective oscillations do not last forever and the electronic motion loses sooner or later its collectivity due to coupling to detailed single-particle motion. This is partially Landau damping (as it is called in plasma physics [LP88], see also Section 4.3.5.1) and partially direct electron emission. The time scale for both processes is about the same and indicated by "s.p. times" in the figure. This initial damping generates "turbulence" in the electron cloud which, in turn, activates electron–electron collisions, which adds further damping and associated internal heating. The thermal energy of the electron cloud can then be transferred to ionic thermalization or released much later in terms of electron evaporation. The excited electron cloud also shakes the ions which react, of course, somewhat slower due to

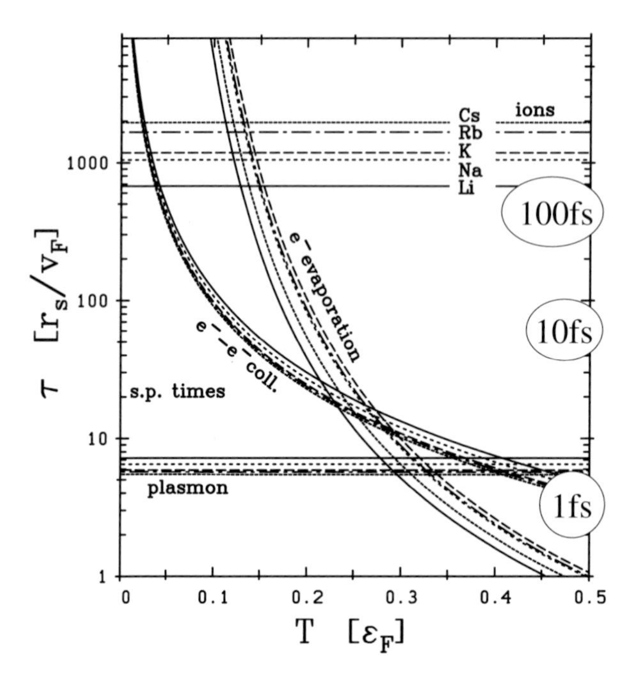


**Figure 1.15:** Time scales for alkaline clusters drawn versus internal excitation in terms of temperature $T$. Times and energies are expressed in natural units of each material according to Table 1.5. Orders of magnitude of the times in fs are also indicated for completeness on the right part of the figure.

their comparatively large mass. A typical time scale is here set by the cycle of ionic vibrations. First effects of ionic motion can already be spotted during the first quarter cycle. Subsequent ionic processes like fragmentation or monomer evaporation usually take much longer (several ionic cycles) but become faster with increasing violence of the excitation, as for example in a Coulomb explosion (see Section 5.5).

The times scales change with the internal state of the cluster. In Figure 1.15 we use temperature as a simple indicator of the excitation regime (keeping in mind that times may depend on the particular excitation mechanism in the early, transient stages of a dynamical process). Temperature has the additional advantage that it is an intensive quantity, independent of system size (unlike an extensive quantity as the internal excitation energy). Figure 1.15 thus shows the various time scales as a function of temperature. Again, we take the Fermi energy $\varepsilon_F$ as a unit for temperatures (the Boltzmann constant $k$ is set to one here). The trends are much different from one time to another. That means the relation between times changes dramatically with the energy of the excitation. For example, electron evaporation is an extremely slow process for low excitations. But it becomes competitive with other electronic mechanisms at high excitations. Electron–electron collisions also depend strongly on temperature and are unimportant for low excitations. The plasmon and single-particle cycles, on the other hand, stay rather robust.

Figure 1.15 also demonstrates that all electronic time scales from the different materials gather close together when expressed in the natural units of the electron cloud. The ionic time scales, on the other hand, are dominated by the independent parameter of ionic mass and thus show a larger spread amongst the materials. But ionic time scales group themselves in a time range well separated from basic electronic scales, as expected in view of the large mass difference between ions and electrons.

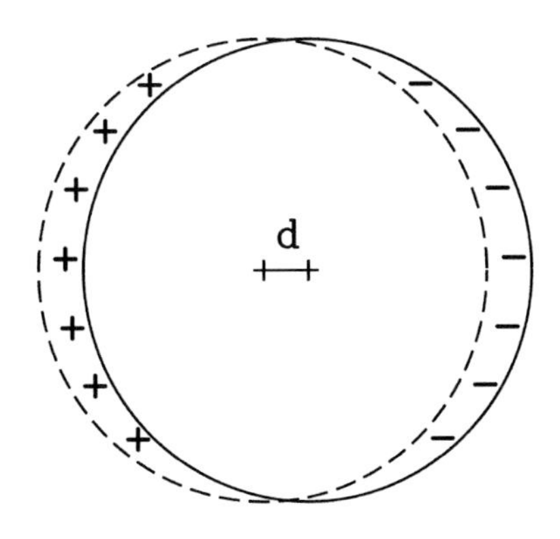


**Figure 1.16:** Schematic representation of Mie surface plasmon. The electrons (represented by a negatively charged spherical cloud) are collectively displaced with respect to the ionic background (represented by a positively charged sphere). Inside most of the system, the original neutrality of the system is preserved. Net charges only appear on the surface of the system, a positive one on one side and a corresponding negative one on the opposite side.

In a dynamical context, the internal time scales of a cluster are to be complemented by the time scale of the excitation processes. Nanosecond lasers are beyond any time shown in Figure 1.15. That is a regime where the frequency hence plays the dominant role. Collisions with highly charged and fast ions may be below any time scale shown (typically in the fs or sub-fs range). Fast ions thus cover the opposite regime where frequencies are unimportant and only forces count, a domain which is also explored by intense laser beams (here for simple reasons of power). Slower ions may be associated with time scales of tens of fs or even more. Similarly, femtosecond (fs) lasers have just the time scales which interfere with the various time scales of metal clusters. It is obvious that this gives rise to a huge variety of accessible dynamical processes which can be triggered by fs laser experiments. We shall come back to these various excitation processes on many occurrences in this book.

## 1.3.3 Optical properties

### 1.3.3.1 The simple Mie picture

As suggested in Section 1.2.2, metal clusters can be viewed as metallic spheres filled with an almost uniform electron gas not too tightly bound to the ionic background. No surprise that a moderate external electric field may easily affect such clusters whose response then provides a fingerprint of crucial cluster properties. This is just the essence of Mie theory of response of a metallic sphere to light [Mie08], for extensive explanations see e.g. [KV93]. Let us briefly outline this simple but generic picture.

We start with an oversimplified model of a spherical metal cluster $X_N$. The ionic background is represented by a homogeneous positively charged sphere with radius $R$ and density $\rho$. Electrons are associated to a negatively charged sphere with the same density $\rho$ such that the total system is neutral in its ground state. For the sake of simplicity, we shall furthermore assume that both electrons and ions move as rigid spheres against each other. In fact, as long as the system is not too much perturbed, electrons do indeed primarily respond in a collective and almost rigid way. Let us now assume that a small uniform external electric field $\mathbf{E}_{\text{ext}}$ is applied to this system. The net effect of $\mathbf{E}_{\text{ext}}$ is to separate slightly electrons from ions. Because ions are much heavier than electrons, they basically remain fixed and the electrons are displaced with respect to ions, as illustrated in Figure 1.16. The separation builds up a strong Coulomb attraction between ions and electrons which provides a restoring force on the displaced electron cloud counteracting $e\mathbf{E}_{\text{ext}}$. Denoting by $\mathbf{d}$ the actual separation of ionic and electronic center of masses the net force acting on the electron cloud (along the direction of $\mathbf{E}_{\text{ext}}$) reads for small $\mathbf{d}$

$$\mathbf{F} = -N\frac{4\pi\rho e^2}{3}\mathbf{d} = N\mathbf{E}_{\text{ext}} \quad . \tag{1.17}$$

Once the external perturbation is switched off, there only remains the attractive force between electrons and ions. This force is proportional to the actual separation $d$ between ions and electrons. The mobile part of the system is the electron cloud having total mass $Nm_{\text{el}}$. It will thus undergo harmonic oscillations around ions, with a frequency

$$\omega_{\text{Mie}}^2 = \frac{4\pi\rho e^2}{3m_{\text{el}}} = \frac{e^2}{m_{\text{el}}r_s^3} \tag{1.18}$$

where $m_{\text{el}}$ is the electron mass. This oscillation frequency is known as the Mie frequency.

When a cluster is irradiated by light the situation is about the same as described above. Optical wavelengths range in the micrometer domain, which is much larger than most cluster sizes. The electrical field is thus locally uniform at the cluster site and provokes the above described collective oscillation of the electron cloud against the ionic background. The typical values of $\omega_{\text{Mie}}$ for simple alkaline metals lie in the visible part of the electromagnetic spectrum, whence the term "optical response" to characterize this phenomenon. For the alkalines, whose $r_s$ were given in table 1.5, the Mie frequencies are: $\hbar\omega_{\text{Mie}}(\text{Li}) = 4.5$ eV, $\hbar\omega_{\text{Mie}}(\text{Na}) = 3.4$ eV, $\hbar\omega_{\text{Mie}}(\text{K}) = 2.5$ eV, $\hbar\omega_{\text{Mie}}(\text{Rb}) = 2.3$ eV, and $\hbar\omega_{\text{Mie}}(\text{Cs}) = 2.1$ eV. These estimates fit fairly well with experimental values as we will see later. The relations are more involved for non-simple metals as, e.g., noble metals. There one needs to take the polarizability of the core electrons into account (see Section 1.3.3.2 and the discussion in Section 4.3.4.1).

#### 1.3.3.2 The dielectric picture

The explicit picture of an oscillating electron cloud led to the expression (1.18) of the Mie frequency simply in terms of bulk density $\rho$. In more involved materials one needs to account for internal polarizability. This can be done at the level of macroscopic models by employing the bulk dielectric constant of the medium $\epsilon$. Note that dielectric media require one to make a distinction between an externally applied electrical field $\mathbf{E}_{\text{ext}}$ and the effective internal field $\mathbf{E}_{\text{int}}$ which emerges when cumulating the additional fields from internal polarization effects. The external field is directly related to the external sources while the internal field describes the net effect on moving charges, accounting for both external sources and induced charges. Both are related at a macroscopic level through the standard relation

$$\mathbf{E}_{\text{int}}(\mathbf{r},\omega) = \epsilon(\mathbf{r},\omega)\mathbf{E}_{\text{ext}}(\mathbf{r},\omega) \quad . \tag{1.19}$$

This relation is given for the dynamical polarizability $\epsilon(\mathbf{r},\omega)$ which is more easily formulated in terms of frequency $\omega$ of the applied field. Transformation to the time domain yields a polarizability with memory kernel depending on all past times. The space dependence can be omitted in homogeneous materials. But it plays a crucial role at interfaces between different materials.

Clusters are finite systems with a large surface area. Let us thus consider a finite piece, namely a spherical sampling, of the bulk material with given dielectricity $\epsilon$. The response of that dielectric sphere to a homogeneous external electrical field can be worked out with standard techniques of electro-dynamics (see e.g. [Jac62, KV93]). The result for this spherically symmetric system is

$$\mathbf{E}_{\text{int}} = \frac{3}{2+\epsilon}\mathbf{E}_{\text{ext}} \quad . \tag{1.20}$$

The Mie resonance frequency is found at the point where the internal field becomes self-supporting, i.e. where we can have a finite $\mathbf{E}_{\text{int}}$ for zero excitation $\mathbf{E}_{ext}$. This defines the resonance frequency $\omega_{\text{Mie}}$ through

$$2 + \epsilon(\omega_{\text{Mie}}) = 0 \quad . \tag{1.21}$$

That definition holds for a much broader class of metals than the simpler relation (1.18). The price is that one has to know the whole bulk $\epsilon(\omega)$.

In the following, we want to show that the definition (1.21) becomes identical with (1.18) for the case of simple metals. To that end, we have first to develop a simple model for the bulk dielectric constant. Let us thus consider an "infinite" piece of metal irradiated by an external light source. The external electric field polarizes the medium. The effect can be estimated at an elementary level by considering the motion of a single electron subject to the net electric field $\mathbf{E}_{\text{int}}$. The electron follows the Newtonian equation of motion

$$m\frac{d^2\mathbf{r}}{dt^2} = -e\mathbf{E}_{\text{int}} \quad . \tag{1.22}$$

For a sinusoidal electric field of frequency $\omega$, the solution is also sinusoidal with the same frequency. The polarization of the medium then simply reads

$$\mathbf{P} = -\rho e\mathbf{r} = -\frac{\rho e^2}{m\omega^2}\mathbf{E}_{\text{int}} \tag{1.23}$$

from which one deduces the total field $\mathbf{E}_{\text{int}} = \mathbf{E}_{ext} + \mathbf{P} = \epsilon\mathbf{E}_{ext}$ and the frequency dependent dielectric function

$$\epsilon(\omega) = 1 - \frac{\omega_{\text{p}}^2}{\omega^2} \quad , \quad \omega_{\text{p}} = \sqrt{\frac{4\pi\rho e^2}{m_{\text{el}}}} = \sqrt{\frac{3e^2}{m_{\text{el}}r_s^3}} \tag{1.24}$$

where $\omega_{\text{p}}$ is the bulk plasma frequency of the medium. The plasma frequency is related to $\epsilon = 0$. It characterizes the point of infinite response in bulk, i.e. an eigenmode of bulk oscillations. The infinite response never occurs in practice. Damping processes take over far before the amplitude explodes. The damping can be taken into account by introducing one relaxation time $\tau$ at the present level of estimate. This generalizes the above dielectric response to the Drude model [AM76], giving $\epsilon = 1 - \omega_{\text{p}}^2\tau/(\omega(\imath + \omega\tau))$ which merges into the form Eq. (1.24) for the limit $\tau \longrightarrow \infty$.

Let us now insert the dielectric function Eq. (1.24) into the condition Eq. (1.21) for a Mie plasmon. This yields with a bit of simple algebra $\omega_{\mathrm{Mie}} = \omega_{\mathrm{p}}/\sqrt{3}$ and thus precisely the frequency (1.18). This bulk-oriented approach thus allows one to recover the simple relation for the Mie frequency when using a spherical dielectric. The macroscopic treatment can easily be extended to more general shapes as, e.g., spheroidal clusters and allows one here to sort a lot of global phenomena. This line is followed to a large extend in [KV93]. In this book, however, we will put more weight on the microscopic treatment as outlined in Chapter 3.

As is clear from the above discussion, in a finite system electron oscillations will dominantly show up at the surface of the system, while in bulk they manifest themselves through electron density oscillations at plasma frequency $\omega_{\mathrm{p}}$, independent from the surface. In fact both mechanisms do exist in finite systems like clusters. In other words, an electric field induces both electron displacement *and* electron density fluctuations. It simply turns out that in a finite system the surface effect dominates over the bulk effect and one thus mostly observes the surface response, theoretically as well as experimentally. This means that photoabsorption spectra of the optical response of metal clusters to light are dominated by the surface plasmon, while a, usually much suppressed, contribution from the so called volume plasmon (corresponding to the bulk-like response) may be spotted in the frequency domain around $\sqrt{3}$ times the surface plasmon.

The existence of a resonance such as the Mie plasmon (or the volume plasmon) has important consequences on the way the system will actually couple to light. To make the point specific let us consider the case of a cluster subject to a laser irradiation, a situation that we shall encounter at many places along this book. In this case, one can easily estimate the amount of absorbed laser power by the system by simply considering the motion of individual electrons subject to the laser field of frequency $\omega_{\mathrm{las}}$ and the damped (with relaxation time $\tau$) plasmon oscillations of frequency $\omega_{\mathrm{Mie}}$ (following, in that respect the simplest Drude model). The instantaneous absorbed power can be expressed as $\mathcal{P}_{\mathrm{abs}} = -e\rho\mathbf{v}\mathbf{E}$ averaged over all possible electron velocities $\mathbf{v}$. Time averaging over one laser cycle, one obtains the well known resonant absorption form

$$< \mathcal{P}_{\mathrm{abs}} >= \mathcal{P}_0 \frac{u^2}{(1-u^2)^2 + (u/\omega_{\mathrm{Mie}}\tau)^2} \qquad (1.25)$$

where $u = \omega_{\mathrm{las}}/\omega_{\mathrm{Mie}}$. The above form obviously exhibits a resonance at $\omega_{\mathrm{las}}/\omega_{\mathrm{Mie}}$. This means that when irradiating a cluster with a laser field one should expect resonant energy absorption when the laser frequency comes close to the plasmon frequency. This resonant behavior will actually be observed and exploited in many situations all along this book (see in particular the discussion in Section 5.3.1.1).

#### 1.3.3.3 Optical response in metal clusters

The Mie model and the dielectric approach, outlined above, show that metals resonantly couple to light, both in bulk as well as in small particles. In finite systems such as clusters, the surface plasmon dominates the response and we shall mostly focus on this aspect in the following. The resonance frequencies provided by the above models, Eqs. (1.18) or (1.24), are independent of cluster size. This does not fully agree with experimentally observed response spectra. Figure 1.17 shows optical absorption spectra of deposited clusters of various sizes.

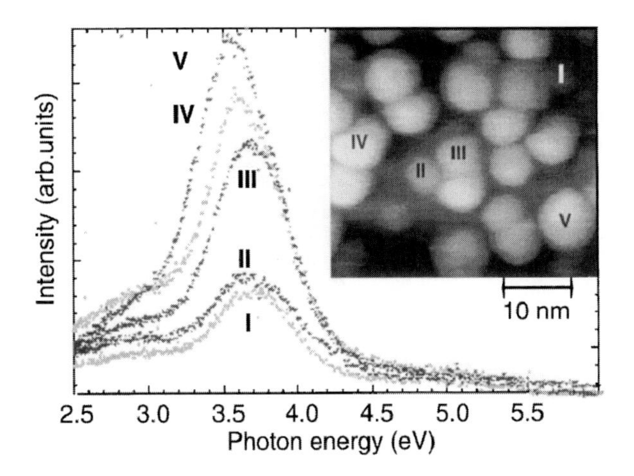


**Figure 1.17:** Photoemission spectra of Ag clusters of various sizes as indicated. The inset shows the corresponding STM image of the various clusters whose spectra have been recorded. Note the dependence of the peak response as a function of cluster size. From [NEF00].

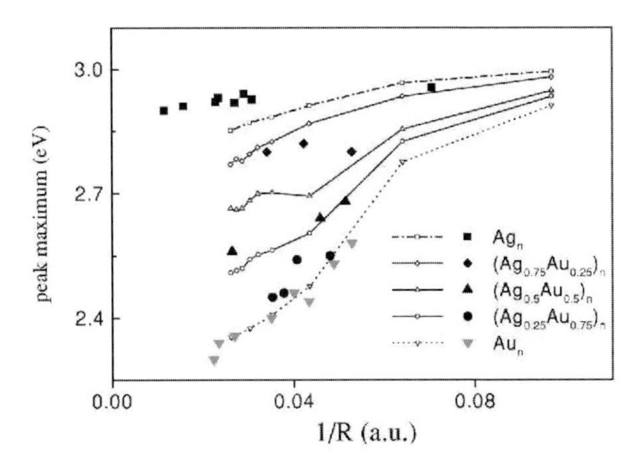


**Figure 1.18:** Resonance frequencies for embedded alloy clusters from Ag and Au with several degrees of mixing. The frequencies are drawn as a function of inverse cluster radius $1/R$. From [GLC+01].

The peak in the plasmon response shows a clear trend as a function of cluster size, which can be roughly estimated from the inset.

More systematic measurements of this effect have been reported in many occurrences. Figure 1.18 shows a systematics of results obtained for metal clusters embedded in a glass matrix. (Such clusters are typically the ones we discussed at the very beginning of this chapter, invoking Roman art and crafts.) The dependence of surface plasmon frequency on cluster radius $R$ is clear. There is a weak linear trend with $1/R$. The bulk limit $1/R \to 0$ tends nicely towards Mie frequency when estimated according to Eq. (1.24) with the bulk dielectric constant. The slope of the trend depends obviously on the material, here on the different degrees of mixing in the alloy. Driving measurements to small clusters shows deviations from the linear trend through finite-size quantum effects. We shall come back in more detail to this point in Section 4.3.

## 1.4 Conclusion

Although clusters have been used for centuries, in particular by craftworkers and in photography, they have become the focus of dedicated scientific investigations only very recently, essentially during the last quarter of the twentieth century. This late interest is, to a large extent, related to the fact that one was not able to produce free clusters, until recently. But since interest in clusters has grown, investigations on these systems have led to major scientific achievements, the most famous being probably the discovery of fullerenes, with all the potential applications it unraveled, in terms of physics and chemistry of carbon at the nanometer scale.

As we have seen throughout this chapter, clusters are specific objects. They are neither big molecules nor finite pieces of bulk. One of their major characteristics is precisely their scalability, which allows them to constitute one almost unique case of systems truly interpolating between individual atom and bulk. This has obviously profound theoretical applications for the many body problem, an aspect which interestingly complements the potential technological applications mentioned just above in the case of Carbon.

In clusters atoms bind together to form the system itself in a way similar to what occurs in simple molecules or in bulk. Indeed, one observes the same different types of bonding, namely ionic, covalent, metallic and van der Waals bonding in clusters. This gives rise to four corresponding classes of clusters, with different binding characteristics, in particular in terms of the robustness of the such formed structures. Of course such a classification primarily provides a guide rather than a rigid classification. Clusters may even exhibit various types of bonding as a function of their size.

Among the four classes of clusters, metal clusters, in which electrons constitute a highly delocalized, quasi free, gas, play a particular role in the field. They have indeed been the focus of a particularly large number of investigations, which allow one to have systematics on several physical properties. They furthermore respond in a particularly simple and robust way to light in terms of the so called surface plasmon, an effect identified in the early twentieth century by G. Mie. This behavior again makes their manipulation, in particular by means of laser fields, especially simple. Metal clusters will thus be considered in a sizable fraction of the discussions in this book. This is particularly true in dynamical questions, which constitute a central topic of this book.

# 2 From clusters to numbers: experimental aspects

The study of clusters requires a proper understanding and management of their production and handling. As for many other physical systems, production conditions strongly limit the level of details accessible in experiments. It is thus of importance to discuss production mechanisms in order to better understand what is measurable and how. It is thus the first aim of this chapter to make a brief presentation of cluster production mechanisms, in terms of cluster sources. On that occasion, we shall address in particular the limitations set by production techniques upon cluster characteristics (size, charge, temperature). Constraints come also from the second step in the experiments, handling of clusters and measurement of observables. The final part of the chapter is devoted to a survey of basic experimental investigations on clusters, both in terms of tools and accessible observables. We shall here illustrate this by several examples from ongoing cluster research.

Cluster production and analysis are intimately linked. Usually, an experimental set up integrates all elements from the source to the collection of data. Such a complete chain of production/measurements/data storage is made possible by the relative compactness of all the elements in the experimental set up. All in all, cluster physics experiments can be performed in modestly sized rooms of a few tens of square meters at most. A schematic picture of a typical experimental set up is shown in Figure 2.1. It demonstrates the case of the widely used supersonic jet source (see Section 2.1.1.1) associated with a mass spectrometer for cluster selection (Section 2.2.1). Let us briefly discuss the various components of this apparatus.

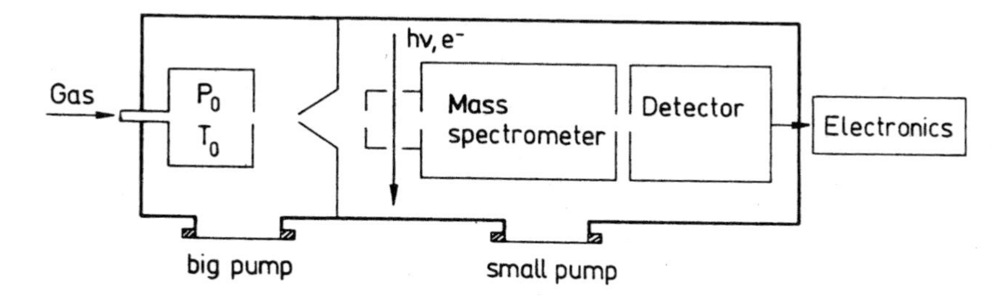


**Figure 2.1:** Schematic view of an experimental setup. The various components of an experimental apparatus are schematically indicated, starting from the key production device, namely the cluster source. Irradiation by a laser, which produces ionized species, allows mass triggering in a mass spectrometer. Finally, properties of the mass-selected clusters are analyzed in a detector.

*Introduction to cluster dynamics.* Paul-Gerhard Reinhard, Eric Suraud

ISBN: 3-527-40345-0

Once produced in the source (left device in Figure 2.1), clusters are ionized either by electron or photon impact. The non-zero net charge thus acquired allows mass selection in the mass spectrometer, an essential step in order to identify the clusters on which experiments are performed. The third compartment in Figure 2.1, labeled "Detector", contains the analysis apparatus itself. As we shall see below, it often employs electromagnetic probing, for example by lasers. The last compartment on the right, sketches electronic devices to store accumulated data. Not shown in the sketch is the very last step of the chain, the data analysis. It is usually done off line. Because this chain is continuous, from production up to data storage, and because it is usually operated as a whole, the production phase is of particular importance as it directly affects experimental outcomes. To a large extent, source managing thus appears as a step in the measurement process. It will be discussed in Section 2.1

As exemplified in Figure 2.1, lasers play a key role in both the handling and the analysis of clusters. We have seen above (in Section 1.2.2) that they are most often driven at sufficiently moderate intensities to remain in a non-destructive regime. Still, this covers a wide range of possibilities and one may wonder what kind of perturbation a laser produces on a cluster. Let us focus on optical lasers, with photon energies of order $\hbar\omega \sim 3$ eV. As we have seen in Chapter 1, such an energy lies close to the range of typical cluster energies. In some cases it may suffice to directly ionize the clusters. The question remains whether the effect is purely energetic or whether such a laser pulse also affects electronic or ionic momenta, and to what extent. Let us consider the example of metal clusters, in which electrons can be viewed as a free Fermi gas characterized by its Fermi momentum $k_{\mathrm{F}}$ (see Section 1.3.2). This corresponds to a typical momentum $p_{\mathrm{F}} = \hbar k_{\mathrm{F}} \sim 0.3 - 0.4\,\mathrm{eV\,fs\,a_0^{-1}}$, see Table 1.5. It is to be compared to the photon momenta delivered by the laser, $p_{\mathrm{las}} = \hbar\omega/c \sim 5 * 10^{-4}\,\mathrm{eV\,fs\,a_0^{-1}}$. The impact of the laser on the electrons is thus purely energetic and does not contribute to their momenta. The impact on ions is even smaller. Let us estimate typical ionic momenta from ionic vibration energies $E_{vib} \approx 10\,\mathrm{meV}$. The associated momenta are thus of order $p \sim \sqrt{2ME_{vib}} \sim 4\,\mathrm{eV\,fs\,a_0^{-1}}$ for Na, i.e. one order of magnitude larger than typical electron momenta and far above photon momenta.

## 2.1 Production of clusters

Cluster production is a key step in cluster studies. As mentioned above, cluster physics, in particular the study of free clusters, really started in the 1970s with the increasing availability of cluster sources. Without proper and controlled means of production, cluster physics was bound to the study of embedded, or at best deposited, systems. In such cases observed phenomena mix in a complex manner the effects of cluster and substrate or matrix. The start of studies on clusters themselves is thus directly related to progress made at the level of production sources. Not surprisingly, many types of sources have been developed to produce clusters of various sizes and properties. Clusters are finite objects, finite pieces of material. In order to produce them, one can thus either aggregate smaller systems (atoms, molecules, small clusters) or break larger systems (bulk typically). Cluster sources hence basically rely either on condensation/aggregation or on break up, and sometimes on both.

One can thus sort cluster production sources into three main classes. In supersonic jets, a gas is expanded into vacuum from a high pressure through a small nozzle. The subsequent adiabatic expansion and cooling leads to cluster formation by condensation. In gas aggregation sources, atoms are injected into a stationary or streaming gas and cluster formation again proceeds via condensation due to cooling of a gas of atoms. In surface sources, on the contrary, clusterization primarily proceeds via break up, in the sense that "proto"-clusters are stripped from a surface by particle or photon impact or by a high electric field, even if some condensation follows. There remains to detail how to activate efficiently these basic production mechanisms. We shall briefly present various types of frequently used cluster sources. One type of source will be discussed in somewhat more detail, namely the supersonic jet sources. This should serve to exemplify the experimental difficulties raised by cluster production, in particular in terms of cluster identification and properties such as size, charge, and temperature. A more detailed discussion of the various sources and their advantages and disadvantages can be found in the review [dH93] or in the books [MI99, Pau00a, Pau00b].

## 2.1.1 Cluster production in supersonic jets: a telling example

### 2.1.1.1 Seeded supersonic nozzle sources

The widely used supersonic jets are among the best understood cluster sources. They thus offer an ideal tool to analyze the difficulties encountered in cluster production and identification. The basic idea in such sources is cluster formation by condensation of an expanding gas of atoms [LNH84]. A highly compressed gas (typical total pressure $P \sim 10$ bar) with atoms of the material to be aggregated is allowed to expand through a small nozzle. The ensuing adiabatic expansion slows down the atoms up to a point at which binding between neighboring atoms becomes energetically favorable. This leads to the successive aggregation of the atoms in clusters.

The expansion mechanism is driven by an inert gas carrier. Supersonic sources are often used for producing metal clusters of low melting point metals (usually alkali metals). A furnace contains melted metal which is heated to produce a metal vapor of pressure around 10–100 mbar. This vapor is mixed with (seeded into) a rare gas introduced into the source at a pressure of several bar. The hot mixture of metal vapor and rare gas is driven through the nozzle and expands after the nozzle. This leads to cluster formation as we shall see below. Seeded supersonic nozzle sources are mostly used to produce intense, cold, and "directed" cluster beams with acceptably narrow speed distributions. They allow the formation of clusters with hundreds, even thousands of atoms per cluster with reasonable abundance. A schematic representation of such a supersonic source, together with other "classical sources" is shown in Figure 2.2. One can see on the scheme the basic elements constituting a supersonic source: furnace (left), injection of inert gas carrier (extreme left), heating system (center), nozzle (right) and skimmer (extreme right) ahead of an expansion vessel.

### 2.1.1.2 Expansion

One starts with an initial gas of atoms at a given temperature $T_0$. Before atoms are expanded through the nozzle their velocities are random; the high pressure implies a small mean free path (much smaller than the nozzle diameter) and thus a highly collisional regime, which can be typically understood in terms of hydrodynamics. The adiabatic expansion through the

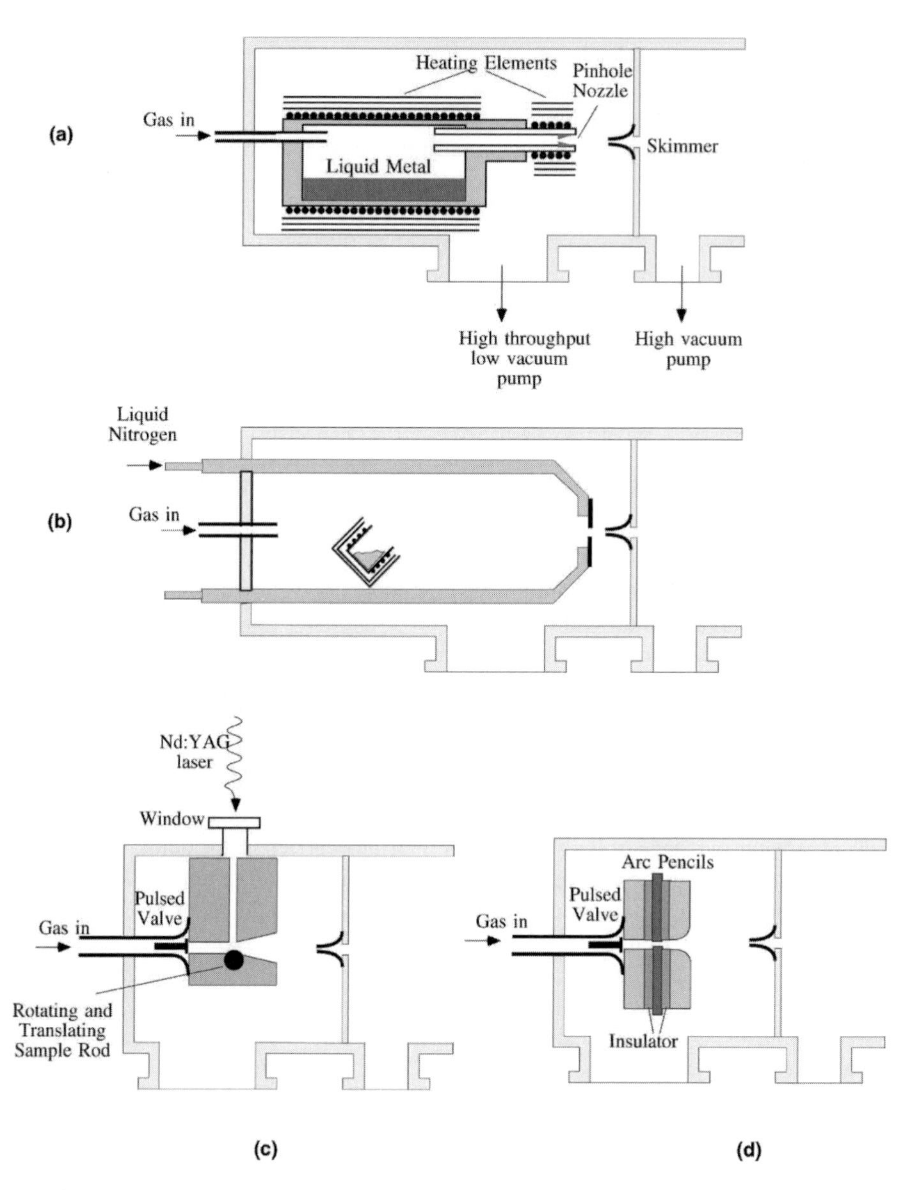


**Figure 2.2:** Basic layout of typical cluster sources. A seeded supersonic nozzle source is represented in (a), a gas aggregation source in (b) and surface sources in (c) and (d). From [Bin01].

nozzle leads to a strong "alignment" of the atom's velocities (the expansion velocity largely overcomes the original thermal velocity of the atoms). The first phase of the expansion nevertheless maintains an "in flight" hydrodynamical behavior in the system, although with a dramatic reduction of the temperature, pressure and density of the gas. It is during this expansion phase, when a proper "window" of density and temperature is reached, that clusterization will occur, as we shall see. With proceeding expansion, one reaches a point beyond which the picture of a continuous medium breaks down, and so does the hydrodynamical approach. From then on, each formed particle will more or less follow its own path. The basic properties of the further expanding gas are essentially frozen beyond this instant of "decoupling" of the system. The velocity profile of the expanding gas, for example, turns out to be essentially Maxwellian from then on, with temperature as given by the one at the end of the "hydrodynamical" phase of the expansion.

#### 2.1.1.3 Cluster formation and size distribution

Cluster formation during the expansion proceeds by condensation. It is a complex process which is not yet fully quantitatively understood. The basic condensation mechanism, though, is simple. When atoms in the jet become sufficiently cold, they can bind together to form a dimer (this corresponds to a temperature smaller than the binding energy of the dimer). These dimers (some of them being possibly already present even in the original vapor) constitute seeds for further clusterization. It should be noted here that the kinematics of the expansion process, with only a small spread of atomic velocities, tends to favor this clustering mechanism by keeping atoms in the vicinity of each other. The actual evolution of the system towards the formation of large or small clusters depends on the thermodynamical properties of the jet itself. When the pressure in the jet is small, cluster growth mostly proceeds on the basis of monomer aggregation, basically leading to low mass clusters. A high pressure $P$ in the jet, in turn, allows growth of clusters by cluster aggregation which leads to the production of large clusters. Not surprisingly, the initial pressure $P_0$ of the gas of atoms is hence directly linked to the actual size distribution of the produced clusters and thus to the abundance spectrum. This feature is illustrated in Figure 2.3 showing mass spectra of $CO_2$ clusters as obtained under various pressure, and thus clustering, conditions. The initial pressure $P_0$ is varied between about 0.7 and 3 bar, while the initial temperature $T_0$ of the atom gas is kept fixed at 225 K. It is interesting to see how the mass distribution strongly depends on $P_0$. First, note that in all cases one obtains, not surprisingly, a distribution of clusters and not one single species. Both the average value of this size distribution and its width depend on $P_0$ and both decrease with the pressure. We confirm here the above remark on the fact that low pressure provides a regime of weak clustering (small clusters) while strong clustering can be attained for higher pressures. With increasing pressure $P_0$ the width of the mass distribution reaches very high values, comparable to the average cluster size. The case at 3000 mbar pressure is quite telling in this respect, with a peak around size 500 and a width (at half maximum) of order 700.

The example displayed in Figure 2.3 exemplifies a basic difficulty in cluster production, namely the fact that one produces a very large variety of clusters in the jet. And it is obvious that clusters with very different sizes will exhibit different physical properties, see the many examples throughout this book. In other words, it is likely that one cannot exploit the cluster beam as is directly formed in a source. A triggering phase is unavoidable in order to select the cluster sizes one aims at studying (see also Figure 2.1). And it is clear as well, that the quality of the production management will directly affect the quality of the measurements. Indeed, if by properly tuning source parameters one can attain an acceptable control on the sizes of the produced clusters, the mass selection will probably become more successful in terms of yield. This allows more accurate measurements on well mass-selected clusters.

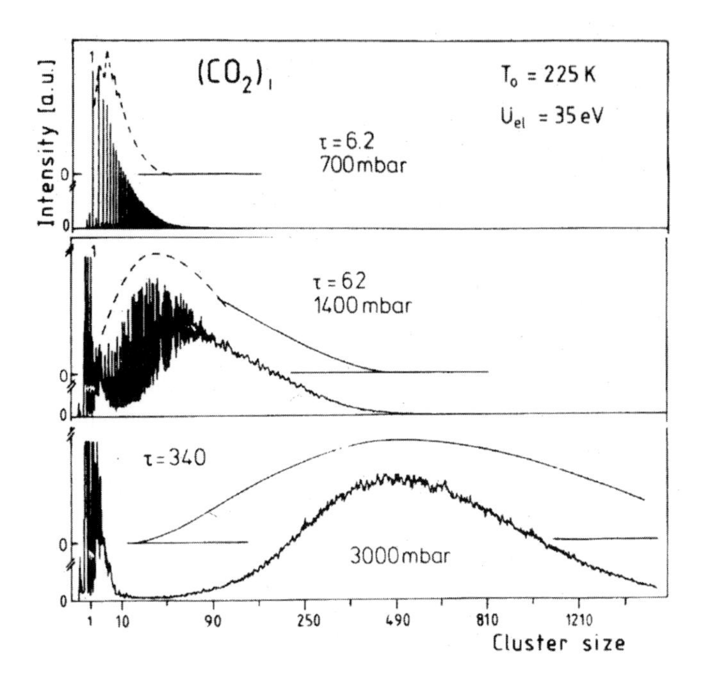


**Figure 2.3:** Mass spectrum of $CO_2$ clusters. Under weak clustering conditions, namely at low pressure (upper panel) a roughly exponential mass spectrum is obtained. Higher pressures (going down the panels) leads to better clustering conditions: clusters can then grow by cluster–cluster collisions. This finally leads to a sizeable, although very broad, peak around a finite value of the average cluster size. Full and dashed lines represent theoretical evaluations of mass spectra. After [Hab94].

#### 2.1.1.4 Temperature effects and handling of supersonic jet sources

The poor mass selection of primary clusters (just behind the source) is however not the single difficulty to overcome. The whole production mechanism involves temperature. The expanding gas cools down; its temperature thus depends on the expansion stage. Clustering hence takes place at different temperatures corresponding to different instants of the expansion. Even more delicate is a proper control of the formation heat involved in the clustering process. In other words, clusters are formed at finite temperature. And this temperature results from the decreasing global temperature of the collectively expanding system and from the formation heat in each aggregation mechanism, making it specific to the history of each cluster. The problem looks even more complex when realizing that clusters in the jet may also cool down, for example by collisions inside the jet, or by monomer evaporation (basically due to the fact that clusters have a finite temperature). In order to understand the thermal behavior of the clusters during the aggregation phase, one has thus to account for all these competing processes, for example by employing a statistical model when the temperature is high enough. But all the details of these models, in particular at the level of evaporation rates, are not known, rendering these approaches not fully predictable. The population of clusters formed in the jet will thus exhibit a temperature distribution which is obviously not easy to predict precisely. One might wonder, though, whether this cluster temperature is likely to affect cluster properties. Roughly speaking, the temperatures attained may be comparable to the clusters dissociation energies, which obviously make them crucial, even for the existence of the clusters themselves. We shall see below that even moderate temperatures may have an important impact (Section 2.1.3.1). All that makes a better control over temperature desirable. This can be achieved by mixing the cluster beam for a while with an inert-gas beam of well defined temperature [ESS+95]. Heat exchange by collisions brings the clusters into thermal equilibrium with the cooler-gas. This allows one to tune temperature over a broad range and in particular to cool down the cluster beam, when necessary.

All in all, this brief overview of cluster production in supersonic jets clearly points out the difficulties one faces in the identification and characterization of formed species. The proper tuning of source parameters thus turns out to be an essential ingredient of any experimental program. As we have seen above, there are no definite models to handle these parameters precisely, although one understands basic underlying physical mechanisms. And of course, experience in the handling of sources becomes essential. In the case of supersonic jets, there thus exist scaling laws for cluster formation involving the initial pressure $P_0$ and temperature $T_0$ of the atom gas, as well as the diameter $D$ of the nozzle. For example, it is known that cluster size increases with increasing $P_0$ or $D$ and decreasing $T_0$. Usually the distribution of formed clusters is broad. It should also be noted that, in such a condensation process, impurities in the gas play a big role. They may constitute condensation germs and lead to an increase of average cluster sizes.

### 2.1.2 More cluster sources

There are many types of cluster sources, which actually all have their own merits. In Section 2.1.1, we have seen how clusters can be formed in the widely used supersonic jets. We want here to consider some other types of sources involving various clusterization mechanisms. Again, we skip the details and keep the discussion at a qualitative level, considering only the most common types of sources.

#### 2.1.2.1 Gas aggregation sources

Gas aggregation sources provide a simple and efficient means to produce large clusters. They are, so to say, laboratory copies of a smoking fire or of cloud and fog formation in nature. The basic production mechanism is simple and well known [SMR80]. A liquid or a solid is evaporated into a colder gas which cools down the evaporated atoms or molecules until condensation starts. Condensation then roughly proceeds as in supersonic jets. The dominant clustering mechanism is here successive atom addition. The major difference with supersonic jets lies in the kinematics of the condensation process which is directional in a jet and not so in an aggregation source. As a consequence there is no "automatic" stopping of growth in an aggregation source contrarily to the case of supersonic jets in which the expansion stops clusterization at some stage. All in all, the condensation process is thus more involved in aggregation sources than in supersonic jets. There have, nevertheless, been quite successful applications of aggregation sources. The most famous example of such (so called) smoke sources are the facilities used to produce fullerenes ($C_{60}$, $C_{70}$,...). But gas aggregation sources have also been used to produce metal clusters, both with alkalines or more refractory metals [dH93]. The intensities of the cluster beams extracted from the gas aggregation sources are nevertheless much lower than from supersonic jets. A typical layout of a gas aggregation source is shown in panel (b) of Figure 2.2. The cold inert gas injected from the left causes the heated vapor (center) to become supersaturated, which allows cluster production. The cluster beam is then ejected through the right aperture.

#### 2.1.2.2 Surface sources

The principle of surface sources is primarily different from both supersonic jets and aggregation sources in the sense that initial cluster formation mostly results from break up. The idea is to remove finite pieces of material from a solid surface. Nevertheless the ablation phase is usually complemented by a further clustering phase which again proceeds via condensation. The first "erosion" phase can be achieved by heavy particle impact, for example a $Xe^+$ at a few keV kinetic energy, in the so called Surface Erosion Sources or sputtering sources. By hitting the surface the projectile ejects atoms, molecules or clusters. When the charged projectile is replaced by photons, typically from a laser with intensity $I \gtrsim 10^8$ W cm$^{-2}$ one speaks of a Laser Evaporation Source (LES) [DDPS81]. Extraction can also be achieved by a high-current pulsed arc discharge in the so called Pulse Arc Cluster Ion Source (PACIS) [SLF$^+$91]. The handling of such sources is technically quite demanding. Because the formation process is violent, clusters are usually formed at very high temperature. In order to overcome this temperature problem, such sources have often been coupled to supersonic jets or aggregation sources to cool down the formed clusters. A typical setup, as illustrated in Figure 2.2 (panels (c) and (d)) will thus associate an "ablation" compartment (central part) to a gas carrier injected from the left, as in a gas aggregation or a supersonic jet device. The cluster beam is finally ejected to the right.

#### 2.1.2.3 Pick up sources

Pick up sources are typically used to produce mixed clusters composed of various materials. The idea is to mix clusters as formed in supersonic jets with a jet of other molecules. This allows, for example, attachment of the latter molecules on clusters formed in the jet. Jet mixing can be done at one location (two orthogonal jets) or "in the flight" (two parallel co-expanding jets). Pick up sources are also used to produce relatively cold jets of charged clusters by using a jet of charged particles (electrons, ions).

#### 2.1.2.4 Embedded and deposited clusters

Up to now we have only discussed how to produce free clusters. However the field of embedded and deposited clusters is of great practical importance and it allows several studies which are not easily possible with free clusters. The substrate fixes the clusters and thus gives much more time for growth. This, in turn, allows one to collect much larger clusters when wanted. Moreover, one can achieve a higher density of cluster. The high density of scatterers thus enables measurements with weak signals, as e.g. the second-harmonic generation (SHG), see e.g. Figure 5.2, or Raman spectroscopy, see e.g. Figure 5.15. Last not least, it is also much easier to control cluster temperature via the substrate.

There are two basically different methods to produce embedded clusters or clusters deposited on a substrate. The first method is growth from supply of atoms or ions. An example of deposited clusters is the growth of Na clusters on an insulating substrate by exposing the substrate to a Na vapor, see e.g. [KWSR99]. An example of embedded clusters is the growth of Ag clusters in glass by diffusion from a surrounding molten Ag salt and $\mathrm{Na}^+ \leftrightarrow \mathrm{Ag}^+$ ion exchange, see e.g. [SKBG00]. The average cluster size can be controlled to some extent by the growth conditions, such as temperature, pressure and time. The method is applicable only if the combination of materials tends to clustering of the vapor atoms or ions respectively. In the case of deposited clusters, a high mobility of the deposited clusters on the surface remains, which may perturb long time measurements by drift and reactions. This can be avoided by sputtering defects onto the substrate before cluster growth, see e.g. [MPT+00].

The alternative method of embedded or deposited cluster production relies on producing free clusters first and depositing them in a second phase onto the substrate or into the matrix. Much depends here on the kinetic energy of the clusters impinging on the material. One distinguishes three regimes [Bin01]: low ($E_{\mathrm{kin}} < 0.1\,\mathrm{eV}$), medium ($0.1\,\mathrm{eV} < E_{\mathrm{kin}} < 10\,\mathrm{eV}$), and high energy impact ($10\,\mathrm{eV} < E_{\mathrm{kin}}$). These produce different results. The low impact deposit places the clusters more or less gently on the substrate. The clusters remain mobile and may rearrange by diffusion along the surface and reactions. The medium impact case pins small defects into the surface which fixes the deposited clusters; however, with the danger of heavily perturbing the clusters themselves. The high impact deposit causes severe damage producing a new composite rather than a deposited cluster.

A problem with embedded or deposited clusters is mass selection. The growth from vapor produces a broad distribution of masses and shapes which can be controlled only roughly through the growth conditions. The explicit deposit of free clusters allows one, in principle, to deal with a beam of mass selected clusters. This holds, however, only before deposit. The actual process of attachment changes the clusters and produces again some spread in masses. There have however come up recently techniques for a size and shape selection of deposited clusters by intense laser fields, see [WBGT99] and Section 5.3.2.

Another problem shows up in connection with the theoretical description of embedded and deposited clusters. The environment has significant impact on the cluster properties. Models for that are yet in a developing stage, see e.g. Section 3.4.4. On the other hand, the many conceivable combinations for cluster and environment generate a huge variety of scenarios whose exploration is an extremely interesting task, see e.g. a demonstration in Figure 5.10. After all, this rich and demanding field is very open for future research. Therefore one will find many further examples for studies with embedded/deposited clusters in Chapter 5.

### 2.1.3 Which clusters for which physics

The above brief overview of the most commonly used types of cluster sources has shown that cluster production is a delicate but essential step of any experimental program. Clusters are usually formed in ill defined conditions (poorly known size and/or temperature). This calls for the best possible handling of source parameters. After that, one needs a post processing (mass selection mostly), before one can use the beam for measurements. In this section, we will briefly review what can be achieved in this respect of post processing, namely which clusters can be selected and under which conditions.

#### 2.1.3.1 Sizes and temperatures

We have noted in many places above that clusters are formed in a mixture of sizes and temperatures. We have insisted upon the fact that the uncertainties on these quantities (size, temperature) should be reduced as far as possible, in order to obtain clear, well defined, signals from the experiments performed on the produced clusters. And we shall see in the next section that mass selection can be, fortunately, achieved to a reasonably high degree of accuracy. Before discussing this question, it is interesting to illustrate the impact of size and temperature on a well defined and widely studied observable: the optical response of metal clusters (see Section 1.3.3.1). We shall come back, in more detail, to the sensitivity of the optical response on various parameters (cluster size, shape, temperature…) in Section 4.3, because the optical response represents a key observable in the physics of metal clusters. But for the time being we shall content ourselves with a qualitative analysis just aiming at pointing out the importance of size and temperature effects in the interpretation of optical response in metal clusters.

The impact of size on the optical response has already been discussed in Figure 1.17 where we have shown that the "color" of clusters strongly depends on size. Of course, in Figure 1.17 we have considered huge variations of size between a few units and several thousands. And in spite of the uncertainties in the sizes of the clusters produced by sources, it is likely that one will hardly observe such large variations of sizes in the studied samples used for one experiment. Still, because the mass distribution may be very broad (see Figure 2.3), one may observe a very broad optical response "peak" (see Figure 1.17) spreading over several "colors". And this is, for example, exactly what one observes in artcraft using embedded clusters in glass (see the discussion of Figure 1.17 in Chapter 1).

The question of the impact of temperature on the optical response of metal clusters has not yet been discussed in this book. We thus briefly explain the basic mechanisms at work here. The typical temperatures considered are in the range of room temperature ($\sim$ 300 K). Such "low" temperatures leave electrons totally unaffected (compare 300K $\sim$ 0.03 eV with a Fermi energy of about 3 eV, see also the temperature scale in Figure 1.15). Electrons can thus be considered as being practically at zero temperature. Ions, in turn, are strongly affected by such a temperature and acquire, accordingly, sizeable thermal excitation. Of course, ions are much more massive than electrons and thus move at a much slower pace. When exciting a surface plasmon in a metal cluster (see, e.g., Section 1.3.2), electrons have thus more than enough time to oscillate several times before ions noticeably move. But in this linear regime of excitation the Mie plasmon lasts very long, for sure long enough to allow ions to move significantly and the cluster thus explores a variety of shapes and/or spatial extensions. The actually measured optical response thus piles up the individual optical responses of a huge set of clusters of the same species but at many different shapes, each of which providing its own specific optical response. The net effect of temperature is thus a substantial broadening of the optical peak which increases with temperature (see Figures 4.17 and 4.23 as well as Section 4.3.5).

Both mix of sizes and shapes (through temperature) thus tend to spread the expected optical response peak. The problem raised by such a spreading is that it contains physical effects which can hardly be disentangled. The interpretation of an experimental optical response spectrum may thus easily become intractable with the result that the experimental data then bring only limited physical insight to cluster properties. On the contrary, as we shall see, when properly handled, the optical response of metal clusters may provide highly detailed information on the structure of these systems (see Section 4.3). This, once again, shows how important it is to produce clusters in the cleanest possible manner.

#### 2.1.3.2 Neutral versus charged clusters

Clusters, as produced in cluster sources, are often neutral. This neutrality raises two problems. First, one may be interested in studying (positively or negatively) charged clusters. Second, as extensively discussed above, the mass distribution of produced clusters is so broad that one does not know which neutral cluster one is handling. And it is obvious that only charged clusters are easy to manipulate with electromagnetic fields as it is in general necessary for mass selection.

Creating charged clusters from neutral ones is an easy task. The most common ways are laser or charged particle impact. Lasers strip electrons from clusters and thus allow an easy and tunable production of positively charged clusters. Impact with a positively charged particle (atomic cations for example) will produce a positively charged cluster as well, typically after a charge exchange between the cluster and the projectile particle. In turn, an electron beam allows one to form negatively charged species by electron capture on the clusters. The latter process remains, however, rather limited for simple binding reasons. The binding energy of anions decreases quickly with increasing (negative) charge. The thus formed cluster anions are consequently rather "fragile" with vanishingly small ionization potentials (see for example Figure 2.17). On the contrary, clusters with one positive charge are especially stable, as the extra charge tends to bind the remaining electrons even more. These clusters are thus particularly easy to manipulate because of their charge and high binding energy. Clusters can even accommodate much more than one positive charge. Of course a too highly charged cluster will become unstable with respect to fission (Section 2.3.11), but it turns out that the maximum charge which a cluster can sustain increases with cluster size. For example, in Na, a cluster with more than 27 atoms may carry two positive charges, while remaining stable. Similarly $Na^{3+}_{n>65}$, $Na^{4+}_{n>123}$, $\ldots$ are stable [NBF+97]. A similar "limiting" trend holds for negatively charged clusters. But the mechanism for instability is here different. It is spontaneous electron emission which limits the negative charge state. For example, the IP of $Na_n^{2-}$ clusters (at temperature $T = 0$) crosses the stability line (i.e. IP = 0) at around $n = 35$. In both cases (negatively or positively charged clusters), once charged, the clusters become easy to manipulate with electromagnetic fields. In particular, they can be sorted in mass spectrometers (Section 2.2.1) so that one can almost perfectly determine on which cluster experiments will be performed.

The case of neutral clusters is a bit more involved. It is nevertheless important to be able to produce and mass selected neutral clusters as well. Beyond the mere tuning of source parameters, which, of course, helps pre-selecting masses (Section 2.2.1), there are basically two ways of producing beams of mass selected neutral clusters. The first approach is of "mechanical" nature. It consists in deflecting a neutral cluster beam by an atomic beam (neutral atoms). Collisions between neutral clusters and atoms allow one to sort the clusters according to their mass. Indeed, the lighter the cluster, the more deflected it will be, which allows mass separation by angular sorting. Note that the same deflection technique also provides a selection in terms of velocities. This approach, however, is not free from technical difficulties. In particular, one has to avoid collisional dissociation in atom–cluster encounters. Low mass and velocity atoms are thus here preferably used. A good focusing of the original cluster beam is also useful to prevent angular and velocity spreading of the deflected clusters. It is equally important to control, as much as possible, the amount of energy deposited in the clusters during the collisions. Finally, one has often to suffer from the relatively small intensity of the available cluster beams.

The second, possibly more "natural", method to produce mass selected neutral clusters is of electromagnetic nature. It consists in re-neutralizing formerly ionized clusters. Starting from initially neutral clusters, a first phase of ionization allows one to mass select the formed charged clusters in a mass spectrometer. The mass selected cluster ions are then re-neutralized to provide mass selected neutral clusters. If the cluster ions are negatively charged re-neutralization can be achieved for example by photodetachment of the extra electron through a laser beam. It can also be obtained by collisional detachment or charge exchange in the course of collisions with a beam of positively charged ions. If the cluster ions are positively charged, re-neutralization can be achieved, for example, by charge exchange with a beam of neutral atoms. Of course, in all these re-neutralization techniques, caution is necessary to avoid (or at least minimize) the possible fragmentation of the resulting neutral clusters, due to electronic excess energy or inelastic effects. Fortunately enough, such fragmentation processes, which affect the size distribution of clusters, are usually small.

## 2.2 Basic experimental tools

We have seen in the previous section how clusters can be produced and which experimental limitations of cluster production techniques may raise, in particular at the level of cluster identification. The purpose of this section is to present the basic techniques which are used to perform experiments on clusters. Detailed examples of measurements will be given in Section 2.3. Here, we want to focus on the basic tools used in measurements, not on the measurements themselves. As pointed out in Section 2.1 a major defect of cluster sources is that they produce clusters with a mix of sizes. The use of mass spectrometers to mass select clusters before experiments is thus a crucial step of the experimental process (see Figure 2.1). We shall thus start our discussions with a brief presentation of the most widely used mass spectrometers. These mass spectrometers serve for cluster selection *before* experiments, as well as for cluster identification *as* or *after* measurements. In a next step we shall also discuss the most commonly used spectroscopic techniques in cluster physics, focusing on optical and, to some extent, on electron spectroscopy.

### 2.2.1 Mass spectrometers

Mass spectrometers constitute a key device in cluster physics for cluster identification before, or after, measurements. Roughly speaking there are two main classes of machines involving either time dependent or time independent electromagnetic fields. Typical devices with time

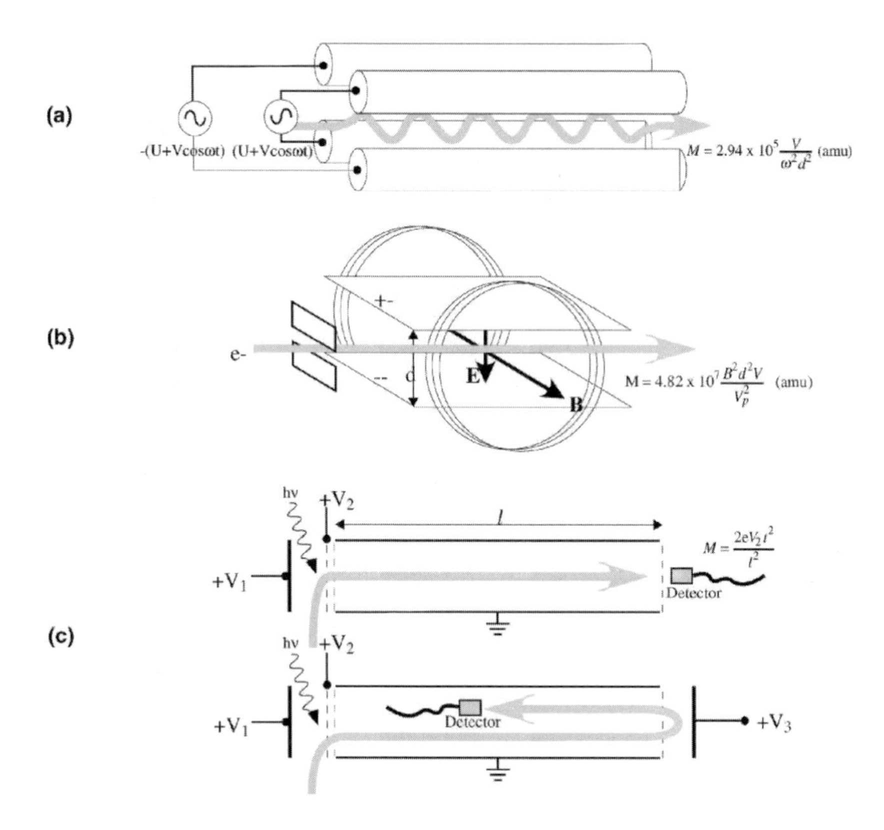


**Figure 2.4:** Schematic representation of basic mass spectrometers used to trigger charged clusters. From top to bottom: quadrupole (a); Wien filter (b) and Time of Flight (c) without (upper part) and with (lower part) reflectron.

dependent fields are quadrupoles, Ion Cyclotron Resonance systems and the widely used Time Of Flight (TOF) set-up, which we shall particularly discuss because of its simplicity and universal use. A TOF is sketched in Figure 2.4. The details will be discussed later in the corresponding sections.

Wien filters and instruments with magnetic and electric sectors are, in turn, typical examples of devices built with time independent electromagnetic fields. We shall show, for illustration, an experimental setup built with a (static) magnetic mass spectrometer. Other more "exotic" mass identification techniques also exist but we shall only mention a few of them for completeness (as e.g. particle traps which are a means to collect and to select clusters at once, see Section 2.2.1.5).

#### 2.2.1.1 Simple devices

The simplest conceivable mass spectrometer is the so called Wien filter, see panel (b) in Figure 2.4. Crossed homogeneous static electric $E$ and magnetic $B$ fields, acting perpendicular to the cluster beam, deviate charged particles and sort them according to their charge over mass ratio $q/m$. This is exactly the basic ingredient of an oscilloscope. Such devices require initially accelerated particles (potential $V$) and are tuned to *one* given $q/m$ ratio for a given combination of electric and magnetic fields. Indeed, for a particle of velocity $v = \sqrt{2qV/m}$ the net force due to $\mathcal{E}$ and $B$ vanishes for $\mathcal{E} = Bv/c$. The particles with $q/m = (\mathcal{E}c/B)^2/2V$ are thus undeflected by the system. It is an advantage of this Wien filter (whence the term "filter") to preserve a straight line trajectory for the selected charged clusters. However, Wien filters suffer from limited resolution ($\Delta m/m \sim 10^{-2}$) due to the dispersion of cluster velocities on entrance, and thus usually require high electric and magnetic fields to counterbalance this dispersion problem. The practical attainable mass range is typically 1–5000 atomic mass units (amu).

Quadrupole filters exploit a proper combination of static (ac) and time dependent (dc) electric fields to select clusters with a given $q/m$ ratio. A typical configuration is sketched in panel (a) of Figure 2.4. It is composed of four rods constituting two pairs of opposite poles. One pair is held at potential $U + V\cos(\omega t)$, the other at opposite potential. In such a system charged particles acquire an oscillatory motion along the poles. Transmission by the filter is achieved only if the amplitude of these oscillations is stable, which occurs irrespective of the particle velocity for a critical ratio $U/V \leq 0.168$. For ratios smaller than that, the device passes specific masses tuned by the value of $V$. Mass resolution attained this way is better than in Wien filters, of order $10^{-3}$ typically and masses up to about 8000 amu can be analyzed. The resolution depends on how close to the critical ratio the filter operates. Resolution also decreases when $V$ is decreased. For vanishing $V$ one actually recovers a simple, non filtering, ion guide.

#### 2.2.1.2 The "universal" TOF

The probably most widely used mass spectrometer in cluster physics remains the Time Of Flight (TOF) set-up, see panel (c) in Figure 2.4. It allows one to access a large mass range, while remaining compact and cheap. Furthermore, we shall see that it is capable of reaching relatively high resolution. Because it is used in many experiments at various degrees of sophistication, we shall discuss its mechanism in some detail.

The basic idea of a TOF spectrometer is to record the arrival time of the ions which were initially accelerated and then left to propagate into free space before detection. Panel (c) in Figure 2.4 illustrates how such a system works. The originally neutral clusters are first ionized by a laser field and then accelerated between the plates of a capacitor. They thus acquire an energy $E = mv^2/2 = qV$, for a given potential $V$ in the capacitor. From then on, they propagate in the field-free region of the TOF with a constant velocity $v = \sqrt{2Vq/m}$ which is thus uniquely related to the $q/m$ ratio. A typical ionizing laser pulse in a TOF lasts for about 10 ns and its intensity can be tuned so that $q$ is essentially one electron charge. Under such conditions, and for a given value of $V$, the ion velocity scales as $v \sim 1/\sqrt{m}$. Until the bound of the TOF, the ions have traveled over a (fixed) distance $D$, characteristic of the TOF, which

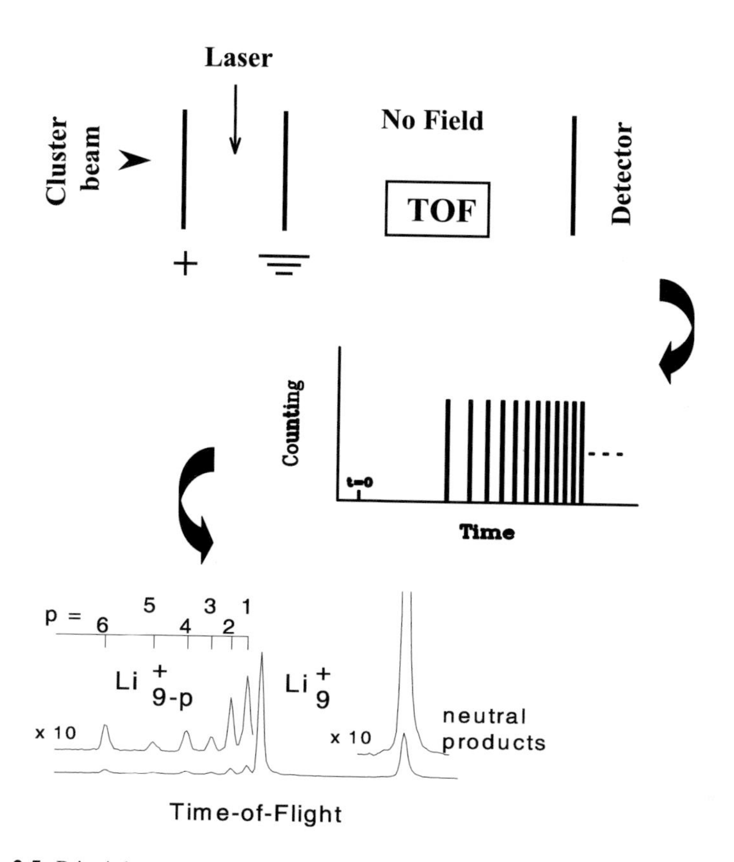


**Figure 2.5:** Principle of TOF analysis. An experimental setup is schematically represented (upper part). The typical TOF spectrum (recording arrival times proportional to the square root of masses) is shown in the middle part. An example of an experimental TOF spectrum is finally displayed in the lower part of the figure. After [BCCL01].

represents the field free region. For an ion of velocity $v$ it takes a time $\Delta t \sim 1/v \sim \sqrt{m}$ to reach the bound of the TOF, which thus allows mass detection. A simple count of the number of hits (one hit per ion reaching the bounds of the TOF) as a function of time provides a mass spectrum. The typical value of $\Delta t$ is $\Delta t \sim 100\,\mu\text{s} - 1$ ms. This is much longer than the duration of laser pulses actually used in such devices. The laser can thus be considered as providing a "zero time" $\delta(t)$ pulse and there is no need to renormalize the delay signal $\Delta t$ with any laser pulse correction.

Figure 2.5 serves to illustrate the TOF analysis. The upper part repeats a short sketch of the set-up. The middle part symbolizes the collected data in terms of a huge sample of recorded flight times. From these one can deduce the final mass spectra as sketched in the lower part. The example shows a simple and small chain of singly charged small $\text{Li}_n^+$ clusters as indicated (shown in terms of the number of removed monomers $p$).

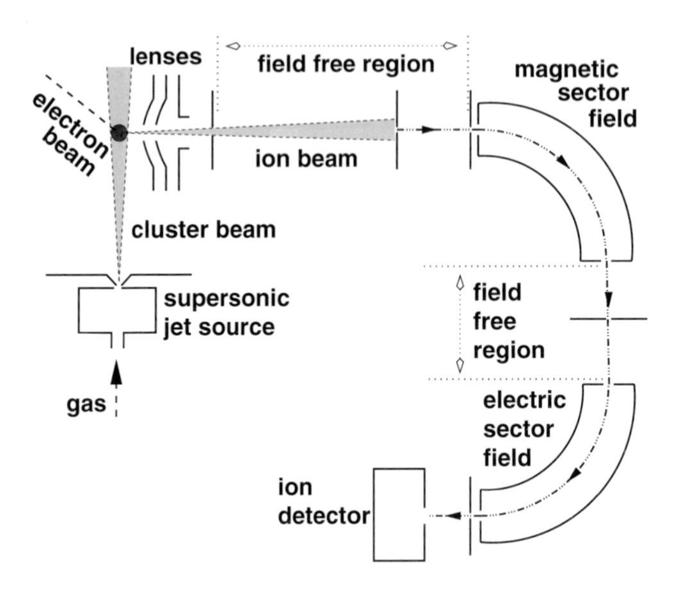


**Figure 2.6:** Schematic drawing of a double focusing mass spectrometer. In this setup clusters are produced by supersonic expansion (left part), ionized by electron impact and then analyzed in the spectrometer which is composed of two field free regions, one magnetic and one electric sector. See text for details. After [Hab94].

A key quantity in a TOF is obviously the extension $D$ of the free-field region: the larger $D$, the larger the proportionality coefficient between $\Delta t$ and $\sqrt{m}$ and hence the better the mass resolution. There is, however, a better solution than increasing $D$ in a simple "linear" way. This is attained in what is called a reflectron. This is a TOF in which at the end of the free path an electromagnetic field reverses the velocities of ions so that they can basically travel twice the distance $D$ (see panel (c) in Figure 2.4). But the gain here is more than a mere doubling of $D$. Reflecting the particles at the TOF extremity allows one to compensate initial velocity fluctuations for a given mass because rapid ions take longer than slower ones to reverse their velocity. This finally results in narrower peaks in the mass spectra and thus a gain in mass resolution. In a reflectron, mass resolution of about $10^{-5}$ can be reached while it stays more around $10^{-3} - 10^{-4}$ in a simple TOF. Because of their high performances in terms of resolution and because they are relatively simple devices, TOF and reflectrons constitute basic tools in many cluster experiments.

#### 2.2.1.3 Magnetic sector devices

Figure 2.6 gives an example of an experimental setup using another type of mass spectrometer, namely a magnetic sector mass spectrometer. In that case, clusters are primarily produced in a supersonic jet source, then ionized by electron impact and accelerated by electromagnetic lenses. The thus produced charged beam expands in a first field-free region before being collimated. After collimation, clusters which basically have the same kinetic energy enter the magnetic sector with an homogeneous static field. The slit at the end of this magnetic sector will thus select only *one* momentum $\mathbf{p} = m\mathbf{v}$, $v = ||\mathbf{v}||$ being constant in this magnetic sector. Clusters are then injected into an electric field ($\mathcal{E}$) sector. The latter field can be chosen to refocus the beam. Indeed, starting from a beam of given $\mathbf{p} = m\mathbf{v}$, $\mathcal{E}$ changes $\mathbf{v}$ and thus allows one to select the mass $m$. The net result is a high resolving power in terms of mass distributions, with a very good suppression of background. Furthermore, the two field-free regions, before the electric and magnetic sectors (see Figure 2.6) allow one to perform further experiments on the cluster beam, for example with lasers or particle beams.

#### 2.2.1.4 Even more spectrometers

As another example of a high resolution mass spectrometer let us also mention the ICR (Ion Cyclotron Resonance) system. It basically consists of a cyclotron dee. Cluster ions are confined by a combination of a static quadrupole electric field and an uniform magnetic field $B$. Because of the magnetic field, ion cluster orbits have a component at angular velocity $\Omega = qB/m$, which, if measured, gives access to the $q/m$ ratio. The measurement of the cyclotron frequencies $\Omega$ are performed by exciting the ions by a radio frequence pulse. The rotating ions induce a current on the plates confining the system, which constitute a sensitive antenna. The induced current is recorded, amplified and Fourier analyzed to extract the rotation frequencies $\Omega$, which provide a direct access to masses $m$. Such a device allows a spectrometry of high resolution of order $10^{-6}$. It is a non-destructive measurement which furthermore provides an access to all masses at once. It is however a relatively expensive setup which requires high magnetic fields $B$ and a proper handling of radio frequency technology.

There are, of course, other techniques to mass select clusters, either by combining the above described tools or by using other basic physical phenomena. For example, molecular beam chromatography is based on a separation of molecules/clusters due to their different mobilities in a gas, as a function of their mass and/or shapes. Such methods have long been used to separate neutral molecules. But we shall not further discuss these techniques here. We refer the reader to [Hab94] for more details on these very specialized techniques. It should, nevertheless, be noted that mass separation of neutral clusters is a question which has long been addressed and which remains to a large extent open, although some researches along that line are still conducted. The chromatography is here one of the few options.

#### 2.2.1.5 Clusters in a trap

Traps are electromagnetic devices which allow one to store particles for incredibly long times. Since their invention (for reviews see [Deh90, Pau90]), particle traps have found widespread applications in various fields of physics: they have been used, e.g., for high precision measurements of magnetic moments; combined with advanced cooling techniques, they are the perfect laboratory for producing and studying Bose condensates [AEM$^+$95, DMA$^+$95, DGPS99]. Their large scale versions are the storage rings with which crucial experiments in nuclear and particle physics become possible. Equally broad is the choice of different constructions for a trap. There are traps relying on static electromagnetic fields, or those working with particularly tuned time-dependent fields, and even the ponderomotive force of laser light is sometimes exploited to keep atoms confined. Whatever construction, a few typical features of a trap ought to be mentioned. There is high mass selectivity. In fact, only one sort of particles (defined by the $q/m$ ratio) can be stored at a time. One can accumulate and handle these particles for very long times. And one can manipulate many features of the stored ensemble as, e.g., temperature.

It is not surprising then that traps are also used to store clusters. Charged clusters can be handled very well in a Penning trap. It consists of a strong magnetic field which keeps the ions circulating, thus confined in the direction orthogonal to the field. Additional electrical fields act like "caps" to prevent escape along the direction of the magnetic field. A detailed description of the trap and its basic experimental capabilities is found in [SKLW99]. Cluster traps allow many precision measurements on clusters which would be much harder to do in free beams. There are, e.g., the detailed measurements of dissociation energies of metal clusters [VHHS01]. Traps can be combined with ionization through a laser beam which gives access to photo-absorption spectra, ionization potentials and related observables. The long storage time allows one to measure very long relaxation times as exploited, e.g., in the measurement of photo-decay lifetime in V clusters, see Figure 5.34 in Section 5.3.4.3. There are drawbacks, too. The clusters are kept in a cage. There is not much space left for further devices in the trap. This hinders all sorts of scattering experiments with well collimated beams. Storage rings may be a solution here. But their price exceeds the typical (low) dimensions which one usually encounters in cluster physics.

#### 2.2.1.6 Beyond spectrometers

We have discussed the production of clusters in Section 2.1 and we are at the end of this Section 2.2.1 on mass selection. This means we are at a stage at which we know how one can handle a well defined type of clusters. The next question is what kind of signal one is able to measure and with what kind of experimental apparatus. This will be addressed in the remaining two subsections of this section for the examples of optical spectroscopy (Section 2.2.2) and of photo-electron spectroscopy (Section 2.2.3) which constitute two generic means of investigation. Both methods are preferred tools to measure electronic properties. There are, of course, several other methods around (a few more of them will be addressed in Chapter 5). And other observables may need methods outside the range of optical photons, e.g. infrared spectroscopy for the study of vibrational modes [Ell98]. Before proceeding, we ought to mention that we have already discussed some tools of analysis. The one large example is the TOF which is not only the doorway to cluster selection but also an important part of measurements. In a similar fashion, traps combine charge/mass selection with measurements.

### 2.2.2 Optical spectroscopy

Optical photons have energies between about 2 and 3 eV, which is typically in the range of excitations of clusters (see Figure 1.17 as an example). Optical spectroscopy thus constitutes an important tool of investigation in cluster physics. Optical methods can be sorted in two main classes, depending on the "fate" of the excited cluster. In *photo-absorption* methods the cluster under consideration may be preserved. There is no necessary dissociation or fragmentation of the system. On the contrary, in *photo-depletion* methods the system is promoted to a dissociative state and thus eventually destroyed after the measurement. We shall successively discuss both these types of methods illustrating them on simple examples of standard measurements. We ought to mention that the most general notion is "photo-absorption" here general. Both experimental techniques, photo-absorption as well as photo-depletion, are used to determine the cross-section for photo-absorption which is the feature of interest to be studied for the given cluster.

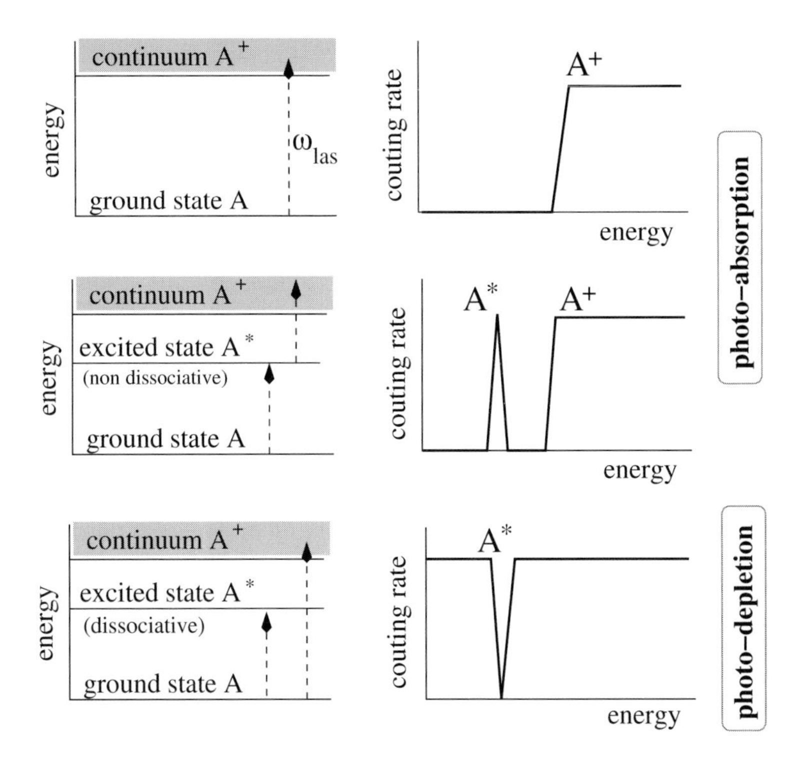


**Figure 2.7:** Schematic view of absorption and depletion spectroscopy. The left part of the figure describes the various situations in the spectral representation. The right column shows the corresponding observed strengths. Several situations are considered: typical IP measurement (top), access to excited (non dissociative) level, access to excited (dissociative) level. The two top cases provide photoabsorption signals while the bottom one corresponds to a photodepletion measurement.

To help in the discussion, both methods are schematically represented in Figure 2.7. We consider a cluster $X$ with schematic excitation spectrum as given in Figure 2.7. The ground state is labeled $A$ (having energy $E_{GS}$) and possible bound excited state by $A^*$ (with its energy $E^*$), ionization continuum being denoted $A^+$ (with energy $E^+$). It should be noted that we do not specify here the nature of the excited state $A^*$. It may correspond to a single electron excitation, as well as to a collective state of the electronic cloud, or to any other choice appropriate for the given cluster.

#### 2.2.2.1 Photoabsorption

Let us first consider the simplest situation of a photoabsorption experiment on a cluster which has no bound excited state, i.e. no further state between ground state and ionization threshold $E_{\text{IP}} = E^+ - E_{\text{GS}}$. The experimentally measured quantity after laser impact is the number of ionized clusters $X^+$. The cluster is irradiated by a weak laser of tunable frequency $\omega$. As long as $\hbar\omega < E_{\text{IP}}$, no electron is stripped from the cluster and no signal is recorded for $X^+$. But as soon as $\hbar\omega > E_{\text{IP}}$, electrons are released and $X^+$ clusters are observed (in a

TOF for example). The threshold frequency $\hbar\omega = E_{\mathrm{IP}}$ is nothing but the ionization potential (IP) of the $X$ clusters, see upper panel in Figure 2.7. Such a simple optical device consisting out of a tunable light source and a mass spectrometer for cluster identification allows a direct measurement of the cluster IP, provided it lies in the accessible range of frequencies $\hbar\omega$ of the laser source. The threshold can be somewhat weakened by multi-photon processes. One has thus to use low intensities to suppress them.

Such a simple situation without intermediate excited state is rare. More frequent is the case in which one (or several) bound excited states $A^*$ exist. We first consider the case for which there is only one non-dissociative state and we see how to access it. Because this state is non-dissociative the method will not immediately destroy the cluster and thus be of photoabsorption type. Again, we irradiate the cluster with a laser with tunable frequency $\omega_1$, and we detect $X^+$ clusters when formed. As long as the cluster absorbs only one photon, the patterns are exactly the same as in the previous case. For $\hbar\omega < E^+ - E_{GS}$, no $X^+$ signal is observed. A large $X^+$ signal is observed otherwise. Of course, when the laser frequency matches the intermediate excitation energy $E^* - E_{GS}$, the $A^*$ state becomes populated. But because this state is non-dissociative it does not lead to ionization and thus $X^+$ is neither formed nor observed. Let us now assume that the cluster simultaneously absorbs a second photon of fixed frequency $\omega_2$ such that $\hbar\omega_2 > E^+ - E^*$ while still residing in the excited state $A^*$. The absorption of this second photon is sufficient to ionize the cluster so that $A^* \to A^+$. The final ionization signal $X^+$ as a function of laser frequency $\omega_1$ thus exhibits a peak at $\hbar\omega_1 = E^* - E_{GS}$, in addition to the "continuum" signal for $\hbar\omega_1 > E^+ - E_{GS}$ (see middle panel in Figure 2.7). Such two photons processes for ionizing clusters are known as Resonant Two Photon Ionizations (R2PI) and give access to excited states of clusters, provided they are sufficiently stable against dissociation. The stability is required to sustain the excited state for a sufficiently long time in order to give the second photon its chance to finally ionize the cluster.

#### 2.2.2.2 Photodepletion

When one wants to study dissociative excited states one has to modify the procedure to the photodepletion method. It is clear that one cannot access a dissociative state as in the former example by first populating it and later on ionizing it, just because the state may have dissociated before the actual probe photon is absorbed. The photodepletion technique provides a solution to overcome this problem. For this purpose one uses two laser beams, a first one with tunable frequency $\omega$ and a second one with fixed frequency $\hbar\omega_0 > E_{\mathrm{IP}}$, i.e. above direct ionization threshold. The clusters are first irradiated by the laser with tunable frequency $\omega$. When $\hbar\omega = E^* - E_{GS}$, the clusters with the $A^*$ excited (dissociative) state are promoted to that state. But such states are usually short lived and the clusters residing in state $A^*$ will thus soon dissociate and disappear. The set of clusters is then irradiated by the second laser with fixed frequency $\omega_0$ above IP. This laser is for sure able to ionize all present clusters. But clusters which were formerly excited in $A^*$ state will not appear in the ionization signal because they have already disappeared by dissociation. If one plots the number of $X^+$ clusters after this second laser irradiation, as a function of the first laser frequency $\hbar\omega$, one will then observe a dip in the signal, located at $E^* - E_{GS}$, see lowest panel in Figure 2.7. This depletion in the ionization signal thus labels the energy of the excited state $A^*$. A typical example of application of this photodepletion technique is the measurement of optical response in metal clusters (see Figure 1.17). The Mie plasmon usually has energies above the threshold for monomer evaporation and thus corresponds to a dissociative state of a cluster.

Both photoabsorption and photodepletion techniques thus provide complementing information on clusters as they allow one to access both non-dissociative and dissociative excited states in clusters. They are thus widely used in various experiments, as we shall see further below (e.g. Section 2.3).

### 2.2.3 Photoelectron spectroscopy

Photoelectron spectroscopy is based on the measurement of the kinetic energy of an electron $\varepsilon_{\text{kin}}$ emitted after irradiation by a laser with frequency $\hbar\omega_{\text{las}}$ (hence the term photoelectrons often used in this context). This gives access to the original single-electron energy $\varepsilon_\alpha$ inside the cluster through the simple relation

$$\varepsilon_{\text{kin}} = n_{\text{phot}}\hbar\omega_{\text{las}} + \varepsilon_\alpha \tag{2.1}$$

where $n_{\text{phot}}$ is the number of photons involved in the process. The case is exemplified in Figure 2.8. The two electron levels, $\alpha = 1$ and 2, are mapped into the kinetic energy spectrum. Three photons are required at least in the given example. A higher number of photons produces further copies of the single-electron spectrum at larger $\varepsilon_{\text{kin}}$ and with much lower yield. The minimum number of needed photons can vary depending on IP and photon frequency. There are several techniques to measure the kinetic energies of the emitted electrons. One widely used one is TOF as done for the clusters themselves. The much larger charge to mass ratio requires careful guiding of the electron flow, e.g., by a magnetic mirror [dH93]. Another interesting technique is photo-imaging spectroscopy. A static electrical field is used to map the distribution of electron velocities onto definite positions on a detection screen, for

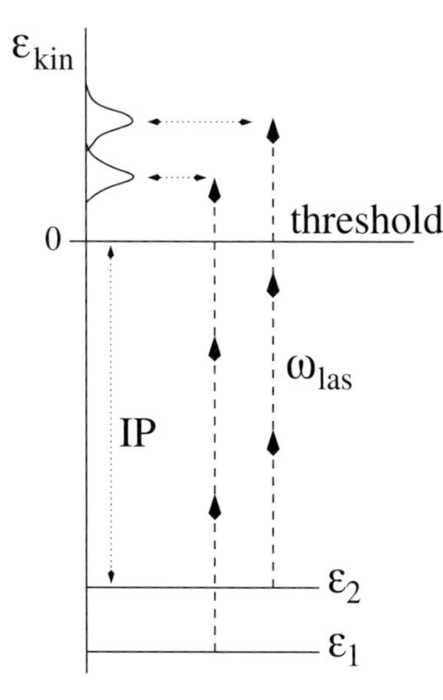


**Figure 2.8:** Schematic view of photoelectron spectroscopy. The system has two single-electron states $\varepsilon_\alpha$. The IP is then $-\varepsilon_2$ while the emission threshold is taken as reference of zero energy. The kinetic energies for free electrons are necessarily positive and go from threshold on upwards. The photons are indicated by dashed lines with arrows. At least three photons are required for onset of emission here. Each initial state, $\varepsilon_1$ or $\varepsilon_2$, produces its own peak in the kinetic energy distribution, as indicated.

a review see [BPHH96]. This allows simultaneous resolution of kinetic energy spectra and angular distributions, see Figure 5.21.

Such photoelectron spectroscopy experiments were performed on negatively charged clusters in the early 1990s. It is indeed easy to strip, with optical photons in a mere one-photon process (i.e. $n_{\rm phot} = 1$), one electron from such clusters because the IP for anions is usually very small, typically in the range of a few eV. More recently, the photoelectron spectroscopy has also been applied to neutral fullerenes [CHH$^+$00] and to positively charged metal clusters [WavI02]. In such cases one may either use high frequency photons or rely on multi-photon processes to extract electrons. We shall re-discuss this question in more detail in Section 5.3.1 as an example for actual trends in cluster dynamics. But the principle remains as sketched in Figure 2.8.

The simple photoelectron spectroscopy technique can also be refined by performing the so called ZEKE (Zero Electron Kinetic Energy) measurements, which has been applied to neutral clusters, in particular in [CTCS90]. The idea is here to tune laser frequencies so that electrons are stripped with almost vanishing kinetic energy. (Two lasers may be used simultaneously to overcome limitations due to the energy of optical photons.) Of course some electrons may be extracted with larger kinetic energies. To get rid of these, one waits a while (typically 1 $\mu$s) so that fast electrons have had enough time to leave the system. In turn, only electrons with kinetic energy close to zero can then be found in the vicinity of the cluster, because of their vanishingly small velocity. A low voltage pulse is then applied: it accelerates these low energy electrons and extracts them from the system. The time delay thus allows one to select the electrons with (almost) zero kinetic energy. One scans the laser frequency and draws the yield as a function of them. This ZEKE technique provides a very good resolution in photoelectron spectra ($\sim 0.1$ meV) as compared to standard photoelectron spectroscopy, for which the resolution is typically 20–50 meV.

A comparison of ZEKE with standard photoelectron spectroscopy is shown in Figure 2.9. In this figure, one actually compares three approaches, all involving one-photon processes. The first one, the so called photoionization efficiency is nothing but a scan of laser frequency and an accumulation of emerging ion signal. This is a standard way to access IP by photoabsorption (Section 2.2.2.1). The ionization threshold is indeed distinguishable but somewhat spread. The second technique is standard photoelectron spectroscopy which provides a structure with a Full-Width at Half Maximum (FWHM) of about $300\,\mathrm{cm}^{-1} \approx 40\,\mathrm{meV}$. In contrast, the ZEKE signal reaches a resolution of order 5 cm$^{-1} \approx 0.6\,\mathrm{meV}$, even resolving vibrational levels.

In spite of its high potential applications electron spectroscopy is not yet routinely used in cluster physics. On the other hand, it provides a unique access to single electron levels in clusters which provides key information on cluster structure and dynamics. Thus this field will be important in future cluster research, for promising examples see Chapter 5, Figure 5.1 and Section 5.2.3.

## 2.3 Examples of measurements

After our brief survey of cluster production methods and basic experimental tools, we can now discuss a few examples of measurements typically performed on clusters. We do not plan to

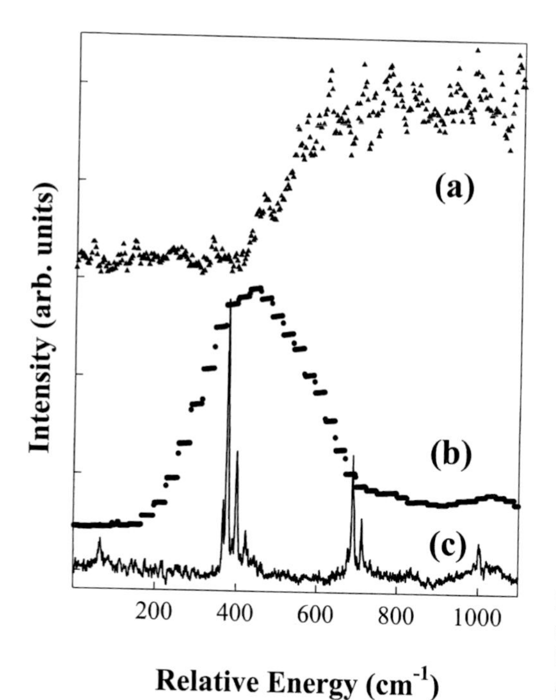


**Figure 2.9:** Spectra of triniobium monoxide obtained from a) photoionization efficiency, b) standard photoelectron spectroscopy and c) ZEKE spectroscopy. From [YH00].

provide here an exhaustive catalog of possible experiments. We would rather like to present a few typical examples, covering widely performed experiments but also more specific examples in order to hint the many explored (and to be explored) facets of cluster physics. More examples will be discussed in Chapter 5. In particular, we shall try to illustrate on selected examples the various types of probes used in cluster physics, in terms of static and time dependent electric and magnetic fields. It should finally be remembered that such experiments lie at the end of the experimental chain (see Figure 2.1) with all its implied limitations in terms of cluster production and selection. These limitations, although not necessarily critical, should be kept in mind in the interpretation of all the experimental results shown below.

### 2.3.1 Abundances

Basic observables in cluster physics are abundance spectra. One collects a protocol of the number of clusters of a given type/size. This can be simply achieved by a mass spectrometer installed after the cluster source. Such an observable looks at first glance too simple to be informative. At second glance, abundances tell a lot about cluster structure. There exist, e.g., often some "special" clusters which are more abundant than others in the sequence. An overabundance of a given species indicates that this species is especially stable which presumably reflects a peculiar arrangement of electrons and/or ions in the system.

The situation is somewhat similar to the case of isotopic abundances in the universe. As is well known, measured abundances of elements in the galaxy show that some specific nuclei are much more abundant than the neighboring nuclei. This overabundance is known to reflect shell closure in these specific nuclei, which explains the particular stability of these systems. The isotopic abundances provide a typical example of a simple observable which allows one to extract the famous sequence of magic numbers in nuclear physics. These magic numbers (2, 8, 20, 28, 50, 82, 126, ...) just correspond to shell closures for neutrons and/or protons, see Section 4.4.2.

Similar patterns have been observed in clusters. In particular in metal clusters, where electrons are quasi free (see Section 1.3) and behave in a way similar to nucleons in nuclei, marked abundance variations were observed in the mid 1980s. This at once suggested electronic shell closure effects (see Section 4.2.1.1), similar to the nuclear case. The effects were soon confirmed by simple jellium model calculations (Section 3.2.3). In the case of metal clusters one can furthermore follow these shell closure effects up to very large cluster sizes (up to several thousands of atoms), at variance with the nuclear case in which basically not more than about 300 nucleons can form a nucleus. When studying these shell oscillations in metal clusters on such a very large scale one observes that the intensity of the shell effects (namely of the overabundances) is modulated with size. It is large in small clusters, decreases up to electron numbers of order 1000 to increase again in larger clusters. This modulation of the intensity of shell effects is known as supershells and was predicted in 1971 by Balian and Bloch on the basis of an analysis of periodic orbits in a closed container (see also Figure 4.5 in Section 4.2.1.1). Clusters are the first physical systems to have given access to this subtle quantum effect because their size can be varied almost completely at will.

Metal clusters are not the only clusters to exhibit pronounced spikes in the abundance spectra. Because clusters are constituted of electrons as well as ions one may also expect geometric shell closures at the level of ions, see Section 4.2.1.2. Ionic shell effects are the only ones in rare gas clusters where electronic shell effects are absent because electrons remain basically bound to their parent atom. Both (electronic and geometric) sorts can be present in other materials. And it may actually be a conjunction of both which will make a peculiar structure especially stable and hence abundant. This is illustrated in Figure 2.10 for the famous case of carbon clusters. As is clear from the abundance spectrum, $C_{60}$ exhibits a particular stability. But this spectrum turns out to be richer than that, as it shows that other structures are also particularly abundant, as for example $C_{70}$.

In spite of their apparent simplicity, abundance spectra thus constitute an important tool of investigation of cluster properties, practically a doorway to structure analysis. As a final remark, it should be noted that only a few cluster species have been studied in this respect. While simple alkaline clusters have been explored in detail over the years, the abundance spectrum of not so different material such as Mg (next to Na) has been explored only very recently [DDTMB01]. And the pattern exhibited by the abundance spectrum of such a material is not yet fully understood, see Figure 4.6.

## 2.3.2 Ionization potentials

Ionization potentials (IP) constitute another basic measured quantity in clusters. We have seen in Section 2.2.2 how such a quantity can be simply attained experimentally using a laser and a mass spectrometer. The IP gives an insight into electronic structure as it is physically associated with the energy of the least bound electron (called the HOMO). One has however to be cautious here and to define more precisely what electron emission means in detail. By

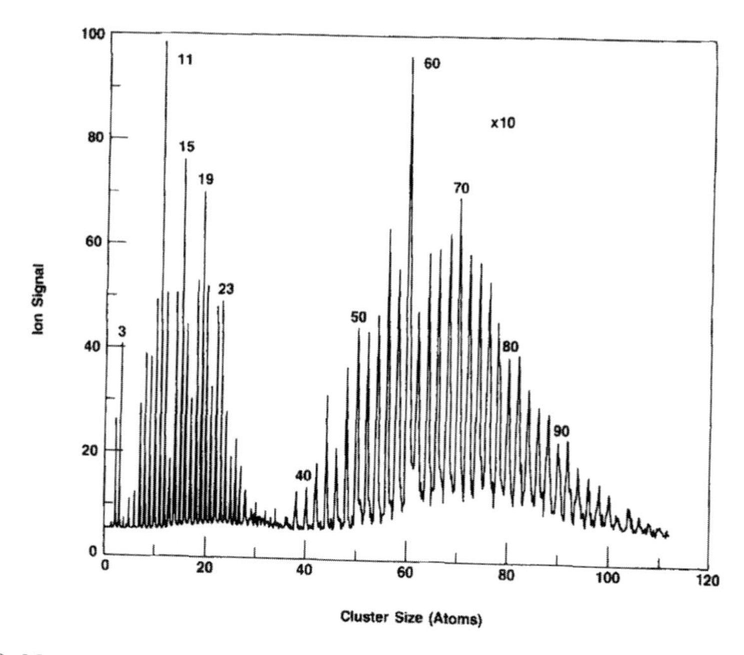


**Figure 2.10:** Mass spectrum of carbon clusters. Clusters have been produced in a supersonic beam by laser vaporization of a carbon target in a pulsed supersonic nozzle operating with a helium carrier gas. Fullerenes correspond to the long distribution of even-numbered carbon clusters starting around $C_{40}$ and up to about $C_{100}$. The $C_{60}$ appears as particularly abundant. This figure corresponds to the first published experiment revealing fullerene cluster distribution [RCK84].

definition, the IP of a system is the minimum energy necessary to extract one electron from the system. However this definition does not specify how the electron is removed from the system. Roughly speaking one may conceive two extreme situations of i) an instantaneous removal or ii) an extremely slow process. These two processes will have different physical consequences. The rapid process i) (usually called "vertical") leaves the ionic part of the cluster basically frozen. Only a difference of electronic energies plays a role. It is a fully non-adiabatic mechanism. On the contrary, in process ii) (usually called "adiabatic") the whole system continuously adapts itself to the situation, namely to the fact that one electron is slowly moving away from the system. Here one counts the energy difference of two fully relaxed configurations. Of course, vertical and adiabatic scenarios lead to different IP values. Actual experimental results will usually lie somewhere in between. An example of measured IP is given in Figure 2.11 in the case of small mixed clusters $Na_nF_{n-1}$. Experimental data, calculated vertical and calculated adiabatic IPs are shown in this figure. The vertical IP's are larger than the adiabatic ones, as expected. The comparison between experimental and theoretical results shows an overall acceptable agreement, in particular for the adiabatic IP's. At second glance, one spots two different versions of adiabatic IP's, the second one called "local adiabatic". This is related to the above mentioned feature of shape isomerism in clusters. Once the electron is stripped off, the ionic configuration will first relax towards the next local

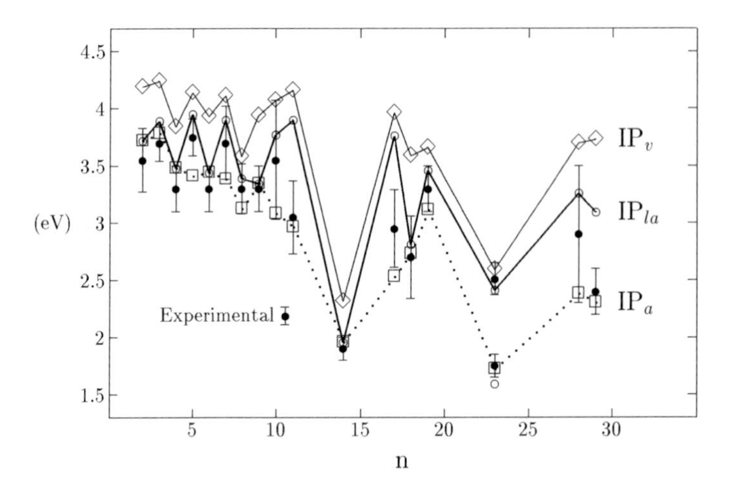


**Figure 2.11:** Comparison between vertical (open diamond), adiabatic (open square) and local adiabatic (open circle) theoretical IP's with experimental IP's in $Na_nF_{n-1}$ clusters. See text for details. From [DGGM$^+$99].

equilibrium point which is not necessarily the absolute minimum. This is the local adiabatic relaxation process. For longer times, thermal motion will carry the system slowly into the deepest minimum, the true ground state configuration. That is then per definition the adiabatic process. Figure 2.2.1 shows visible differences between this intermediate and the final stage. Last not least, it is also interesting to consider the variations of the ionization potential with cluster size. These are connected to the various types of structures taken by $Na_nF_{n-1}$ clusters as illustrated in Figure 1.13. This demonstrates, not surprisingly, that IP's are closely related to the underlying cluster structure and thus provide valuable structural information.

Of course ionization potentials were measured in many types of clusters. They were in particular systematically measured in metal clusters. These exhibit, not surprisingly, variations in the IPs which reflect the strong electronic shell effects already identified in abundance spectra, see Section 2.3.1.

### 2.3.3 Static polarizabilities

The static dipole polarizability $\alpha_D$ of a cluster is a measure of its response to an external static electric field $\mathcal{E}$. Cluster polarizations are measured by applying a static but inhomogeneous field $\mathcal{E}$ to a cluster jet. The clusters are polarized by the field $\mathcal{E}$. They acquire a dipole moment $\mathbf{D} = \alpha_D \mathcal{E}$, due to the displacement of the electronic center of mass as compared to the ionic one. This dipole moment then couples to the static field $\mathcal{E}$ and although clusters remain neutral, the beam is deflected. Indeed, because the electric field is inhomogeneous, it possesses a non-vanishing gradient and clusters are exposed to a force $F = -\nabla(\mathbf{D}\cdot\mathcal{E}) \propto -\alpha_D \nabla\mathcal{E}^2$. This, in turn, leads to a deflection $d \propto \alpha_D \nabla\mathcal{E}^2$ which allows a direct measurement of $\alpha_D$. The deflection depends on the polarizability and/or the cluster size. By measuring the deflection

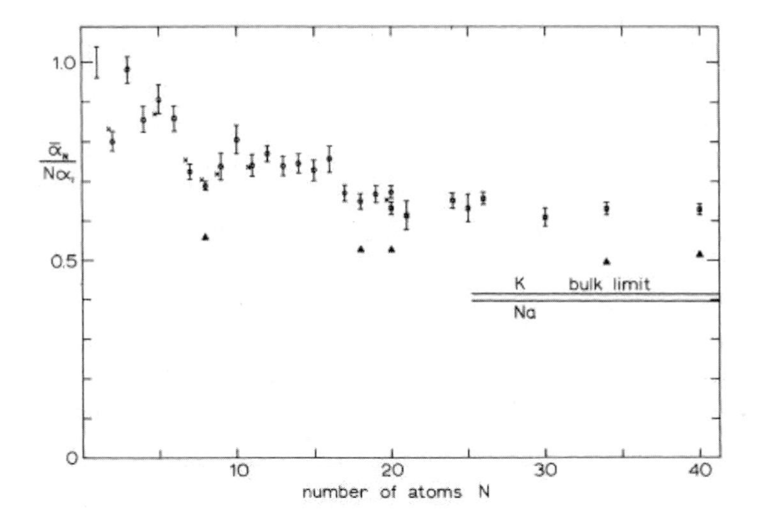


**Figure 2.12:** Static experimental polarizabilities of sodium and potassium clusters. From [KCdHS85].

and mass simultaneously, one obtains the polarizability of the clusters as a function of their mass.

Metal clusters exhibit sizable polarizabilities because in these systems the electron cloud is loosely bound to the ionic background. Measurements have been performed in various alkalines. Figure 2.12 shows the example of $Na_n$ and $K_n$ polarizabilities. The results are plotted as a function of size and scaled to the atomic polarizabilities. Indeed, the larger the cluster, the larger the polarizability. But the total cluster polarizability of course does not reduce to a mere sum of individual atomic polarizabilities because of the electron interactions inside the cluster valence cloud. The experimental results shown in Figure 2.12 exhibit interesting features. First, one can identify shell structure effects. For a few particular system sizes, the values of the polarizability are smaller than the general trend. These effects appear for numbers of valence electrons equal to 2, 8, 20 mostly, which indeed corresponds to known electron shell closure in alkaline clusters (see Section 4.2.1.1). Second, the convergence towards the bulk limit is relatively slow. For example, the actual value of the polarizability for $Na_{40}$ is still far away, about a factor 1.5, above the bulk ($N \to \infty$) value. This is again an example illustrating the fact that clusters are not simply pieces of bulk matter, see the discussion in Section 1.2.1.2.

Thus far we have discussed the static dipole polarizability. The notion of polarizability is often also used for the dynamical polarizability $\alpha_D(\omega)$. This quantity is closely related to the optical response which will be discussed in the next section.

### 2.3.4 Optical response

The optical response is the response of a cluster to a time dependent electromagnetic field. It is, so to the say, the natural extension of the static polarizability (see previous section) to time dependent fields. The quantity to be measured here is the cross section for photoabsorption which is closely related to the dipole strength distribution or dynamical dipole polarizability. The enormous selectivity on the dipole transitions is related to the fact that the photon wavelength in the optical range ($\approx$ 400nm) is usually much larger than the cluster diameter, justifying the dipole approximation for the electromagnetic vector potential $\mathbf{A} = \mathbf{A}_0 \exp(\imath \mathbf{kr}) \longrightarrow \mathbf{A}_0$. The most widely used means to measure photoabsorption cross sections are the photodepletion techniques, explained in Section 2.2.2.2. The photoabsorption cross section is then proportional to the number of photons of a given frequency $\omega$ absorbed by a cluster.

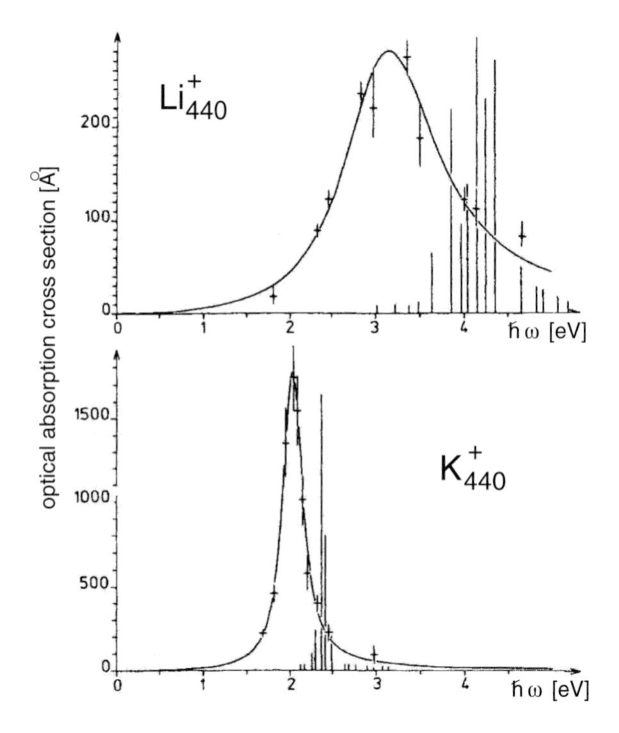


**Figure 2.13:** Optical response in $Li^+_{440}$ and $K^+_{440}$ Experimental points are indicated by crosses (with error bars). The full lines are results of a theoretical linear response calculation. After [BC95].

The peaks of the photoabsorption cross section as a function of photon energy map the spectrum of dipole excitations of the system. There is a wide variety of results depending on the material. The general case shows a fragmented spectrum with a series of many comparable response peaks, see e.g. the result for LiH clusters in Figure 4.12. Metal clusters, however, are distinguished by exhibiting one strong resonance peak usually in the range of optical frequencies. We have already discussed in Section 1.3.3 a simple model for the resonance peak on the basis of Mie theory. We just note here the basic mechanism: the Mie surface plasmon is viewed as a collective oscillation of the valence electron cloud with respect to the ionic background. The optical response is a most basic observable coming next to the abundances. Particularly for metal clusters, it has been among the most studied observables in the past two decades. An example with experimental and theoretical results is shown in Figure 2.13. One sees the pronounced peaks typical of metal clusters. However, in these experimental results the widths are relatively large. There are various physical mechanisms contributing to this width, as discussed in more detail in Section 4.3.5. It is most likely that temperature effects play an important role in the particular case considered here (see also Section 2.1.3.1 for a discussion of the impact of temperature on the optical response). The ionic configurations undergo large thermal fluctuations for high temperatures (near the melting point). The spectrum thus shows the incoherent mix of spectra from very different ionic configurations which, of course, develops a certain width. On the other hand optical response provides more detailed spectra and clear clues on the global geometry of the cluster when temperatures are sufficiently low (see also Section 4.3.2). As the optical response is a key tool of investigation we shall discuss it in much more detail in Section 4.3 of Chapter 4 and in several other places throughout this book.

### 2.3.5 Vibrational spectra

Vibrational spectra are most important observables in molecular physics. They allow one to conclude on molecular symmetries and on the curvature of the Born-Oppenheimer surfaces (Section 3.1.3 and Appendix C.1) in the vicinity of its equilibrium. Vibrational bands have thus been widely studied in molecular physics in many kinds of molecules [Wei78]. The situation is more complex in clusters. These larger systems have an unoverseeable amount of vibrational modes. Although certainly useful, vibrational spectra in clusters are hard to interpret in the standard terms of molecular physics. They are thus not nearly so much studied as optical spectra.

Vibrational spectra have energies in the range of several tens of meV, thus in the deep infrared region. The analogue to optical spectroscopy is thus here infrared spectroscopy [Ell98], for an example from cluster physics see [LZWR95]. It proceeds, to a large extent, similarly. But the much smaller photon energies require special care. A way to deal with the simpler more energetic photons and yet to access the low vibrational energies is Raman spectroscopy [AM76, Ell98]. This is basically inelastic photon scattering. The cross section as a function of the energy loss provides a direct mapping of the low-energy excitation spectrum. An example for that is given in Figure 5.15 where breathing modes of large clusters are investigated. The concentration on collective modes as a first step is one way to approach the very complex vibrational spectra of large clusters (for a recent theoretical analysis see also [RS02]).

Another possible path to the detailed information in vibrational spectra is provided by ZEKE spectroscopy and its high resolution, see Section 2.2.3. An example is given in Figure 2.14 which shows a ZEKE spectrum of the small $V_3$ cluster in an energy range around its ionization potential. The ZEKE measurements give here an IP of 44342 $cm^{-1} \equiv 5.5$ eV corresponding to the main peak in the figure. But one also observes a second peak at 44514 $cm^{-1}$ namely 172 $cm^{-1} \equiv 21$ meV above ground state. By comparison with theoretical predictions of vibrational modes of $V_3$ one may attribute this second peak to the transition from the ground state to the first excited level of a totally symmetric vibration of the ionized $V_3^+$, as depicted in the inset in the figure.

### 2.3.6 Conductivity

The conductivity is a key feature of bulk material. It was rarely a matter of concern in traditional molecular physics. As clusters stay in between molecules and bulk, conductivity is an interesting observable, the more so as electronic circuitry is stepping down to these small length scales. Even single molecules are coming into the focus as micro-switches between

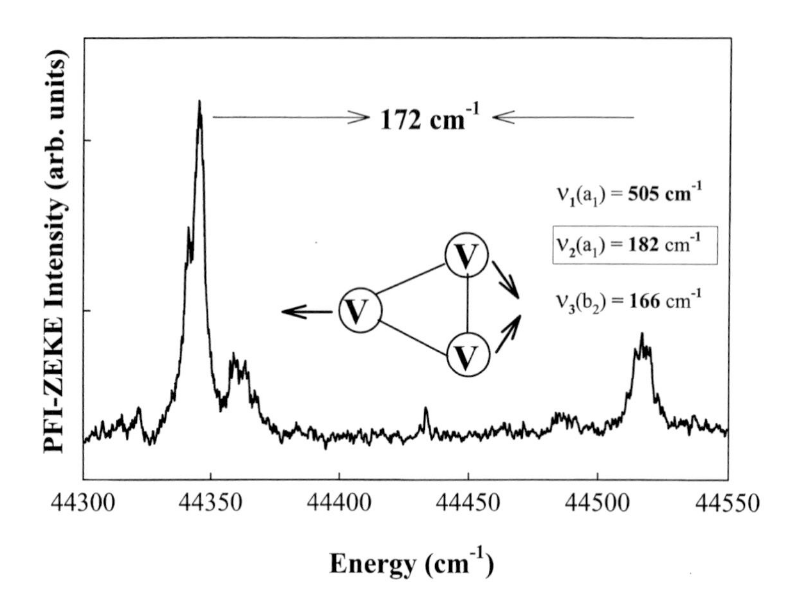


**Figure 2.14:** ZEKE spectrum of $V_3$. The calculated vibration frequencies are indicated on the right. The right peak is attributed to the frequency $\nu_2$. The corresponding vibrational mode is depicted in the inset. From [YH00].

conducting and non-conducting states. In any case, switches at the nano-scale are a technical challenge of high current interest. Clusters are promising candidates in that development and their conductivity is a crucial property. The experimental determination is, in principle, simple. One applies a constant voltage and measures the emerging current. The practical problem is the size of the samples which makes it a bit tricky to establish contacts with the measuring apparatus. Half of the problem is solved for clusters on a surface because the substrate serves at the same time as the one contact medium. The contact to the other side can be established, e.g., by the tip of a tunneling or transmission electron microscope. Other setups pin the molecule or cluster between nano-contacts placed on an insulator for reasons of mechanical stability. In any case, the contacts are an essential part of the measurement and it turns out that the conductivity is a property of the whole setup and not a feature of the cluster as such. This makes conductivity not so useful an observable for investigating cluster properties as such. But it is, of course, a most important feature in applications. There are thus many investigations on the conductivity of clusters and arrays of clusters. For a review of nano-contacts partially addressing clusters see [JBD$^+$00]. The topic is also much discussed in connection with the mesoscopic cousins of clusters, namely carbon nano-tubes [Kai01].

A typical and transparent example for conductivity of a single cluster is shown in Figure 2.15 with the current voltage characteristics of an Au cluster pinned to a Si surface [LLC$^+$00]. At first glance, the characteristics look like many others in electronics. But a closer look reveals step-like pattern rather than a smooth growth. The structure becomes more apparent from the lower panel which shows the differential of the current. The $dI/dV$ shows marked oscillations. At some point (indicated by an arrow) the oscillations even overshoot

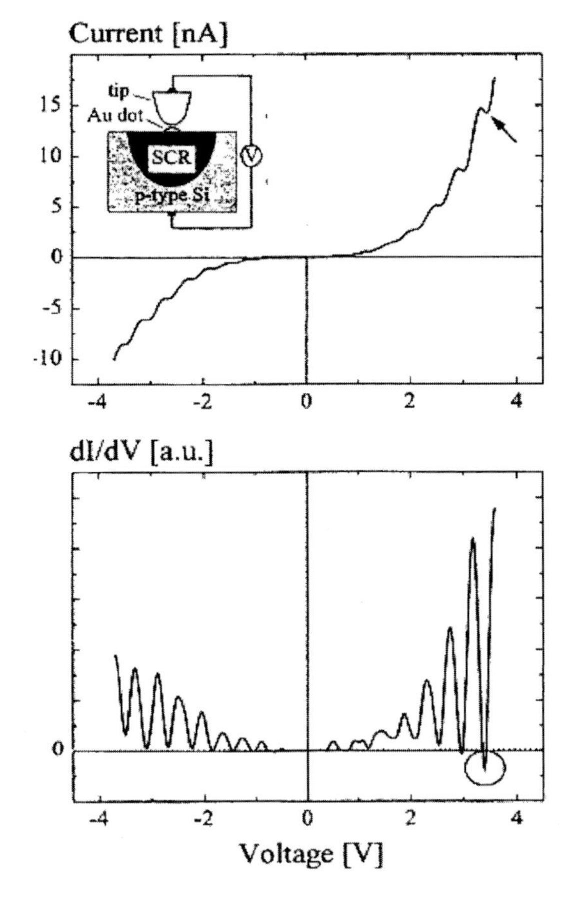


**Figure 2.15:** Current–voltage characteristics for an Au cluster attached to a Si substrate. The upper panel shows the standard $I$–$V$ diagram and the inset sketches the experimental setup. The lower panel shows the differential characteristics $dI/dV$ versus voltage $V$. From [LLC$^+$00].

and produce a negative slope, i.e. a non-monotonous piece in the characteristics as such. The oscillations are an effect of the small size. For clusters with less than 3 nm diameter, the capacitance of the cluster $C$ is so small that the single charging energy $e^2/C$ becomes larger than the temperature $T$. This discretizes flow which, in turn, leads to what is called "Coulomb staircases" in the $I - V$ curve (and subsequently to the oscillations in its derivative). Moreover, there is obviously a suppression of current at low voltage. This is also due to the large charging energy $e^2/C$ and is called Coulomb blockade. The example thus demonstrates that clusters are special conductors due to their small size.

### 2.3.7 Magnetic moments

Magnetic moments characterize the response of a cluster to a static magnetic field. The experimental setup is essentially a Stern-Gerlach device [HW00]. The measurements rely on the existence of an inhomogeneous magnetic filed, similarly to the setup used for measuring static electric polarizabilities (see Section 2.3.3). A cluster jet is injected between two magnets providing a magnetic field $\mathbf{B}$. A cluster with a magnetic moment $\mu$ couples to the external field $\mathbf{B}$

with associated potential energy $E_p = -\mu \cdot \mathbf{B}$. As usual this leads to the appearance of a force $-\nabla E_p$ and a moment $\mu \wedge \mathbf{B}$. In a homogeneous field, only the moment is active and there is no deflection of the beam. As soon as the magnetic field is inhomogeneous, a net force acts on the cluster. This provokes a deflection of the cluster beam. For a given cluster the amplitude of the deflection depends on the magnetic moment and/or the size. The deflections are then recorded and a mass identification is performed by laser ionization and injection in a mass spectrometer, once the beam is outside the region where the magnetic field is active.

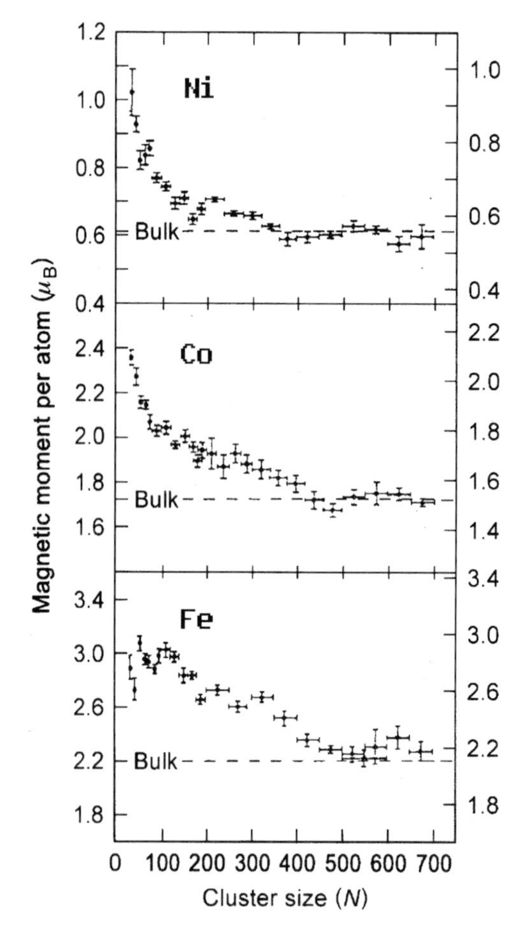


**Figure 2.16:** Average magnetic moments of size-selected clusters of various materials: Ni, Co and Fe, in unit of Bohr's magneton $\mu_B$. The estimated temperature of the clusters is around 100 K. After [BCdH94].

As is well known, the magnetic moment is proportional to the total angular momentum of a system. When no global rotation is present it becomes directly proportional to the spin itself. Not surprisingly these are thus mostly noble metals which exhibit the largest magnetic moments while such effects are very small in simple alkaline systems. An example of measurements is shown in Figure 2.16. In this figure the magnetic moments per atom are displayed as a function of size in three "noble" metals namely Ni, Co and Fe. It is interesting to note the differences and similarities exhibited by the curves for the three materials. First, as is clear from the indicated bulk values, these three materials have somewhat different magnetic characteristics. The bulk magnetic moment per atom is about four times larger in Fe than in Ni and about 1.5 larger than in Co. Nevertheless, the general trend of the magnetic moment with size is the same for the three materials: a decrease towards the bulk value which is reached for sizes around 400-500. However, when looking more into the detail, one can identify different behaviors in the various materials. Particularly in Ni, one observes marked oscillations of the magnetic moment as a function of size. Such oscillations are also visible in Co and, to a lesser extent, in Fe, but at different sizes than in Ni. Finally it should be noted that temperature effects play here a crucial role, similar as in bulk material. In Figure 2.16 the temperature was about 100 K and thus sufficiently low to see pronounced magnetic moments. It is obvious that magnetic moments will fade away with increasing temperature. A most interesting question is also the magnetic "memory", the hysteresis in the magnetic response of a system to an applied static magnetic field. This effect depends much on temperature in bulk material. This remains so for magnetic clusters and one finds additionally a strong dependence on cluster size. This point, facing to actual and future developments is taken up in Section 5.1.5. As in the case of the conductivity, the quest for cluster magnetism is much inspired by its enormous potential for applications.

## 2.3.8 Photoelectron spectroscopy

Photoelectron spectroscopy provides a most direct access to electronic energy levels in clusters, as discussed in Section 2.2.3 where the principle of this technique was presented. The first experiments with photoelectron spectroscopy were performed in the late 1980s on cluster anions [MEL$^+$89]. In that case a cluster beam is first negatively charged by an electron beam through a reaction $\mathrm{X}_n + e^- \rightarrow \mathrm{X}_n^-$ and then ionized by a laser as $\mathrm{X}_n^- + \hbar\omega \rightarrow \mathrm{X}_n + e^-$. The kinetic energy of the emitted electron is then recorded to provide the photoelectron spectrum. The IP of neutral metal clusters are typically 5 eV and even larger for cluster cations. This is above standard laser frequencies. Ionization thus requires at least two photons which reduces the signal and/or requires higher intensity lasers. On the contrary, negatively charged metal clusters have a very small ionization potential, typically in the eV range. Any optical photon will thus be able to strip directly the least bound electron. Deeper lying levels will also be more easily accessible by optical photons than in the neutral or positively charged species. The use of negatively charged clusters thus facilitates the spectroscopy and this is the reason why cluster anions were analyzed first. Meanwhile, there are also several measurements on neutral and positively charged clusters, see Sections 5.2.3 and 5.3.1.3.

Examples of experimental photoelectron spectra in negatively charged metal clusters are shown in Figure 2.17. These measurements have been performed both for alkaline and transition metals. The figure displays results for alkalines. It exhibits several interesting features. As expected, the energy range explored here is relatively modest (0–2 eV) as experiments have been performed with a laser frequency of 2.54 eV. In all materials, when the sizes of the clusters are increased one observes more and more complex spectra, which reflects the fact that the density of electronic states increases with cluster size. One can also note that the resolution of the peaks decreases with increasing cluster size. This reflects, to some extent, the densification of the electron spectrum and increasing correlations, but there is probably, on top of this physical effect, an experimental degradation of the resolution itself. Nevertheless, the highest occupied levels are very clearly identified with narrow structures well separated from deeper bound levels.

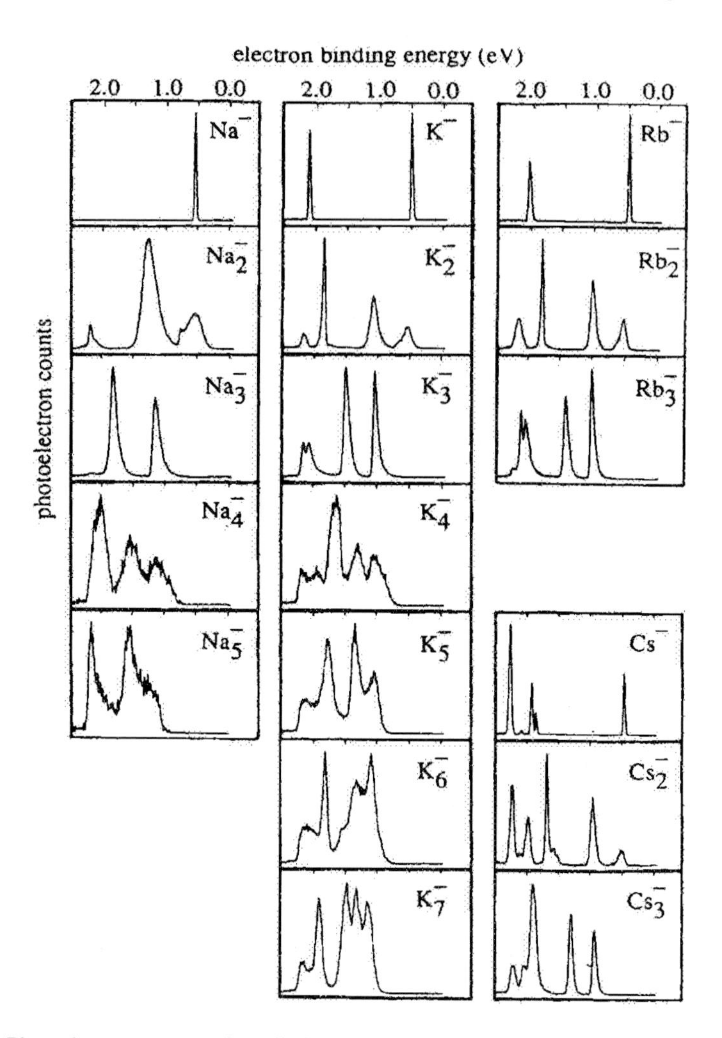


**Figure 2.17:** Photoelectron spectra of small alkali cluster anions. From [dH93].

The major interest of photoelectron spectra lies in the access to single electron levels. It should, however, be noted that even in the case of well separated experimental peaks, the link to theory is rather involved, as well as the actual interpretation of the peaks themselves. Indeed, when electrons are emitted the electrons remaining in the cluster rearrange and thus do the levels they occupy. For very slow emission, even the ions have time to relax into a new minimum. The situation is here similar to the case of IP (Section 2.3.2) with a higher degree of complexity, due to the fact that one considers several electron levels and not just the least bound one. This means that a proper interpretation of photoelectron spectra has to account in detail for the actual ionization process, fast or fully adiabatically. And the interpretation of the spectrum will depend on this mechanism. There are again the two extreme cases of an instantaneous and a fully adiabatic emission where only the instantaneous process may map the original electron levels. In slow excitations, there comes a further complication for multi-photon processes. This is the possibility of sequential emission where the first photon occupies a bound excited state of the cluster. Moreover, there is the chance of multiple ionization where each successively emitted electron will, in fact, probe the ionization potential of $X_n$, $X_n^+$, $X_n^{++}$ etc, and not only the electron levels of $X_n$. A proper interpretation needs backing from theoretical simulations of the photo-ionization which model the details of the excitation mechanism, e.g. the full pulse profile in case of lasers. Short pulses and large photon energies (to ensure one-photon processes) minimize these uncertainties. Free-electron lasers and synchrotron radiation are here promising sources with a broad choice of energies delivering very short pulses in future developments, for an example see Section 5.2.3.

### 2.3.9 Heat capacity

We have seen above that temperature plays an important role in cluster physics. Clusters are produced at finite temperature in most sources. And we have seen that temperature may affect several physical observables measured on clusters. Thermodynamics of clusters is thus an important issue both for itself and because of the impact of temperature on our access to cluster properties. In this context, measurements of cluster heat capacities represent an important challenge. In particular, typical experimental conditions often imply temperatures of several hundreds of K, a temperature range in which one expects, by interpolation from the bulk, the solid/liquid (melting/fusion) phase transition. The heat capacity is then a particularly well suited observable to track such a melting transition. Basic thermodynamics tells us that the heat capacity $C_{\rm heat}$ links internal energy and temperature variations $\Delta E_{\rm int} = C_{\rm heat}\Delta T$. At melting, fusion latent heat leads to an energy variation $\Delta E_{\rm int}$ at fixed temperature ($\Delta T = 0$) which implies that $C_{\rm heat} \to \infty$ at the transition temperature. In the bulk, where standard thermodynamics applies, melting transition can thus be searched by tracking a divergence ($C_{\rm heat} \to \infty$) of the heat capacity. In a finite system like a cluster this divergence is smeared. But one still expects a strong peak of $C_{\rm heat}(T)$.

The melting transition in terms of heat capacity is illustrated in Figure 2.18 for the case of $Na_{139}^+$. The figure displays three panels which deserve some explanations. In all three panels, experimental data are denoted by dots while solid lines correspond to the bulk and dashed lines to a bulk "adapted" to the finite $Na_{139}^+$ cluster. The top panel gives the internal energy from which, by derivation, one extracts the heat capacity $C_{\rm heat} = \delta E_{\rm int}/\delta T$. In the case of bulk, of course, the transition is sharp, with a finite energy step at bulk sodium melting temperature $T_{\rm bulk} = 370$ K. On the contrary, experimental data on $Na_{139}^+$ show a smooth internal energy variation taking place between about 250 and 300 K. The corresponding heat capacities (bulk and $Na_{139}^+$) are displayed in the bottom panel of Figure 2.18. In the bulk, $C_{\rm heat}$ diverges at $T = 370$ K, while the $Na_{139}^+$ heat capacity only exhibits a marked increase (0.1 versus 0.04 for the background) at the transition. This experimental result thus exhibits two interesting behaviors: i) a smeared melting transition in the finite system, as expected, and ii) a much smaller melting temperature in the finite system than in the bulk.

It would be very interesting to make similar heat capacity measurements in other clusters. And yet there are only a few results presently available for medium size sodium clusters. This may be due to the fact that such measurements are rather involved. We briefly explain how this first experiment on $Na_{139}^+$ was performed in order to appreciate the inherent difficulties.

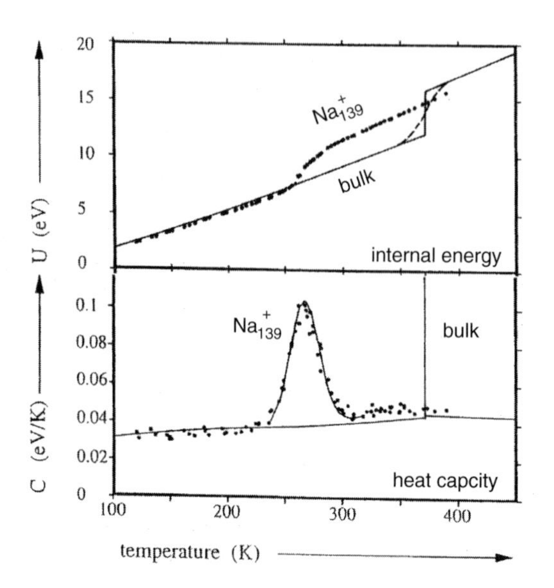


**Figure 2.18:** Total internal energy $E_{\text{int}}(T)$ of a $\text{Na}_{139}^+$ cluster (a) as a function of temperature. The heat capacity estimated from bulk entropy (full line) or smoothed for finite size effects (dashed line) is shown in panel b. The experimental heat capacity is shown with points in panel c. After [SKK+97].

The measurements were performed by analyzing photofragmentation spectra. At a given temperature, the cluster is fragmented by a laser $\text{Na}_N^+(T) + n_{\text{phot}}\hbar\omega \rightarrow \text{Na}_{N-x_{\text{mon}}}^+ + x_{\text{mon}}Na$ and the number $x_{\text{mon}}$ of monomers is recorded. This number $x_{\text{mon}}$ directly depends on how much energy $U = n_{\text{phot}}\hbar\omega$ the laser has deposited. One then changes the cluster temperature $T$ and the laser frequency and simultaneously monitors both, so that the same number $x_{\text{mon}}$ of monomers is always measured. This means, practically, that one has tuned the internal energy $E_{\text{int}}$ and the temperature $T$ of the cluster so that $E_{\text{int}}(T + \delta T) = E_{\text{int}}(T) + \delta E_{\text{int}}$, where $\delta E_{\text{int}} = C_{\text{heat}}\delta T$ provides the seeked heat capacity. As is obvious from this brief outline, such experiments require a high degree of accuracy, in particular in the experimental capability to monitor the cluster temperature. But the results are worth the effort. As this topic is still of much future interest, we continue the discussion in Section 5.1.3.

### 2.3.10 Dissociation energies

The IP is the key signal for electronic binding and thus electronic structure. The analogue for the atomic structure of a cluster is the dissociation energy, i.e. the minimum energy required to remove one atom (monomer) from the cluster. The main decay channel for the monomer is by thermal evaporation according to the Weisskopf estimate [Wei37]. The dissociation energy could be measured as reaction threshold for monomer emission by scanning laser frequency in conjunction with subsequent mass selection. One has to take care to control the effect of the initial temperature. Such measurements are reported, e.g., in [BCC+90]. A well cooled and mass selected beam of K clusters is excited with photons of well defined frequency. By letting the ensemble evolve a further while and finally mass selecting the fragments, one can deduce the reaction rates for dissociation which allow one to conclude on the underling dissociation energies with the help of simple rate models. These experiments on free cluster beams are tedious. The well controllable conditions and long life-time of cluster ensembles in a trap allow a very precise and model-free determination of dissociation energies [VHHS01]. A result of

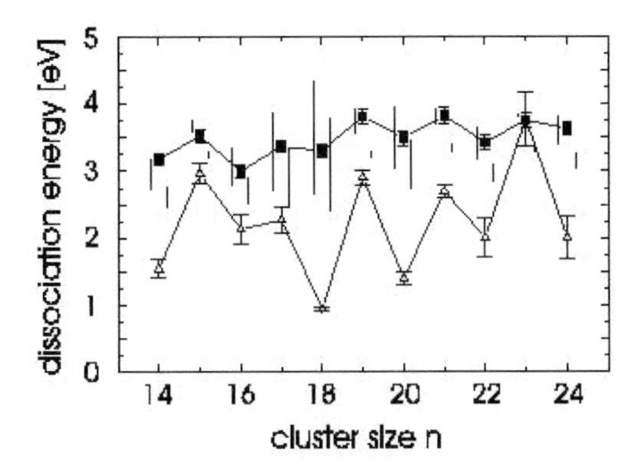


**Figure 2.19:** Dissociation energies in $Au_n^+$ clusters as a function of cluster size determined by time resolved photodissociation. Full squares show the values from the experimental analysis and open triangles from Arrhenius plots. The vertical lines show results from rate equations with RRKM or Weisskopf evaporation rates. From [VHHS01].

such an analysis for Au cluster cations is shown in Figure 2.19. The experimental results are compared with more traditional forms of analysis (Arrhenius plots and rate equations). The differences hint that some of the assumptions underlying the simpler approaches are probably not valid for noble metals. For a more detailed discussion see [VHHS01].

### 2.3.11 Limit of stability

We have pointed out at several places that clusters are distinguished by their scalability, i.e. by the fact that they exist for all sizes between atom and bulk. There should be no limit here. But there is a limit in the net charge which a cluster can carry. Cluster anions are restricted in charge due to electronic stability. The IP decreases quickly when adding further electrons which limits the amount of negative charge. Small clusters can thus mostly appear as simply anions. The capacity to attach further electrons grows slowly with increasing system size giving rise to double anions for heavier systems and eventually more for huge clusters. Cluster cations are more likely because they are even more stable with respect to the electronic binding. But here emerges an instability due to increasing Coulomb pressure between ions which eventually leads to fission or explosion. Looking at it from a different perspective, we find that for a given positive net charge $+q$ stable clusters can only exist from a certain size $N_{\rm app}(q)$ on. This is called the appearance size, the lowest size at which a cluster $X_N^{+q}$ appears in the spectrometer. First systematic measurements of appearance sizes were done by multiple ionization with nanosecond lasers.

A summary of results for Na clusters is presented in Figure 2.20. The results are compared with a simple model of charged liquid drops. It estimates the fission stability in terms of the fissility which is the ratio of Coulomb energy to twice the surface energy, for details see [NBF$^+$97] and references therein. The Coulomb energy drives fission through quadrupole deformations away from the spherical shape. The surface tension counteracts and tries to maintain a stable spherical cluster. Increasing Coulomb energy (here by charging the cluster) eventually leads to fission instability. Experimental results follow nicely along one fissility line ($Z^2/N = 0.13$) which confirms that view of the counteracting effects on the stability. It turned out in more detailed investigations that the appearance size depends sensitively on the initial cluster temperature and the actual ionization mechanism. This will be discussed more extensively in Section 5.4.1

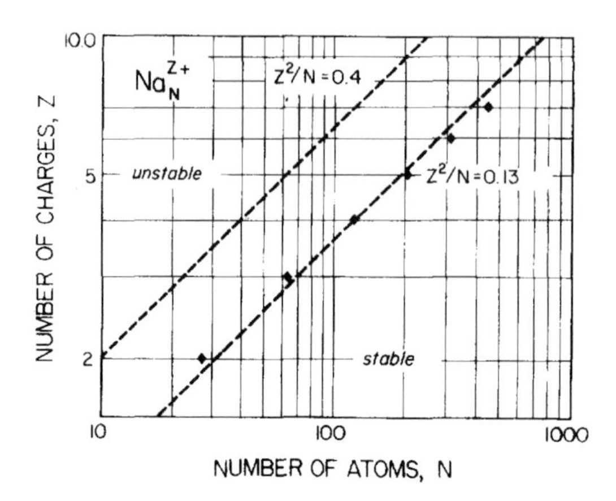


**Figure 2.20:** The dots show experimentally observed appearance sizes in multiply charged sodium clusters. The dashed lines correspond to fissilities $X \sim 1$ $(Z^2/N = 0.4)$ and $X \sim 0.3$ $(Z^2/N = 0.13)$. From [NBF⁺97].

### 2.3.12 Femtosecond spectroscopy

Nowadays a very fashionable means of analysis is femtosecond spectroscopy in terms of pump and probe experiments. This is a very versatile tool allowing time-resolved tracking of molecular dynamics and chemical reactions [Zew94]. The concept is simple. A first short laser pulse ("pump") excites the system out of equilibrium. This triggers a dynamical evolution of the system. With a certain time delay, a second short pulse ("probe") is exerted on the system. It can produce a reaction, or not, depending whereto the system has evolved in the meantime. The measurement is repeated several times with systematically varied delay time between pump and probe. Scanning the reaction products as a function of the delay time then yields valuable information on the underlying dynamics. A clear and much celebrated example is the tracking of vibrational modes in dimer molecules, see e.g. [GS95]. Clusters are much more complex systems than these. There are thus incredibly plentiful conceivable scenarios for femtosecond spectroscopy. Cluster physics is just getting first (already very exciting) steps into that field. We postpone the discussion of more detailed examples to Section 5.3.4 in Chapter 5.

## 2.4 Conclusion

In this chapter we have discussed basic aspects of experiments performed on clusters. The first stages of almost any experiment deal with proper cluster production and identification. We have thus provided detailed discussions on both these aspects before presenting actual measurements of various types of observables.

The first part of the chapter was devoted to a presentation of the most frequently used cluster sources. We have seen that cluster production is a complicated process involving either condensation of an atomic vapor or fragmentation of bulk samples. Whatever method is used to produce the clusters, one usually ends up with an ensemble of objects at finite temperature. Before attacking any experimental measurements it is thus crucial to define, as much as possible, the initial conditions and at least to mass select the clusters on which one aims at performing experiments. Tuning of source parameters helps to preselect the produced clusters. But it is usually not sufficient to provide a clean and homogeneous set of clusters.

The second part of the chapter was thus devoted to a brief presentation of basic tools to properly sort clusters before experiments. Thereby we have paid particular attention to mass spectrometers. We have presented the most frequently used devices, discussing in particular detail the case of Time Of Flight (TOF) mass spectrometers which, because of their simplicity and efficiency, are universally used in cluster physics. As a complement to the presentation on mass spectrometers (which are also used in many experiments for final identification purpose) we have also discussed the widely used optical spectroscopy as a tool to investigate several cluster properties by means of lasers.

In the last part of the chapter we have presented several examples of measurements in clusters, not aiming here at an exhaustive survey. The presentation widely used observables, such as abundances, ionization potentials (IP), and optical response, as well as more special quantities such as heat capacity, or photoelectron spectra. These examples serve to show what kind of information one may extract from cluster experiments. We have also addressed for several of these examples the problem that several sources of pollution may affect experimental data, such as an insufficiently well known cluster size and/or temperature. In spite of this necessary caution in the interpretation of experimental data, we have shown the richness and diversity of experiments performed on clusters.

# 3 The cluster many-body problem: a theoretical perspective

Clusters are complex many-body systems in between atom or small molecules and bulk. Their theoretical description thus requires dedicated methods, in part inspired from atomic, molecular or bulk many-body techniques, but "adapted" to the finite nature of the system. This finiteness of clusters indeed imposes severe constraints. Even small clusters are composed of a large number of degrees of freedom. Remember that each atom contains possibly a large number of electrons (even a simple atom like Na has 11 electrons !). And one has to account for both intra and inter atom (and electrons) interactions, and this at a quantal level. The number of *a priori* active degrees of freedom may thus quickly reach tens or hundreds of units. Furthermore, because a cluster is composed of several atoms, basic atomic symmetry properties are lost, in particular the simplifying central nature of the driving force imposed by ions. In turn, finite objects like clusters cannot directly be treated with the approximations used in bulk materials, in which simple periodic boundary conditions can be used. Such approximations allow one to reduce the study of an "infinite" piece of material to a finite (usually small) sample with simple boundaries. In clusters such "boundary conditions" are more involved as they are so to say generated by the system itself.

Although atomic, molecular or bulk techniques may provide useful tools for attacking the theoretical description of clusters, it is essential to realize that the cluster many-body problem has its own, highly non-trivial, specificity. In this chapter we will thus present the theoretical methods developed to describe clusters, focusing, in particular, on the aspects relevant for dynamical problems. We shall also, as much as possible, base our discussions on microscopic quantum foundations. We shall try to cover most of the various approaches presently used in cluster physics, some of them more briefly and the time-dependent density-functional theory a bit more extensively. The emphasis on density-functional theory is motivated by the fact that this is the most widely used method for describing truly dynamical scenarios while keeping a high degree of microscopic sophistication.

Clusters are composed of ions and electrons. The theoretical description of these two types of particles however does not raise the same difficulties and specific approximation schemes can be worked out for each species. We shall thus successively study these approximations in the course of this chapter. After discussing, on a qualitative basis, the differences between electrons and ions we shall quickly see how to treat ions and then address the key question of the treatment of the electron-ion interactions. Indeed in many-atom systems an account of all electrons (most of which remaining safely bound to their parent atom) is hardly conceivable and one has thus to properly build up an approximate description of interactions between truly active electrons (valence electrons) and ions. Once this question discussed we shall present the various methods developed to describe the electronic cloud as a set of interacting

*Introduction to cluster dynamics.* Paul-Gerhard Reinhard, Eric Suraud

ISBN: 3-527-40345-0

fermions. The treatment of electrons in clusters, just as in molecules or bulk, represents a key and difficult problem, which requires elaborated many-body techniques. Finally, and in particular in view of potential applications in truly dynamical problems, we shall discuss how to put all these pieces (ions, electrons) together in order to reach a proper modeling of clusters. This can be done either by including electronic and ionic degrees of freedom separately or in the framework of model Hamiltonians in which microscopic details can be hidden.

## 3.1 Ions and electrons

### 3.1.1 An example of true cluster dynamics

This chapter is devoted to a brief presentation of the various theoretical approaches employed in describing cluster structure and dynamics up to the non-linear regime. To illustrate the goal, we show in Figure 3.1 key observables of the time evolution of a ${Na_{41}}^+$ cluster in a pump and probe scenario, in which the dynamics of a primarily excited cluster (pump pulse) is analyzed with help of a second laser pulse (probe pulse). In such a situation the first task is to optimize the cluster structure. The result for the ground state structure is indicated at the bottom left corner of the figure. Note that the systematics of ground state structures carries already a world of interesting information, as e.g. mass abundances, radii or shapes. Many investigations concentrate on this stage. The next step is to consider the dynamics of the cluster as it could be triggered by various excitations. Figure 3.1 shows the dynamics in response to strong fs laser pulses. We first apply a pump pulse of 50 fs duration. The electronic dipole moment of electrons with respect to ions (middle panel) responds immediately. With

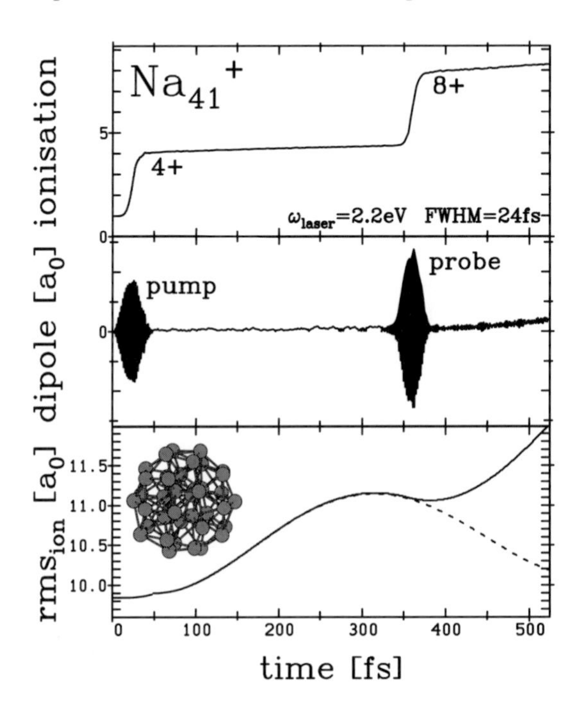


**Figure 3.1:** Typical pursuit of of a pump and probe scenario. Test case is ${Na_{41}}^+$ described with TDLDA (Section 3.3.3) and local pseudo-potentials (Section 3.2.2). Pump and probe pulses have the same laser parameters: frequency $\hbar\omega = 2.2\,\mathrm{eV}$, intensity $1.1 * 10^{12}\,\mathrm{W/cm^2}$, and the profile is a $\sin^2$ pulse with FWHM= 24 fs. The probe pulse is applied with an offset of 340 fs. Lower panel: time evolution of the ionic radius, including probe pulse (full line) and for pure pump (dashed); the inset shows the ionic configuration of the ground state ${Na_{41}}^+$ cluster. Middle panel: electronic dipole moment. Upper panel: time evolution of net ionization at the cluster site.

only few fs delay, electron emission takes place and levels off soon after the end of the laser signal (upper panel). The thus enhanced net cluster charge drives oscillations of the ionic radius (full line, lower panel) which could continue for long. In a second step, the cluster is perturbed by a probe pulse. The same pattern appears again: immediate response of the electronic dipole, subsequent direct electron emission, and finally an ionic reaction which is here a violent Coulomb explosion. A detailed description of such processes is extremely demanding. It simultaneously requires a resolution of the electronic time scale ($\sim$ 1fs), the ability to treat the electron continuum, endurance to continue up to the much longer ionic time scales ($\sim$ 100–1000 fs), and a coupled propagation of ions and electrons. Only few methods can handle that. Most applications are content with less. For example, studies of optical absorption can be done with frozen ionic configuration and need to cover only $\sim$ 100 fs. Studies of ionic vibrations without electron excitation can travel through a sequence of instantaneous electronic ground states with a large time step at ionic scale. However, notwithstanding these different dynamical scenarios, all approaches need to deal with three basic aspects: handling of the involved many-body problem for the electrons, proper description of the ions, and interaction between ions and electrons. In the following, we shall address these three aspects, first separately, and combine them finally.

## 3.1.2 The full many-body problem

### 3.1.2.1 Starting Hamiltonian

A cluster represents a complex many-body system consisting of $N$ atoms with their nuclei and all electrons. A perfect treatment would require one to solve the Schrödinger equation for all electrons and nuclei (called more generally ions in the following). This is an extremely demanding task, at the edge of present day computing facilities, even for small systems. Practical calculations make more or less dramatic approximations. We will present in this chapter the various, most commonly used approaches. A most basic and most widely used approximation is to treat the nuclei, or ionic cores respectively, as classical particles (Section 3.1.3). We will assume this throughout the whole chapter. The starting point is then the Hamiltonian

$$\hat{H} = \hat{H}_{\mathrm{el}} + \hat{H}_{\mathrm{coupl}} + \hat{H}_{\mathrm{ion}} + U_{\mathrm{ext}} \tag{3.1a}$$

$$\hat{H}_{\mathrm{el}} = \sum_i \frac{\hat{\mathbf{p}}_i^2}{2m_{\mathrm{el}}} + \frac{1}{2}\sum_{i\neq j} \frac{e^2}{|\mathbf{r}_i - \mathbf{r}_j|} \tag{3.1b}$$

$$\hat{H}_{\mathrm{coupl}} = \sum_{iI} \frac{Z_I e^2}{|\mathbf{r}_i - \mathbf{R}_I|} \tag{3.1c}$$

$$\hat{H}_{\mathrm{ion}} = \sum_I \frac{\mathbf{P}_I^2}{2M_I} + \frac{1}{2}\sum_{I\neq J} \frac{Z_I Z_J e^2}{|\mathbf{R}_I - \mathbf{R}_J|} \quad . \tag{3.1d}$$

The $\mathbf{r}_i$ and $\hat{\mathbf{p}}_i$ represent electronic position and momentum operator while $\mathbf{R}_I$ and $\mathbf{P}_I$ stand for the same quantities for the nuclear centers (or ions). The separation of the above Hamiltonian hints already that very different constituents are coming together in molecules and clusters, the light and fast electrons together with the heavy and slow ions. Different levels of approaches apply naturally for these two different species. We will in this chapter discuss the three pieces of the Hamiltonian step by step: Section 3.1.3 explains briefly the (simple) approximations for treating the purely ionic part $\hat{H}_{\text{ion}}$, Section 3.2 deals with the coupling of ions to electrons described in $\hat{H}_{\text{coupl}}$, Section 3.3 attacks the electronic many-body problem posed by $\hat{H}_{\text{el}}$, and finally the pieces are put together in Section 3.4. The external potential $U_{\text{ext}}$ comprises the external potentials coming from the the excitation mechanisms (ionic Coulomb field, laser field). This is specified in more detail in Section 4.1.1 of Chapter 4. At the present stage, it suffices to know that $U_{\text{ext}}(\mathbf{r}, t)$ is a local one-body potential acting on electrons (and possibly on ions).

Magnetic effects and spin-orbit force are omitted in the Hamiltonian (3.1) to keep the presentation simple. Note that the spin-orbit force plays a crucial role in magnetic materials, as for example Cr or Fe clusters (for an example see [KB99]), and in heavy elements. Cluster magnetism is a rich field with its own interest and often studied explicitely, as exemplified in Section 5.1.5. Magnetic effects and response in strong external magnetic fields is also an important issue in the study of quantum dots [RM02].

#### 3.1.2.2 A few words on notations

Ionic variables are denoted by capital letters, as e.g. $\mathbf{R}_I$ for the ionic position and $I$ for labeling the ions. Electronic variables, on the other hand, use lower case, as e.g. $\mathbf{r}_i$ for position or $\hat{\mathbf{p}}_i$ for the momentum operator of electron number $i$. The operator hat is usually omitted for the position $\mathbf{r}_i$ because we tacitly assume that we are working in the local representation in most cases, because that is the natural frame for local energy-density functionals. The same holds for any local operator, as e.g. a local potential $U(\mathbf{r})$.

A crucial distinction is to be made for the electronic wavefunctions. The electron cloud of a many-Fermion system is to be associated with an antisymmetrized many-body wavefunction. These are denoted by the edgy bra's $\langle\Psi|$ and ket's $|\Psi\rangle$. The objects which are actually handled in the most widely used mean-field theories are single-electron wavefunctions $|\varphi_\alpha)$ for one electron in a single particle state $\alpha$. These states are distinguished by round bra's and ket's. This distinction is accompanied by conventions for energy and Hamiltonian. Energies related to the whole $N_{\text{el}}$-body system are denoted by capital $E_{\text{label}}$ while single particle energies are written as $\varepsilon_\alpha$. A full $N_{\text{el}}$-body Hamiltonian will read $\hat{H}_{\text{label}}$ while the mean-field Hamiltonian is $\hat{h}_{\text{label}}$.

An extensive table of symbols with associated explanations is given in Appendix A.2.

### 3.1.3 Approximations for the ions as such

A fully quantum mechanical treatment of ions and electrons at the same level is extremely cumbersome and rarely done. One takes advantage of the fact that ions are much heavier than electrons, e.g. $M_{\text{I}}/m_{\text{el}}$= 2000 for H, 46000 for Na or 420000 for Pb. This allows one to separate the problem. There exist basically two different routes of approximation: the (adiabatic) Born-Oppenheimer approximation or a classical treatment of the ionic dynamics.

The Born-Oppenheimer (BO) approximation is a widely used standard in molecular physics and quantum chemistry. Although we will not use it much throughout this book, let us ponder briefly on the Born-Oppenheimer approach. The mass difference causes also the large difference in time scales as seen for example in Figure 3.1 (see also Figure 1.15). The electrons can thus follow almost immediately any given change in ionic configuration. That allows one to decouple electronic from ionic motion provided that electrons start and stay close to (instantaneous) equilibrium. This is the essence of the adiabatic approximation where the fully coupled many-electron–ion wavefunction $\Psi$ is approximated by the adiabatic separation

$$\Psi(\mathbf{r};\mathbf{R}) = \psi_{\text{ion}}(\mathbf{R})\phi_{\mathbf{R}}(\mathbf{r}) \quad . \tag{3.2}$$

In the compact notation here $\mathbf{R}$ stands for the set of ionic coordinates and $\mathbf{r}$ for the electronic ones. The electronic state $\phi$ still depends parametrically on the actual ionic positions. The idea beyond this ansatz is that the electronic state adjusts almost instantaneously to the very slowly changing ionic configuration $\mathbf{R}$. One applies the full Schrödinger equations with Hamiltonian (3.1) to that ansatz. This reads

$$\left(\frac{\hat{P}^2}{2M} + V_{\text{ion}}(\mathbf{R}) + \hat{H}_{\text{el}} + \hat{H}_{\text{coupl}}(\mathbf{r},\mathbf{R})\right)\Psi = E\Psi \quad , \tag{3.3}$$

where $V_{\text{ion}}$ embraces ion-ion interactions, $\hat{H}_{\text{el}}$ electron-electron ones (including also electron kinetic energy) and $H_{\text{coupl}}$ labels the coupling between electrons and ions. The underlying assumption that $\phi$ only weakly depends on $\mathbf{R}$ allows one to neglect all gradient terms $\nabla_{\mathbf{R}}\phi$ and $\nabla^2_{\mathbf{R}}\phi$. The thus decoupled electronic and ionic problem leaves for the ionic Schrödinger equation the effective potential

$$V_{\text{BO}}(\mathbf{R}) = V_{\text{ion}}(\mathbf{R}) + \langle\phi_{\mathbf{R}}|\hat{H}_{\text{el}} + \hat{H}_{\text{coupl}}|\phi_{\text{R}}\rangle \tag{3.4}$$

while electronic wavefunctions are reduced to the eigensolutions of the following Schrödinger equation

$$(\hat{H}_{\text{el}} + \hat{H}_{\text{coupl}})|\phi_{\text{R}}\rangle = E_{\text{el}}|\phi_{\text{R}}\rangle \tag{3.5}$$

at given ionic configuration $\mathbf{R}$. For more details of the derivation see Appendix C.1. It is interesting to note that the BO separation can be formulated easily with quantum mechanical ions. That is gratifying to know because quantum effects in molecular ionic motion are crucial at various places. They are for example required to obtain the observed discrete vibrational and rotational spectra. The dynamics of vibrational wavepackets is actually explicitely explored in pump-and-probe experiments [GS95]. One has even produced quantum mechanical interference patterns for scattering of rather large molecules [GST$^+$00]. All these processes deal with slow ionic motion and can be very well handled with the BO (or adiabatic) approximation. The scheme separates the solution into two steps: first, the electronic state is determined for frozen ionic positions (Eq. (3.5)), and second, the thus derived effective ionic Hamiltonian is used to determine the static and dynamical features of molecular motion (Eq. (3.4)). The heavy work is involved in the solution of the electronic problem for fixed ionic positions and most approximations are employed here. All methods discussed in Sections 3.2 and 3.3 apply here as well.

However, the restriction to truly adiabatic processes is too narrow for modern scenarios in cluster dynamics, as e.g. excitation by intense lasers or collisions with fast ions. Fast excitation processes drive the electron cloud quickly far off equilibrium, i.e. far off a BO surface. A fully coupled (non-adiabatic) treatment of electrons and ions is required. Thus we do not really want to separate the problem in order to maintain flexibility for truly dynamical processes. The molecular motion is highly excited in those cases and that justifies a classical treatment of ionic motion. The assumption may be a bit critical for the lightest species of hydrogen molecules. But most practical calculations of fully dynamical processes up to now use classical ions. Thus we prefer the second route dealing with classical ionic molecular dynamics (MD). The simplification is justified by the huge mass difference. The Compton wavelength is $\propto M^{-1/2}$ and thus very short for the ions. This allows one to treat them as classical particles in most applications. This simplifies the ionic part while allowing to keep the ion–electron coupling without restrictions. Again, the heavy workload remains for the electronic part. The detailed classical equations of motion are given later in Section 3.4.1, see Eq. (3.37), because they are better understood once the description of electrons is settled.

As a final remark, we note that the handling of the electronic problem is the same in the BO approximation and in the case of coupling to full ionic MD. The difference consists in how the electronic information is used in relation to the ions. The BO approximation relies on adiabatic decoupling (relative slowness) and allows a quantum mechanical treatment of the ions. On the other hand, the classical treatment of ions allows fully coupled propagation (see Section 3.4.1). The combination of both limiting assumptions leads to a simple classical Born-Oppenheimer MD, which will be discussed in Section 3.4.2.

## 3.2 Approximation chain for the ion–electron coupling

### 3.2.1 Core and valence electrons

The starting point Eq. (3.1) consists in naked nuclei as ions and all electrons for each atom in detail. The atomic electrons are grouped into shells of extremely different energy and length scales, see the examples given in Figure 1.3. The deeply lying electronic states are so tightly bound that they are insensitive to varying molecular environment. They can thus be lumped together with the nucleus into one inert object, the so called ionic core. The aim is hence to eliminate these core electrons completely from the treatment and to handle only the least bound states, the valence electrons (the ones actually taking part in bonding, see Section 1.1.2). For the sake of simplicity, let us confine the more general discussion around Figure 1.3 to metals. Even here we have to distinguish different cases. Two typical metals are depicted in Figure 3.2. The leftmost panel shows the atomic level sequence for the noble metal Cu and the middle panel for the simple metal Na. Note that there are several lower states ($1s$ for Na and $1s, 2s, 2p$ for Cu) which are so deeply bound that they fall far below the bounds of the plot. For the remaining states one still sees a clear distinction. The $3s, 3p$ states for Cu and the $2s, 2p$ states for Na are far away from the least bound states (which constitute the Fermi surface of the system) and can be safely considered as core states. There remains one occupied valence state, the $3s$, for Na and the group $3d, 4s$ for Cu. The distinction between core and valence electrons is, of course, a matter of decision and may differ under different circumstances. For example, one may put the $3d$ state in Cu into the core states for low-energy situation where the $4s$-$3d$ gap of about 10 eV is sufficiently large. On the other hand, highly dynamical processes (involving high energies and/or frequencies) may require one to include amongst the active electrons also some of these deeper core levels ($3d$ states in the case of Cu).

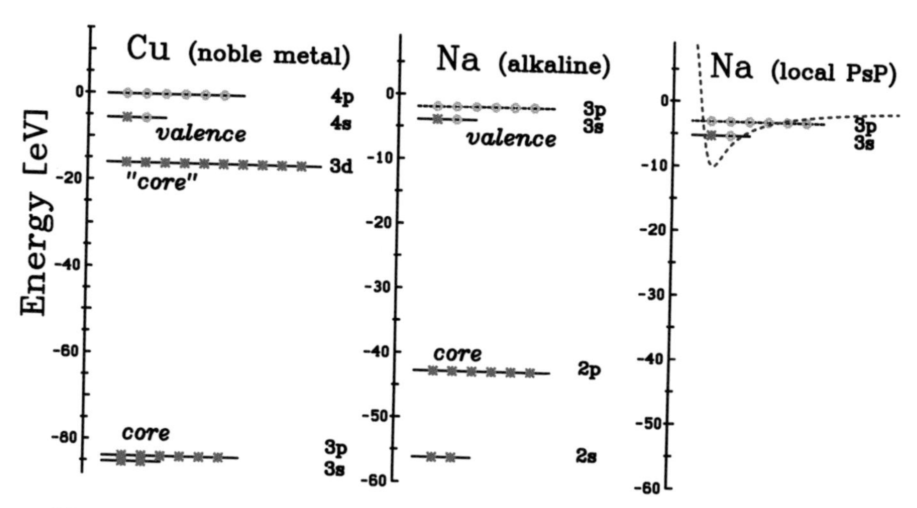


**Figure 3.2:** Comparison of the level sequence in a noble metal (Cu, left panel) with a simple metal (Na, middle panel). Single electron energies are indicated by horizontal bars. The naming of the states (atomic convention) is given to the right near each level line. Occupied states are indicated by stars and unoccupied ones by open circles. Deeply lying states are not shown. The rightmost panel shows the spectrum for the Na atom in a pseudo-potential approach. The deepest state corresponds to the $3s$ state of the full atom. All results have been obtained from LDA using the functional of [PW92] complemented by a self-energy correction [PZ81].

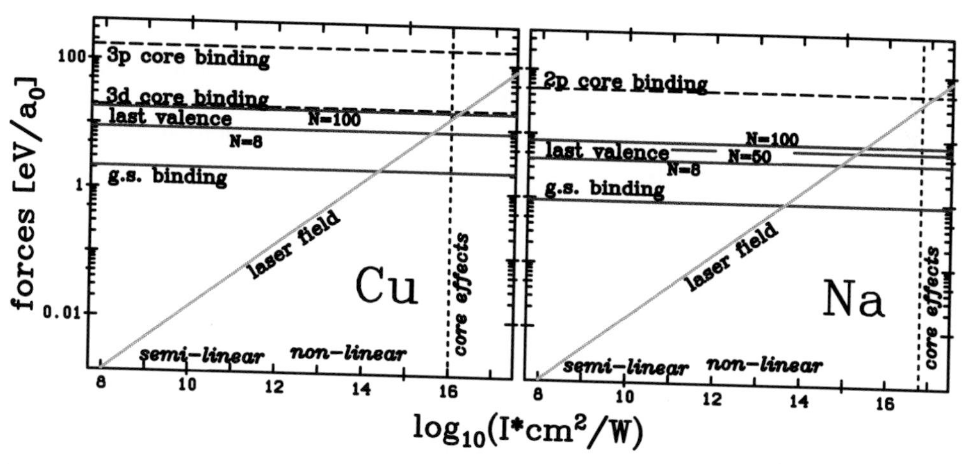


**Figure 3.3:** Typical forces in Cu (left panel) and Na clusters (right panel) compared with the external force of a laser and drawn versus laser intensity $I$. The binding forces at the surface of a neutral cluster and an 8+-charged cluster are shown as well as the binding force for the core electrons (for energies see figure 3.2).

A complementing view of these relations between core and valence electrons is given in Figure 3.3 for Na and Cu, in terms of forces. Binding of the valence electrons in a cluster or binding of the various electrons in an atom can be characterized by the peak force exerted on the electron in the surface zone of its density distribution. The figure compares the binding forces for the HOMO level in $Na_9^+$, for $Na_9^{8+}$ where all valence electrons except one are removed, and for the $2p$ core state in the Na atom. The case of Na shows a clean separation of scales. One can strip off all valence electrons from the cluster without having to fear much interference from the core state. The figure also compares these forces with the external force exerted by a laser beam. The $x$ axis scans the laser intensity and the straight line with positive slope shows the relation between intensity and the peak force from the electrical field on one single charge. This has to be viewed in relation to the force lines discussed above. Most interesting is the comparison with the binding force for $Na_9^+$ which is a typical binding for many Na clusters. A critical point is reached at an intensity around $I = 10^{13}\mathrm{W/cm}^2$ where the laser force becomes comparable to binding. For lasers above that intensity there is no barrier any more to remove an electron at once. Details of cluster structure, electron shells or resonant response are unimportant here and the more so for larger intensities. That is the regime of strong laser impact where details of the material become less important and where one basically studies features of a nano-plasma [DTS+97a]. More sensitivity to cluster properties is achieved for lower intensities. Far below we have the linear regime where mainly optical response is explored. With increasing intensities one reaches a semi-linear regime where multi-photon processes (e.g. second harmonic generation, multi-photon emission) become likely. A typical range for that is $I = 10^8 - 10^{12}\mathrm{W/cm}^2$ depending somewhat on the actual laser frequency, and material, of course. Details of clusters are explored in the linear as well as in the semi-linear regime (Section 5.3). The semi-linear regime is more demanding concerning its theoretical description. It requires a fully fledged dynamical treatment.

Figure 3.3 shows the binding forces to strip off the last valence electron for different cluster sizes $N$. It is obvious that this binding increases with $N$ due to the growth of the attractive Coulomb force. This is the point to mark the difference between the simple metal Na and the noble metal Cu. For Cu, the binding of the last valence electron in larger clusters come close to core binding while it remains still a long way off for Na. This means that the dynamics of the uppermost core electrons needs to be taken into account when dealing with noble metals. This holds already for the linear regime, see Section 4.3.4.1, and the more so for the semi-linear regime. In all other respects, the Figure 3.3 looks much similar for Cu and Na.

## 3.2.2 Pseudo-potentials

### 3.2.2.1 Formal foundation of pseudo-potentials (PsP)

Once the decision is made, one has a grouping into inert core electrons and reactive valence electrons. One then wants to develop a scheme where only valence electrons are treated explicitely. This is achieved by pseudo-potentials (PsP). We give here a brief introduction of PsP. We start from the basic (still exact) ideas and make quickly a big leap to the practitioner's view. For the basic steps we follow the original and pedagogical presentation of [Sza85]. Let us thus assume that we have one isolated atom with $N_c$ core electrons in single-electron states $\{|\varphi_1\rangle, \ldots, |\varphi_{N_c}\rangle\}$ and one valence electron in state $|\varphi_v\rangle$. The states are supposed to be generated by some one-body theory, such as for example Hartree-Fock or Kohn-Sham as will be discussed in Section 3.3. In any case, the single-electron states fulfill the one electron equations

$$\hat{h}|\varphi_\alpha) = \varepsilon_\alpha|\varphi_\alpha) \tag{3.6}$$

where the one-body Hamiltonian $\hat{h}$ may itself depend on all occupied states. The set of states is ortho-normalized. In particular, we have $(\varphi_v|\varphi_\alpha) = 0$ for $\alpha \in \{1, \ldots, N_c\}$. Now we want to represent the valence state by a wavefunction $\psi$ which should be free from explicit reference to the core states, as we want to explicitely "hide" core states from the description. This state $\psi$ may then differ from the original valence state $\varphi_v$. In particular, it need not to be orthonormal to the core states. But it is essential that $\psi$ carries at the end the same physics, in particular in terms of energy. One should thus be able to reconstruct $\varphi_v$ from $\psi$ by orthogonalizing on the core states

$$|\varphi_v) = |\psi) - \sum_{c=1}^{N_c} |\varphi_c)(\varphi_c|\psi) \quad . \tag{3.7}$$

Inserting $\varphi_v$ as a function of $\psi$ into Eq. (3.6) and regrouping terms yields an equation for $\psi$

$$\hat{h}|\psi) + \hat{V}_{\mathrm{PsP}}|\psi) = \varepsilon_v|\psi) \tag{3.8}$$

$$\hat{V}_{\mathrm{PsP}} = \sum_{c=1}^{N_c} |\varphi_c)\,(\varepsilon_v - \varepsilon_c)\,(\varphi_c| \quad . \tag{3.9}$$

Note that $\hat{h}$ depends only on the core states when acting on $\psi$ or $\varphi$ respectively. The pseudo-potential $\hat{V}_{\mathrm{PsP}}$ naturally emerges as an additional term which takes care of the now hidden orthogonalization and which ensures that $\psi$ possesses the right energy $\varepsilon_v$ of the original valence state $\varphi_v$, but for the complemented $\hat{h} + \hat{V}_{\mathrm{PSP}}$ Hamiltonian. The core states entering $\hat{V}_{\mathrm{PSP}}$ provide a repulsive term which repels $\phi$ outside the core region, as would "naturally" occur for a true valence wavefunction, but this now occurs for a pseudo wavefunction, possibly without nodes. The pseudo potential $\hat{V}_{\mathrm{PSP}}$ is a non-local and energy dependent operator. This form, which actually suffers from some formal difficulties, was first developed in [Sze55]. One may recast it on purely formal grounds into a local potential [PK59, PK60]. In any case, these strictly formal expressions are not so useful for practical applications. But it is an invaluable tool to study typical patterns of PsP which then can be built into more phenomenological models. For a few interesting examples see [Sza85]. In practice, one takes these formal ideas to motivate an ansatz for the PsP and then adjust the free parameters in the ansatz to desired observables, as e.g. atomic spectra or asymptotic profiles of electron wavefunctions. Before coming to the practical part, we mention briefly a basic feature of PsP which can be learned from formal grounds: they are ambiguous. Let $\psi$ be a solution of Eq. (3.8), then any other wavefunction $|\psi) \longrightarrow |\psi) + \sum_c \gamma_c|\varphi_c)$ fulfills this equation as well. We thus have the freedom of adding arbitrary portions of occupied states. That can be exploited to match additional requirement as, e.g., smooth local form or desired number of nodes etc. This shows that there is some versatility in the adjustment of a PsP, see next subsection and Appendix D.

#### 3.2.2.2 Practical pseudo-potentials

There is a long way from these still exact grounds to a practical, efficient, and reliable PsP. And there is a broad range of choices derived from more or less strict paths. Formally the simplest choice are the local pseudo-potentials. Early proposals are given, e.g., in [AH65, AL67]. And soft shapes are preferable for numerical handling. A particularly soft parameterization is provided by a combination of two error functions [KBR99]

$$V_{\rm PsP}(\mathbf{r}) = \sum_{i=1}^{2} c_i \frac{\mathrm{erf}(\mathbf{r}/\sigma_i)}{|(\mathbf{r})|} \qquad \text{with} \quad \mathrm{erf}(x) = \sqrt{\frac{2}{\pi}} \int_0^x e^{-y^2} dy \tag{3.10}$$

where the $\sigma_i$ are widths and the strength parameters have to line up to the total charge of the ionic core $c_1 + c_2 = Z_{\rm ion}$ (see Figure 3.2 for an example in the case of Na). A few examples for appropriate parameters of that form of local pseudo-potentials are given in more detail in Appendix D.3.

Local potentials are bound to deliver a sequence of states which has $l = 0$ as lowest eigenstate. This is appropriate for alkalines and earth-alkalines where an $ns$ state is the wanted valence state. But it is impossible to accommodate other materials for which active shells start at higher $l$, as e.g. for the example of Cu in Figure 3.2. This (and often other reasons) calls for the more elaborate and more flexible non-local pseudo-potentials for which there exists a very broad selection. In some of them one parameterizes the non-locality through projectors on good angular momentum $\hat{\Pi}_L = \sum_M \int d^2\Omega' Y_{LM}(\Omega) Y^*_{LM}(\Omega') \ldots$ in the form

$$\hat{V}_{\rm PsP}(\mathbf{r}) = V_{\rm loc}(\mathbf{r}) + \sum_L V_L(\mathbf{r}) \hat{\Pi}_L \tag{3.11}$$

where each angular momentum channel $L$ possesses its own especially tuned potential $V_L$. A typical example can be found in [BHS82]. Others use a technically simpler form in terms of sums of dual products

$$\hat{V}_{\rm PsP} = V_{\rm loc}(\mathbf{r}) + \sum_c |\varphi_c) h_c (\varphi_c| \tag{3.12}$$

where $c$ runs over an appropriate set of projected states and $h_c$ is the corresponding weight. Again, a numerically particularly robust form employs Gaussians times a polynomial for the $h_c$ [GTH96]. All PsPs are of course defined with respect to a given ionic center. The projection, if any, has to be done with respect to that center, a procedure which can become rather cumbersome in the multi-centered case of molecules and clusters. Local PsPs are simpler in that respect. On the other hand, non-local forms are more flexible and there exist PsPs where this flexibility has been exploited to provide a particularly soft numerical representation allowing, e.g., coarse grids in the calculation [TM91].

Figure 3.4 demonstrates the effect of pseudo-potentials on the description of the wavefunctions. It compares the exact wavefunction for the $3s$ state in the Na atom (obtained from an all electron calculation) with the corresponding wavefunction as obtained from a local pseudo-potential. The nodal structure in the interior of the atom is much different. But this is bearable because it is unimportant for the description of long range features and in particular bonding (Section 1.1.2). What counts is that the two wavefunctions nicely agree outside the region of the core states. And that is visibly well fulfilled for the example here, and for any well fitted pseudo-potential. For completeness the lower panel of Figure 3.4 displays the local pseudo potential used in this case as well as the Coulomb field as exerted by the net charge (one !) of

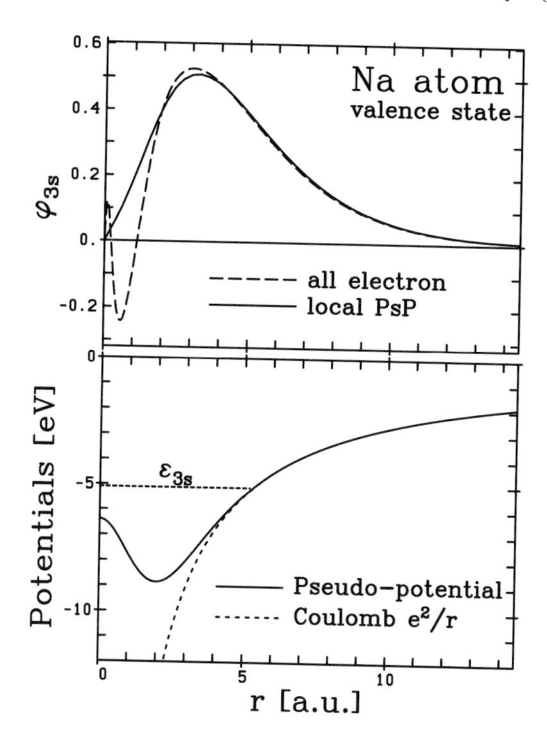


**Figure 3.4:** The upper part shows the effect of the pseudo-potential on the wavefunction. Test case is the Na atom. The dashed line shows the exact wavefunction of the $3s$ state from an all-electrons calculation with LDA and SIC. The full line shows the ground state in the local pseudo-potential 3.10 with parameters as given in Table D.1. The lower part of the figure compares the actual (local) pseudo-potential (full line) with the Coulomb field corresponding to one charge, namely the charge of the sodium core. The asymptotic behaviors of the potentials of course perfectly match.

the sodium core. Both potential asymptotically match, but strongly differ in the interior. In particular the singularity at ionic site is removed by the smooth pseudo potential.

The correct asymptotics of the pseudo-wavefunction is one crucial condition. Indeed, one wants more. The goal is transferability, which means that one and the same pseudo-potential is also appropriate to describe molecular binding up to the bulk. This requires that an atomic calculation with the PsP should reproduce as well as possible the energy levels of the states near the Fermi surface, including the first excited states. Most widely used are fitting procedures which rely on atomic calculations, such as the so-called norm conserving pseudo-potentials (see Appendix D.1). One starts from a reliable all-electrons calculation for the atom including several excited states; one selects the relevant atomic states (valence and a few excited states) which one wants to describe by the pseudo-potential; one chooses an appropriate form for the pseudo-potential, and then one adjusts its free parameters to accommodate the relevant states optimally. The careful fit to sufficiently many atomic states arranges automatically the polarizability which then yields correct binding properties from molecule to bulk. Beyond the reproduction of electronic single particle energies, key constraints in the fit can also be an exact matching of the (possibly nodeless) pseudo wavefunctions with the atomic wavefunctions beyond a given cutoff, which ensures proper screening behaviors (see Appendix D for details).

An alternative is to adjust the atomic valence states together with key bulk properties (equilibrium energy and density, interstitial density), as done e.g. in [AL67, KBR99]. The pseudo-potential then "interpolates" between atom and bulk. It is found that this strategy then also provides automatically a good description of atomic excited states and polarizability. Thus both fitting schemes are equivalent and lead to the same result. An example for the

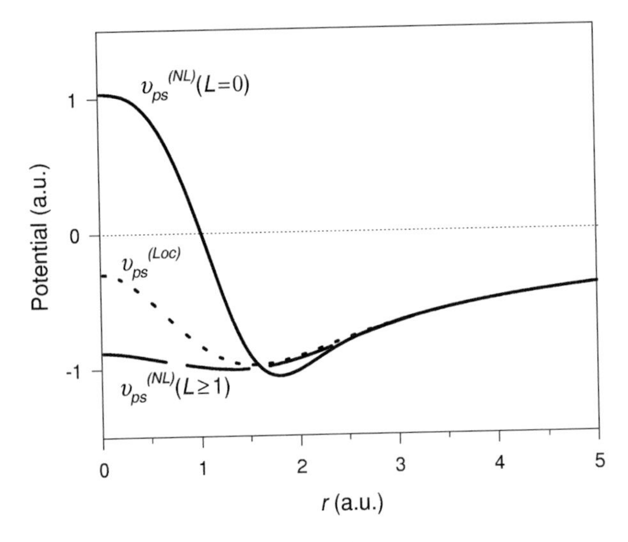


**Figure 3.5:** Demonstration of non-local pseudo-potentials for the example of Mg and the $L$-dependent parameterization of [BHS82]. The two different branches $L = 0$ and $L > 0$ are shown and exhibit very different behaviors, in particular at short distance. For comparison a purely local approximation is also shown, which somewhat "interpolates" between the two branches of the non-local pseudo potential. From [SRS02].

performance of a simple local PsP in atomic spectra is shown in Figure 3.2. Compare the middle panel showing the full atomic spectrum of Na with the right panel showing the spectrum obtained from the local PsP of [KBR99]. The (last occupied) $3s$ and the (first unoccupied) $3p$ level are well reproduced. In particular, the $3s \rightarrow 3p$ dipole transition comes out perfectly fine. This guarantees that the polarization interaction between two (or more) Na atoms is correctly described which promises transferability, i.e. an appropriate description also of molecules and clusters up to the bulk. As countercheck, one can fit only the last occupied state together with bulk properties and one finds that the transitions across the Fermi surface are automatically well described [KBR98, KBR99].

From a technical point of view, local PsPs are most simple to use. But there is only a small subset of favorable cases which allows one a description with local PsPs. Simple alkaline metals as Na, K, Cs, or Rb belong to that group. But Li, as an exception, is better treated with a non-local PsP [BG93]. The next column with valence two, as e.g. Mg, is a bit at the edge. In most cases, a smoother and more robust representation is obtained with non-local PsPs. Non-local PsP are compulsory to model the situation in noble metals where an active $d$-shell is energetically below the $s$-shell. Figure 3.5 exemplifies a non-local pseudo-potential, here for the $L$-dependent form Eq. (3.11) as it is used, e.g., in [BHS82]. The local profiles in the two different $L$-channels look very different. It is obvious that such an ansatz is extremely flexible and allows one to adjust any energetic relation between the various valence states (think e.g. of the above mentioned "unnatural" ordering in noble metals).

#### 3.2.2.3 Core polarization effects

Thus far, the motivation for a PsP employed the working assumption that the eliminated core states are perfectly inert. At second glance, one realizes that an external charge applies a Coulomb force on the core which, in turn, reacts with a small polarization. The polarization effect lowers the energy of the combined system and this contributes to the effective potential. Those polarization potentials are a leading ingredient in effective rare gas atom-atom potentials, e.g. for Ar–Ar [AC77] or He–He [AP78]. They also add to the electron–ion PsP. In most cases, they are implicitly contained in the adjustment of the effective PsP. One has to keep in mind that such an adjustment then fixes the static polarization properties. An explicit handling of polarization is advisable if the frequency of core vibrations is not sufficiently well separated from the dynamical processes under consideration. An example is the description of noble metals (Cu, Ag, Au) with only one valence electron per atom. We see from Figure 3.2 that, for example, the $d$ shell in Cu is not so far away from the $4s$ shell. The same holds, for example, in Ag and Au, commonly used materials in cluster physics. By far most PsP do not count $d$ electrons as core electrons in such systems. If one deliberately wants to eliminate them, one should thus treat their dynamical polarization explicitely. Only this guarantees an appropriate description of the optical response of noble metal clusters. Such effects have for example been clearly identified in the case of Ag clusters [SR97].

#### 3.2.2.4 Symmetry averaged approaches

The total potential created by the ionic background is the sum over each ion PsP:

$$U_{\text{back}}(\mathbf{r}) = \sum_I V_{\text{PsP}}(|\mathbf{r} - \mathbf{R}_I|) \tag{3.13}$$

In general, it breaks all spatial symmetries and one then needs to perform a fully three-dimensional (3D) calculation for the electrons. The 3D treatment is very expensive and limits the cluster size for large scale studies. On the other hand, one knows that many clusters are close to a more or less symmetric shape. For example, clusters with magic electron shell closures have nearly spherical shapes (Section 4.2.1). This suggests that a spherically averaged ionic background potential provides a reliable approximation to the exact ionic potential in such cases. The result is the spherically averaged pseudo-potential scheme (SAPS) which cooperates with local PsP and which consists in the approximation

$$U_{\text{back}}^{(\text{SAPS})}(r) = \int d^2\Omega U_{\text{back}}(r\mathbf{e}_\Omega) = \sum_I \int d^2\Omega V_{\text{PsP}}(|r\mathbf{e}_\Omega - \mathbf{R}_I|) \quad , \tag{3.14}$$

where $r = |\mathbf{r}|$ and $\mathbf{e}_\Omega$ is the unit vector for space angle $\boldsymbol{\Omega} \equiv (\vartheta, \phi)$ in the center of mass of the cluster. The SAPS approximation has proven to provide an efficient and reliable approach for nearly spherical clusters of simple metals [BG93, BLMA92, MF91]. The spherical averaging is formulated in Eq. (3.14) for a local PsP. (Here it can be performed analytically with the soft form involving error functions and Gaussians.) One can equally well develop a spherical average on a non-local PsP [ASB$^+$95]. This gives access to materials like Li which are better treated with a non-local PsP.

Spherical clusters are a minority. The majority of clusters is deformed. But most of them remain at least close to cylindrical (or axial) symmetry. The success of SAPS has inspired an analogous averaging procedure for that situation, the cylindrically averaged pseudo-potential scheme (CAPS) [MR94, MR95a]. It consists in approximating the ionic background potential through the cylindrical average

$$U_{\text{back}}^{(\text{CAPS})}(r, z) = \int d\phi U_{\text{back}}(r\cos(\phi), r\sin(\phi), z) \quad , \quad r = \sqrt{x^2 + y^2} \quad . \tag{3.15}$$

CAPS can describe reliably well a broad range of clusters [SPMR98]. It allows one to compute much larger clusters than in a full 3D approach and becomes increasingly valid with increasing cluster size. Its efficiency makes it a welcome tool for first explorations of new dynamical scenarios [CRS97, URS97] also including coupled electronic and ionic dynamics [RCK+99].

It is to be remarked that the ion–ion interactions, third term in Eq. (3.1), remain treated fully three-dimensional, in SAPS as well as in CAPS. The approximation consists only in forcing the mean fields acting on the electrons to become spherically or axially symmetric. This holds for static structure optimization as well as for true ionic dynamics.

### 3.2.3 Jellium approach to the ionic background

#### 3.2.3.1 Steep jellium

Simple metals have valence electrons with long mean free path throughout. The relevant electrons move with momenta near Fermi momentum (Section 1.3.1) which corresponds to a spatial resolution of the order of the Wigner-Seitz radius $r_s$. The fine details of the ionic background are thus seen by the electrons only in an average manner. This motivates the jellium approximation in which the ionic background is smeared out to a constant positive background charge. It is a standard approach in the theory of bulk metals [AM76]. And the adaptation to a finite cluster is straightforward. One carves from the bulk a finite element of constant positive charge. The volume is chosen such that its total charge coincides with the given ionic charge. The simplest geometry is a sphere with a radius given by $R_{\rm jel} = r_s\, N_{\rm ion}^{1/3}$ ($N_{\rm ion}$ being the number of ions). The jellium density

$$\rho_{\rm jel,steep}(\mathbf{r}) = \rho_0 \theta\left(R_{\rm jel} - |\mathbf{r}|\right) \tag{3.16a}$$

exhibits a sharp transition from bulk density $\rho_0 = 3/(4\pi r_s^3)$ to zero at the surface. The simple form allows one to derive an equally simple explicit expression for the associated Coulomb potential, namely

$$U_{\rm jel,steep}(\mathbf{r}) = \begin{cases} -\frac{Ze^2}{r} & \text{for} \quad r > R_{\rm jel} \\ \frac{m_{\rm el}}{2}\omega_{\rm Mie}^2(r^2 - R_{\rm jel}^2) - \frac{Ze^2}{R_{\rm jel}} \quad , \quad \omega_{\rm Mie}^2 = \frac{4\pi\rho_0}{3m_{\rm el}} & \text{for} \quad r \leq R_{\rm jel} \end{cases} \tag{3.16b}$$

where the Mie frequency $\omega_{\rm Mie}$ appears in the curvature for the inside region. This steep jellium model served very well to outline the basic properties of metal clusters, in particular concerning gross shell structure (Section 4.2.1.1), see e.g. the early papers of [Bec84, Eka84]. However, electronic shell effects in finite systems call for deformation, see e.g. Section 4.2.2.2. This can be accommodated by a straightforward generalization to ellipsoidal drops of jellium [EP91], as done in Eq. (3.17b).

#### 3.2.3.2 Soft jellium

The jellium model with steep surface provides a pertinent zeroth order account of cluster properties. However, it is found that the Mie plasmon resonance energy comes out notoriously too high, about 10% above experimental ones. That deficiency can be cured by using a soft surface profile [RBA91]. Versatile and easy to handle in this respect is a Woods-Saxon profile for the jellium density

$$\rho_{\text{jel}}(\mathbf{r}) = \frac{3}{4\pi r_s^3}\left[1+\exp\left(\frac{|\mathbf{r}|-R(\vartheta,\phi)}{\sigma_{\text{jel}}}\right)\right]^{-1} , \tag{3.17a}$$

$$\text{with} \quad R(\vartheta,\phi) = R_{\text{jel}}\left(1+\sum_{lm}\alpha_{lm}Y_{lm}(\vartheta,\phi)\right) , \tag{3.17b}$$

and $R_{\text{jel}}$ defined by the normalization to total particle number

$$\int d\mathbf{r}\,\rho_{\text{jel}} = N_{\text{ion}} . \tag{3.17c}$$

The central density is determined by the bulk density $\rho_0 = 3/(4\pi r_s^3)$. To give an impression about the surface width, the transition from 90% to 10% bulk density is achieved within $4\sigma_{\text{jel}}$, while the steep jellium model Eq. (3.16a) is recovered in the limit $\sigma_{\text{jel}} \longrightarrow 0$. Possible deformations are parameterized in $R(\vartheta,\phi)$ through the coefficients in front of the spherical harmonics with the parameters $\alpha_{lm}$ determining the amount of deformation. For example, axially symmetric ellipsoids are tuned by $\alpha_{20}$, positive values producing prolate and negative values oblate shapes. Octupole (pear-like) shapes are generated by $\alpha_{30}$ and can have a considerable influence in metal cluster spectroscopy [MR95b]. Next come hexadecapoles $\alpha_{40}$ which play a role for fine-tuning the shape [MHM+95]. Higher moments are rarely considered. Triaxial shapes are produced by moments with $m \neq 0$. This rich palette of shapes suffices for most applications. An exception is cluster fission where the shapes undergo dramatic changes. A good coverage of initial and final shapes requires specially tuned parameterizing, see e.g. [GAB+94]. After all, there remains the cluster radius $R_{\text{jel}}$. It is fixed by the others parameters, predominantly by the bulk density. It becomes simply $R_{\text{jel}} = r_s N_{\text{ion}}^{1/3}$ for the spherically symmetric steep jellium model. It needs slight readjustment with condition (3.17c) in the case of soft surface profiles and deformation. The leading parameters of the soft jellium model Eq. (3.17) are thus finally the Wigner-Seitz radius $r_s$ and the surface thickness $\sigma_{\text{jel}}$. They are universal for a given material. For example, typical values are $r_s \sim 4\,\mathrm{a}_0$ and $\sigma_{\text{jel}} \sim 0.9\,\mathrm{a}_0$ for Na clusters, $r_s \sim 2.66\,\mathrm{a}_0$ and $\sigma_{\text{jel}} \sim 0.76\,\mathrm{a}_0$ for Mg clusters, or $r_s \sim 3\,\mathrm{a}_0$ and $\sigma_{\text{jel}} \sim 0.78\,\mathrm{a}_0$ for Ag clusters. The deformation parameters $\alpha_{lm}$ depend on the actual cluster and strongly vary with size.

To illustrate the performance of the jellium model, we consider the simple, though telling, case of a Na cluster. We compare in Figure 3.6 densities and Kohn-Sham (KS) potentials (for its definition see Section 3.3.3) from a fully ionic calculation with results from steep and soft jellium models. The ionic model uses spherical averaging (see Section 3.2.2.4) to allow simple visualization of densities and potentials along a radial profile. The KS potential with detailed ions is strongly fluctuating due to huge repulsive peaks at the ionic cores. But note that spherical averaging mellows these fluctuations for large $r$. The effect of a single ion can for example be read off from the sharp dip at $r = 0$. The jellium models provide smoother trends where the remaining fluctuations are solely due to electronic shell effects. The average trends of jellium potentials approximate very well the case with ions, and that

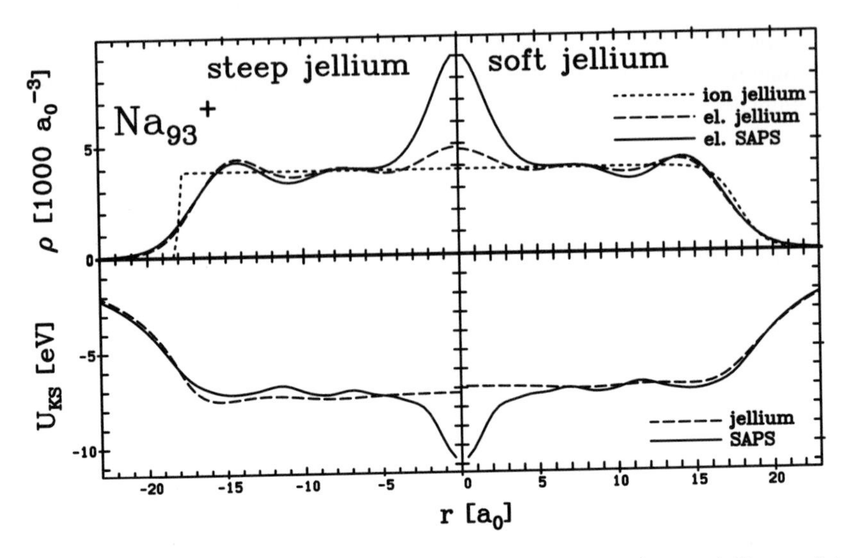


**Figure 3.6:** Comparison of approaches for the ionic background in $\mathrm{Na}_{93}{}^{+}$: steep jellium, soft jellium, and local pseudo-potentials. The lower panel shows the net Kohn-Sham potentials for the three cases as indicated. The upper panel shows the corresponding electron densities. All quantities are spherically averaged and drawn as radial profiles. The (spherical) jellium approaches use a radius of $R = r_s N^{1/3}$ with $r_s = 3.93\,\mathrm{a}_0$. The surface parameter for the soft jellium is $\sigma_{\mathrm{jel}} = 0.8\,\mathrm{a}_0$.

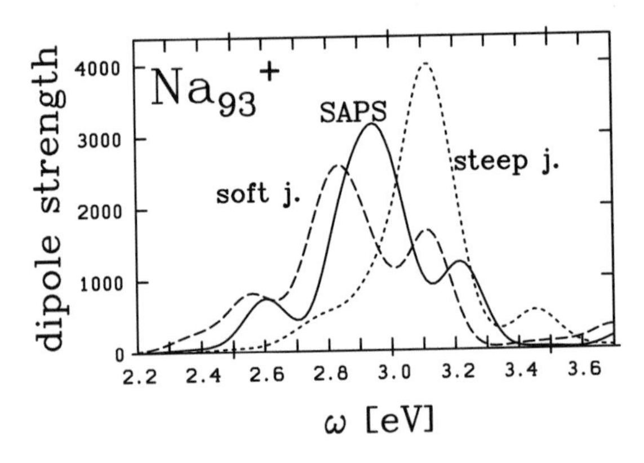


**Figure 3.7:** Optical response of $\mathrm{Na}_{93}{}^{+}$ for the three different approaches to the ionic background as shown in Figure 3.6 .

particularly in the surface region which determines the binding. A similar situation is seen for the densities. Ionic cores print occasional spikes into the electron density. But the gross features are well reproduced by the jellium model. The difference between steep and soft jellium models is hardly visible in these plots. But one can spot that the soft jellium model performs somewhat better and provides the better match in the surface region. The difference becomes more apparent only when looking at sensitive observables. An example is provided by the optical absorption strengths compared in Figure 3.7 (for details of its computation, see Section 4.1.7). The average position of the Mie plasmon agrees fairly well between soft jellium model and fully ionic structure while it is visibly blue shifted for the steep jellium. The strong influence of surface softness on the Mie plasmon can be read off from a simple sum rule estimate (see [Bra93] as well as Section 4.3.1 and Eq. (G.9)) and it is also well known in surface physics [TPL+91]. Taking both figures together, we see that a properly tuned soft jellium can reproduce very well the salient features of electron distribution and dynamical response of simple metal clusters.

#### 3.2.3.3 Relation to PsP

The soft jellium approach can be deduced from the concept of pseudo-potentials. Assume a material for which a local PsP is legitimate, as for example Na. Then let us identify the steep jellium as resulting from smearing a collection of point charges to a smooth background. In order to go one step further, we now associate a local PsP to each point charge. The soft jellium then emerges as a folding of the steep jellium with the given PsP

$$U_{\text{jel}}(\mathbf{r}) = \int d\mathbf{r}'\, U_{\text{back}}(|\mathbf{r}-\mathbf{r}'|)\frac{3}{4\pi r_s^3}\theta\left(R(\vartheta,\varphi)-|\mathbf{r}'|\right) \tag{3.18}$$

where $R(\vartheta,\varphi)$ is again given as in Eqs. (3.17b) and (3.17c). The surface parameter is now replaced by the profile and width given through the PsP. This definition of smoothed ionic background yields a perfect description of magic shell closures and optical response in metal clusters [RWGB94, RGB96a], and that without any free parameters ($r_s$ is taken from bulk and the surface profile comes from the PsP). It is often called a pseudo-jellium model as abbreviation for PsP motivated jellium. The soft jellium model (3.17) should be considered as a simple parameterization of such pseudo-jellium models. The subtle differences in the surface profiles play a minor role in practical calculations.

The same strategy of folding steep jellium with a PsP can be pursued for a non-local PsP. This yields effective Hamiltonians with non-local contributions, as e.g. coordinate-dependent effective mass or angular momentum-dependent potentials [SBGL93, ASB+95, CPCV95]. The thus generalized models are also often called pseudo-jellium, or more specifically pseudo-Hamiltonians. These non-local models extend the range of applicability of simple jellium to a broader range of metals which require non-local PsP, as e.g. Li. The performance of pseudo-Hamiltonians is as good as it was for the local pseudo-jellium. In particular the optical response is very well described.

Let us finally mention that there is a close relation between pseudo-jellium and the symmetry averaged ionic models SAPS and CAPS, introduced in Section 3.2.2.4. One can view that as a hierarchy of averaging. Starting from full 3D one first averages over angular dependences and obtains CAPS or SAPS. Ionic structure is still present there in terms of strong axial or radial fluctuations of the effective KS potential (see for example Figure 3.6). In a final step, one averages over these radial fluctuations and obtains pseudo-jellium, or pseudo-Hamiltonians respectively.

#### 3.2.3.4 Further variants of jellium

The jellium model provides the correct shell effects and optical response for simple metals. The single electron energies are acceptable but the total binding energies are awfully wrong. The predictive value and the range of applicable metals can be substantially extended by mapping ionic structure energies into effective potentials. This leads to the stabilized jellium model where the averaged extra binding through ionic structure is taken from bulk metal [PTS90]. The contribution of ionic structure to the surface energy of a finite cluster is still missing and it can be large depending on the material. This detail has been added in the structure averaged jellium model (SAJM) [MRM94]. These extensions of the jellium model provide better single-electron energies and reliable total binding. The SAJM even allows one to estimate such fine details as the compression of small clusters through surface tension.

## 3.3 Approximation chain for electrons

### 3.3.1 Exact calculations

Many-electron systems are highly correlated. Exact calculations of their properties are extremely involved. On the other hand, the simplicity of the (local) Coulomb force and the steady progress of computational facilities have given access to several exact numerical calculations. We mention here a few of them. Two extremes are accessible: very small molecules and infinite, homogeneous systems.

The infinite systems are extremely useful for developing density functionals in LDA (Section 3.3.3.3). The Hartree-Fock energy can be evaluated here analytically, see appendix F.1. Correlations, although less simple, are at least in reach. Hierarchical expansions of many-electron theory can, to a large extent, still be solved analytically (see e.g. the treatments of [vBH72] and [GL76] on which the respective energy functionals are based). Even a fully exact treatment is numerically feasible. A crucial benchmark is given here by the early results for the homogeneous electron gas at zero temperature [CA80]. The development has gone further to finite temperatures and non-homogeneous systems, as e.g. hydrogen in its various phases [NMC95]. But the first calculations from [CA80] are still the basic input for practically working density functionals, see Section 3.3.3.3. They are also quite telling as such because they shed light onto the composition of energies in an electron cloud. Figure 3.8 shows the various contributions to the binding energy of electron matter with homogeneous and neutralizing positive background (jellium). Kinetic and exchange part can be expressed exactly in analytical form as we will see later. The correlation energy is extremely involved. It is taken from

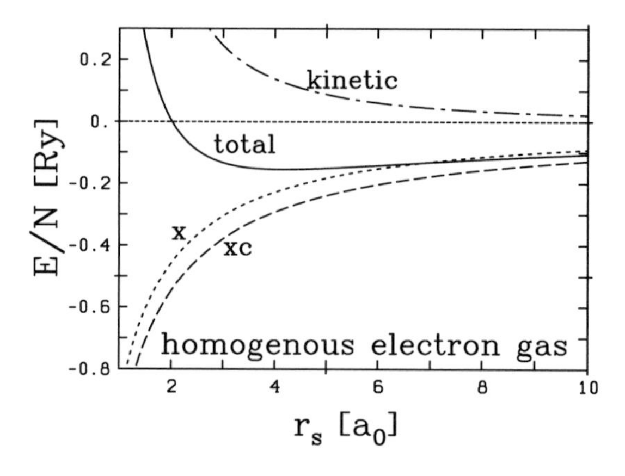


**Figure 3.8:** Binding energy per particle and the various contributions to it for the homogenous electron gas. The total energy, the kinetic energy, the exchange energy (x), and the exchange-correlation energy (xc) are indicated. Results are drawn versus the Wigner-Seitz radius $r_s$ of the gas. The xc-results are taken from [CA80] in the interpolation of [PW92].

the numerical calculations of [CA80]. It is interesting to see that the main balance of binding is found as a compromise between kinetic and exchange energy. The correlation energy adds a smaller, but non-negligible, part. The contribution of the correlation energy relative to the exchange energy increases for dilute media (large $r_s$) and fades away for dense systems.

At the side of small molecules, there exist several exact calculations for structure, see e.g. rather recent calculations for carbon clusters up to fullerenes [NKP+02]. One can meanwhile even find fully dynamical exact calculations for small systems, e.g. for the laser-induced dynamics of the He atom [PTCBF96]. The progress of computer technology will steadily extend the range of applicability for exact calculations of many-electron systems. Yet, exact methods will be limited for a long while. Computations of larger systems and systematic surveys on a broad scale require approximations to make the task manageable. The following sections will discuss a selection of the most widely used approaches in cluster physics.

## 3.3.2 Ab initio approaches

There is a large class of approaches called "ab initio" methods. In a broader sense, it embraces all techniques which can be derived more or less strictly from the initial Hamiltonian Eq. (3.1). For example, the density functional method discussed in Section 3.3.3 is often called "ab initio" although the relation to the initial Hamiltonian is somewhat sophisticated. In a narrower sense, methods are "ab initio" if they employ directly the Hamiltonian (3.1) and deal with it under well defined approximations, in particular at the level of the choice of the Hilbert space in which electronic wavefunctions are constructed. We will sketch in this section two examples for this narrower class. A few more technical details can be found in Appendix C.2.

### 3.3.2.1 Hartree-Fock (HF)

The lowest level "ab initio" method is the Hartree-Fock (HF) approach. It is derived by making an ansatz of independent particle states (Slater determinants)

$$\Phi(\mathbf{r}_1,\ldots,\mathbf{r}_N) = \mathcal{A}\{\varphi_1(\mathbf{r}_1)\ldots\varphi_1(\mathbf{r}_N)\} \tag{3.19}$$

(where $\mathcal{A}$ is the N-body antisymetrisation operator) and by optimizing the single-electron wavefunctions $\varphi_\alpha$ by applying the Ritz variational principle to the expectation value of the Hamiltonian Eq. (3.1), see e.g. [Wei78]. The emerging self-consistent HF equations then read

$$\hat{h}_{\mathrm{HF}}\varphi_\alpha = \varepsilon_\alpha\varphi_\alpha \quad \text{for} \quad \alpha \in \{1,\ldots,N\} \quad , \tag{3.20a}$$

$$\text{with} \quad \hat{h}_{\mathrm{HF}} = \frac{\hat{p}^2}{2m} + U_{\mathrm{back}} + U_{\mathrm{C}}(\mathbf{r}) + \hat{U}_{\mathrm{ex}} \quad , \tag{3.20b}$$

$$U_{\mathrm{C}}(\mathbf{r}) = \int d\mathbf{r}' \, \frac{e^2}{|\mathbf{r}-\mathbf{r}'|}\rho(\mathbf{r}') \quad , \tag{3.20c}$$

$$\rho(\mathbf{r}') = \sum_{\beta=1}^{N} |\varphi_\beta(\mathbf{r}')|^2 \quad , \tag{3.20d}$$

$$\left(\hat{U}_{\text{ex}}\varphi_\alpha\right)(\mathbf{r}) = \int d\mathbf{r}' \underbrace{\frac{e^2}{|\mathbf{r}-\mathbf{r}'|}\rho(\mathbf{r},\mathbf{r}')}_{U_{\text{ex}}(\mathbf{r},\mathbf{r}')}\varphi_\alpha(\mathbf{r}') \quad , \tag{3.20e}$$

$$\rho(\mathbf{r},\mathbf{r}') = \sum_{\beta=1}^{N}\varphi_\beta(\mathbf{r}')\varphi_\beta^+(\mathbf{r}) \quad , \tag{3.20f}$$

where $U_{\text{back}}$ stands for the ionic background potential in any one of the forms as discussed in Section 3.2. The single-electron wavefunctions $\varphi_\alpha$ are, in fact, two-component spinors and the $\varphi_\alpha^+$ represent the hermitian conjugates spinors. Note that the symbol $\rho$ is used in two slightly different contexts, once as density matrix $\rho(\mathbf{r},\mathbf{r}')$ and once as local density $\rho(\mathbf{r})$. The latter is obtained from the density matrix as $\rho(\mathbf{r}) = \text{tr}_\sigma\{\rho(\mathbf{r},\mathbf{r}')\}$ where the trace is taken over the 2×2 spin-matrix. The direct Coulomb term $U_{\text{C}}$ is a local operator which is easy to handle. The exchange term $\hat{U}_{\text{ex}}$ is highly non-local and causes the dominant fraction of the numerical expense. Both (direct and exchange) HF potentials depend themselves on the set of occupied states $\varphi_\alpha$ which are to be determined by the HF equations. This poses a non-linear problem which is usually solved iteratively. The generalization to dynamical processes is time-dependent HF (TDHF) which can be derived from the time-dependent variational principle [KS67] $\delta\int dt\,[\sum_\alpha(\varphi_\alpha|\imath\partial_t|\varphi_\alpha) - E_{\text{total}}] = 0$. The modification is that the stationary HF equations (3.20a) are replaced by

$$\hat{h}_{\text{HF}}\varphi_\alpha = \imath\partial_t\varphi_\alpha \tag{3.21}$$

which determines uniquely the time-evolution $\{\varphi_\alpha(\mathbf{r},t)\} \longrightarrow \{\varphi_\alpha(\mathbf{r},t+\delta t)\}$. The composition of the HF Hamiltonian $\hat{h}_{\text{HF}}$ remains as in the static case above.

The (TD)HF equations are the most that one can do within an independent particle model. Their solution is already a demanding task due to the non-local exchange potential. There exist a few studies using HF in connection with clusters, see e.g. [BG93, MGJ95, LSG02]. The results are agreeable. This corroborates experience from atomic and molecular physics that HF is reliable and performs well. And yet, correlations beyond mean field are considered to be important for a more satisfying agreement. There is a large variety of approaches to deal with correlations. A truly “ab initio” method is discussed in Section 3.3.2.2. But most methods deal with constructing effective interactions. As examples we will discuss here the density-functional approach in Section 3.3.3 and effective atom–atom potentials in Section 3.4.5.4.

#### 3.3.2.2 Configuration interaction (CI)

The HF method optimizes one single Slater state Eq. (3.19). An obvious extension is to consider a correlated state as a sum of Slater states

$$\Psi = \sum_n \Phi_n c_n \quad . \tag{3.22a}$$

For a given set $\{\Phi_n\}$, the superposition coefficients $c_n$ are determined variationally which yields

$$\sum_m \langle\Phi_n|\hat{H}|\Phi_m\rangle c_m = E\sum_m \langle\Phi_n|\Phi_m\rangle c_m \quad . \tag{3.22b}$$

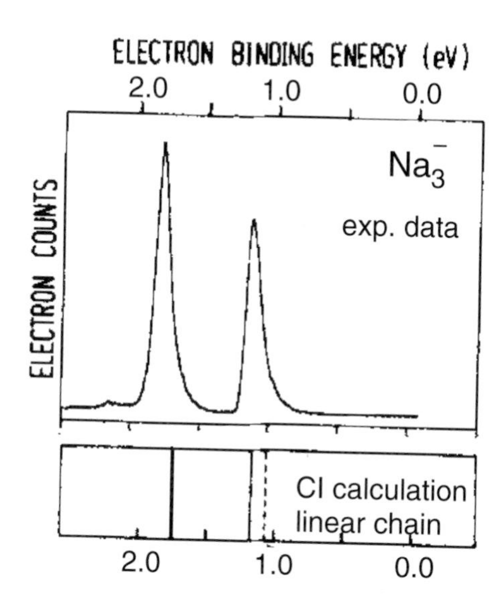


**Figure 3.9:** Photoelectron spectra for the anion cluster $Na_3{}^-$ as measured by [MEL$^+$89]. Lower panel shows single electron spectra from CI calculations [BKFK89].

This is a most general formulation of correlations. Much depends on how one actually chooses the basis. For an extensive discussion on the many variants actually used, see e. g. [YOGI94]. We briefly mention here a few typical examples.

It is most straightforward to compose all determinants from the same set of ortho-normalized single-electron wavefunctions, usually provided by the solution of the HF problem. A natural hierarchy of choices is here to start from the HF state which we denote by $\Phi_0$, then to add one-particle-one-hole ($1ph$) states about $\Phi_0$ up to a given energy, then to add $2ph$ states, and so on until one finds sufficient convergence of the results. Such a set of basis states is orthogonal which simplifies the right-hand side of Eq. (3.22b) to $\langle\Phi_n|\Phi_m\rangle \longrightarrow \delta_{nm}$. This constitutes the configuration interaction (CI) method that is widely used in quantum chemistry and has also found several applications in cluster physics (there are several examples of applications throughout this book). Still, the necessary size of the basis can grow huge and the CI method can become very expensive. On the other hand, the appeal of CI is that it emerges as a systematic improvement beyond HF and that all compromises are concentrated in one well controllable place: the choice of the expansion basis. In fact, carrying the expansion far enough allows to approach an exact calculation and there exist many such large-scale CI calculations in atomic and molecular physics. Figure 3.9 shows a simple example of a CI calculation in cluster physics. Test case is the ground state of $Na_3{}^-$ and the lower part of the figure shows the single electron energies computed from CI. These energies can be measured directly by photeelectron spectroscopy and an experimental result is shown in the upper part of the figure. The agreement between theory and experiment is excellent.

These elaborate CI calculations are also used to compute excitation spectra of clusters. In principle, the diagonalization of the CI Hamiltonian automatically produces also excited states. In practice, an appropriate description requires an even more elaborate choice of the basis set. The necessary extensions depend very much on the actual system (symmetries, density of states). For large clusters, one may even decide to use approximations, as for example a linear response theory built on the CI ground state. A detailed discussion of these subtle points can be found in [KFK91]. There are several applications of CI to compute optical response along these lines. They all stay in the linear domain of small amplitude motion. A truly time-dependent CI for large amplitude processes has, to our knowledge, not yet been attacked.

A more refined strategy to optimize the basis states for the superposition Eq. (3.22a) is multi-configurational HF (MCHF) where not only the base point $\Phi_0$ but also the first selection of excited states is determined by variation. In that case one considers not only the superposition weights $c_n$ as free variational parameters, but the total energy is also varied with respect to all single-particle wavefunctions in all different $\Phi_n$. The states $\Phi_n$ will usually turn out to be non-orthogonal with respect to each other. It is obvious that such an optimized expansion converges much faster. The matrix equation (3.22b) thus becomes much smaller. But the expense is shifted to the mutually linked optimization of the basis set $\{\Phi_n\}$. This elaborate MCHF strategy is rarely used in cluster physics, an interesting exception is an application to $Na_4^+$ in [MYB98]. Most of the CI calculations there are done with the conceptually simpler strategy explained in the previous paragraphs, i.e. with an expansion into a $n\ p-h$ hierarchy about the HF state.

There also exists an extension of CI with MCHF that deals with a continuous set of Slater states. It is called the generator-coordinate method (GCM) and it has found widespread application in nuclear physics [RS80, RG87].

### 3.3.3 Density-functional theory

The goal of density-functional theory (DFT) is to develop self-consistent equations which employ purely local effective potentials for the contributions from exchange and correlation. These potentials are to be expressed in terms of the local electron density $\rho(\mathbf{r})$ of the system. DFT is thus more than HF in that correlations are included. But it is less than HF in that exchange is treated only approximately. In any case, it is much simpler to handle because only local potentials are dealt with. The success of the method depends on a diligent choice of these effective potentials. In turn, a suitable effective potential may allow one to restrict the electronic Hilbert space to a rather simple one. This is at variance with the ab initio methods of Section 3.3.2, in which the price to pay for working with the full Hamiltonian is the construction of huge and complex electronic wavefunctions.

It took decades to develop DFT to today's level of an efficient and reliable approach and developments are still going on. In fact, DFT is a huge field in itself. There exist several books and review articles which are devoted exclusively to that topic, see e.g. [JG89, PY89, DG90, GDP96]. We will only give here a brief account with an emphasis on the necessary practical steps.

#### 3.3.3.1 Basic development

Simplification of electronic calculations by virtue of effective energy-density functionals started very early. Recall the Slater approximation $U_{\text{ex}} \propto \rho^{1/3}(\mathbf{r})$ for exchange [Sla51] which is still part of most density functionals, see Eqs. (3.25). Another long standing energy functional is the Thomas-Fermi approximation for the kinetic energy [Fer28, Tho26] starting from the local energy-density functional (3.32) for the kinetic energy (for more details see Section 3.3.5.4). A sound theoretical foundation was delivered much later. The first crucial step was here the Hohenberg-Kohn theorem [HK64] which states that the ground state energy of a Coulomb system can always be written as a functional of the local density alone. But the theorem does not provide that functional. One has to search appropriate approximations for it. The Thomas-Fermi functional (3.32), for example, constitutes such an approach in terms of local density alone. It also serves as an example that the simplest step can be insufficient. It was found out later that one should better augment the functional Eq. (3.32) by corrections depending on derivatives of the density, $\nabla\rho$ and $\Delta\rho$, leading to the extended Thomas-Fermi theory (ETF) based on the functional Eq. (3.35) [BB97]. The latter provides a pertinent description of average electronic properties over a broad range of system sizes, for applications in clusters see e.g. [Bra93] and references cited therein. But it fails to reproduce the electronic shell structure of atoms and clusters. The escape from that deadlock was the Kohn-Sham scheme which reintroduces the detailed single-electron wavefunctions $\varphi_\alpha$ back into the energy functional and which applies density-functional mapping only to the potential energy [KS65]. This yields detailed, quantized single electron states and with it all quantal shell effects. The self-consistent equations still remain fairly simple because all effective potentials stay local. But the situation for the remaining potential energy is the same as before: an exact density functional is not available and approximations have to be invoked. The direct Coulomb term is immediately a functional of the local density $\rho(\mathbf{r})$. It remains to consider approximations for the exchange and correlation energy.

#### 3.3.3.2 The Kohn-Sham equations

The starting point for the (time-dependent) Kohn-Sham equations is an expression for the total energy which employs a local energy-density functional for the potential energy of the electrons. As we aim at a coupled description together with ions we include them from the beginning. The total electronic energy $E_{\text{total}}$ thus reads

$$\begin{aligned} E_{\text{total,el}} &= E_{\text{kin}}(\{\varphi_\alpha\}) + E_{\text{el,ion}}(\rho, \{\mathbf{R}_I\}) + E_{\text{C}}(\rho) + E_{\text{xc}}(\rho) \\ &\quad + E_{\text{ext}}(\rho, t) \quad , \end{aligned} \tag{3.23a}$$

$$E_{\text{kin}} = \int d\mathbf{r} \sum_\alpha \varphi_\alpha^+ \frac{\hat{p}^2}{2m} \varphi_\alpha \quad , \tag{3.23b}$$

$$E_{\text{el,ion}} = \int d\mathbf{r} \sum_\alpha \varphi_\alpha^+ \hat{V}_{\text{back}} \varphi_\alpha \quad , \tag{3.23c}$$

$$E_{\text{xc}} = \int d\mathbf{r}\, \rho(\mathbf{r}) \epsilon_{\text{xc}}(\rho(\mathbf{r})) \quad , \tag{3.23d}$$

$$E_{\text{ext}} = \int d\mathbf{r}\, \rho(\mathbf{r}) U_{\text{ext}}(\mathbf{r}, t) \quad , \tag{3.23e}$$

$$\rho(\mathbf{r}) = \sum_\alpha \varphi_\alpha^+(\mathbf{r}) \varphi_\alpha(\mathbf{r}) \quad , \tag{3.23f}$$

where $E_{\text{C}}$ is the direct part of the electronic Coulomb energy which is naturally a functional of the local electron density $\rho(\mathbf{r})$. The coupling to the ions is described by the background potential $\hat{V}_{\rm back}$ for which the various approaches were discussed in Section 3.2. Local pseudo-potentials and the jellium model provide here also a purely local term $E_{\rm el,ion} = \int d\mathbf{r}\,\rho(\mathbf{r})U_{\rm back}(\mathbf{r})$. The $E_{\rm ext}$ stands for the external, time-dependent perturbation (laser field, bypassing ion) acting on the electrons as already mentioned in the context of the basic Hamiltonian (3.1) and as will be discussed in more detailed in Section 4.1.1. Stationary calculations and simple excitation spectra (linear response) do not require that part. The $E_{\rm xc}$ is the energy-density functional for electronic exchange and correlations. It is only approximately known, see Section 3.3.3.3. The most general form for $E_{\rm xc}$ in dynamical cases should, in principle, depend on densities over all past times [GDP96]. For simplicity, we tacitly assume here already an instantaneous approximation for $E_{\rm xc}$ which will be discussed also in Section 3.3.3.3. All other terms are self-explanatory. The more detailed expressions are provided in Appendix E.1. The electronic density $\rho$ should be split, in fact, into spin up $\rho_\uparrow$ and spin down $\rho_\downarrow$ densities. This technical detail is also taken care of in Appendix E.1. Here we keep things as simple as possible to concentrate on the essentials.

The Kohn-Sham (KS) equations are derived by variation of the given energy (3.23) with respect to the single-electron wavefunctions $\varphi_\alpha^+$. This yields

$$\hat{h}_{\rm KS}\varphi_\alpha = \varepsilon_\alpha\varphi_\alpha \quad \text{for} \quad \alpha \in \{1,\ldots,N_{\rm el}\} \quad , \tag{3.24a}$$

$$\hat{h}_{\rm KS} = \frac{\hat{p}^2}{2m} + \underbrace{U_{\rm C}(\mathbf{r}) + \hat{U}_{\rm xc} + \hat{U}_{\rm back} + U_{\rm ext}(\mathbf{r},t)}_{U_{\rm KS}} \quad , \tag{3.24b}$$

$$\hat{U}_{\rm xc}(\mathbf{r}) = \epsilon\left(\rho(\mathbf{r})\right) + \rho(\mathbf{r})\frac{\partial\epsilon_{\rm xc}\left(\rho(\mathbf{r})\right)}{\partial\rho} \quad . \tag{3.24c}$$

The external field $U_{\rm ext}$ is, in fact, obsolete for the stationary calculations. But it can play a crucial role for the time-dependent case (3.24d) following below. The Coulomb potential $U_{\rm C}$ is given in the standard manner, see Eq. (E.2b). The exchange-correlation potential involves a construction $\partial\epsilon_{\rm xc}/\partial\rho$ which is the formal derivative of the exchange correlation energy per particle $\epsilon_{\rm xc}$ with respect to the density $\rho$ considered as an independent variable. The time-dependent KS equations analogously read

$$\hat{h}_{\rm KS}\varphi_\alpha = \imath\partial_t\varphi_\alpha \tag{3.24d}$$

where $\hat{h}_{\rm KS}$ is composed in the same manner as above, by virtue of the instantaneous approximation for $E_{\rm xc}$. The Kohn-Sham equation (3.24a) poses a stationary eigenvalue problem. It provides the electronic ground state of a system. The time-dependent Kohn-Sham equation (3.24d) poses an initial value problem. The natural starting point is the ground state as obtained from the stationary Kohn-Sham equation. The stationary Kohn-Sham equation is highly non-linear due to the self-consistent feedback of the local density. It is to be solved by iterative techniques. The time-dependent Kohn-Sham equations pose a standard problem of first order differential equations for which several methods are readily available. For a personal view on the numerical representation of wavefunctions and densities and for iterative as well as propagation schemes, see appendix H.

The KS scheme was invented to maintain shell structure in a density functional approach. Figure 3.10 shows a typical result. The test case is $\mathrm{Na}_{138}$ with soft jellium background Eq. (3.17) and an energy-density functional at LDA level (see Section 3.3.3.3). The left upper panel shows that jellium and electron densities stay indeed very close to each other. That

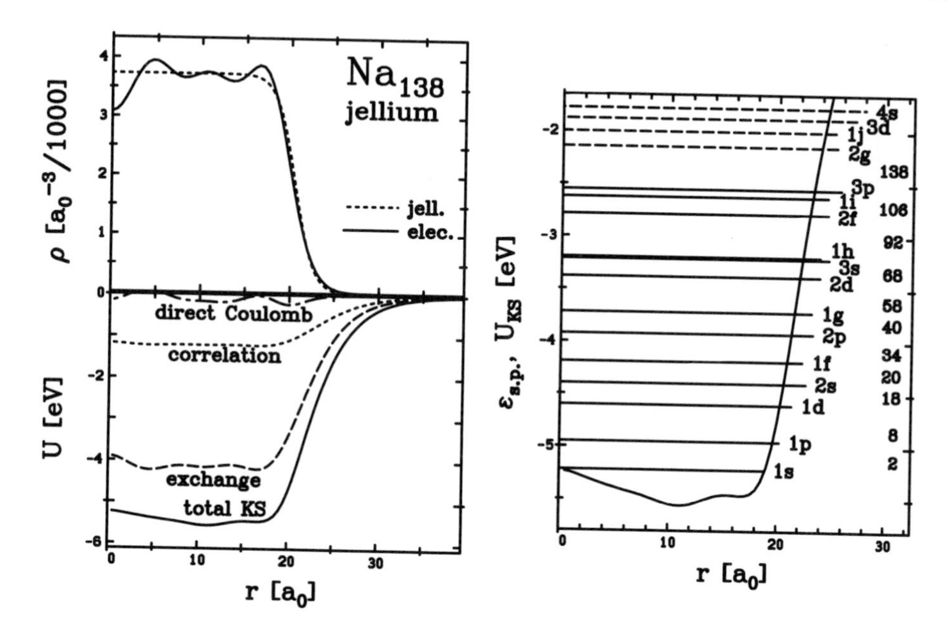


**Figure 3.10:** Kohn-Sham results for $Na_{138}$ computed with the energy functional of [PW92] and soft jellium background ($r_s = 3.93\,a_0$, width $\sigma = 1\,a_0$). Left lower panel: the total Kohn-Sham potential as well as its separate contributions as indicated. Left upper panel: total electron density (full) and jellium background density (dashed). Right panel: sequence of single-particle levels; full lines = occupied, dashed lines= unoccupied.

minimizes the direct Coulomb energy. There remains an irreducible discrepancy in that the electron density has unavoidable fluctuations due to shell effects. One sees the oscillatory pattern of the last filled shell which resides at Fermi momentum $k_F = (9/4\pi)^{1/3} r_s^{-1}$ and which produces a spatial oscillation frequency $\pi/k_F$ ($\approx 2\pi a_0$ in the case of Sodium). The point has also been made in a nuclear physics context; for an extensive discussion in that case see [RFV84]. These oscillations would be wiped out in a Thomas-Fermi approach, see [Bra93] and Section 3.3.5.4. The left lower panel of Figure 3.10 shows the KS potential and its separate contributions. The part called "direct Coulomb" is, in fact, the total Coulomb field from electrons and jellium background. It is extremely small as expected from the close matching of the densities. The Coulomb field may be a bit larger if detailed ionic structure is considered. But even in this case it remains a small part. The largest contribution to cluster binding comes from the Coulomb exchange energy and the correlation energy adds another 20–30% to the binding. The right panel of Figure 3.10 shows the shell structure in terms of the sequence of electron levels. We see the large HOMO-LUMO gap (Highest Occupied Molecular Orbital to Lowest Unoccupied Molecular Orbital) which hints that $N_{el} = 138$ is an electronic shell closure. Metal clusters are saturating systems, i.e. systems which develop along a nearly constant average density (see Section 1.3 and Chapter 4). Accordingly, the potential depth and the Fermi energy are insensitive to the particle number. The Fermi energy is about 3 eV above the bottom of the potential in accordance with the Fermi momentum $k_F = 0.5/a_0$ (for Na).

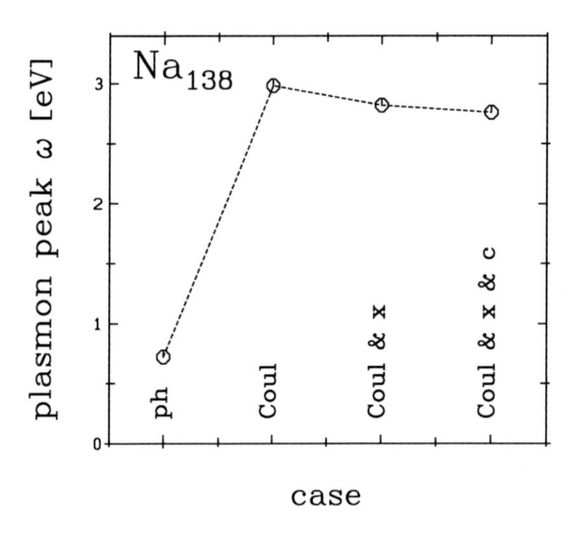


**Figure 3.11:** Position of Mie plasmon peak in $Na_{138}$ for various stages of the energy functional in the response calculation: pure particle-hole (ph) with no interaction, calculation with only direct Coulomb (Coul), with Coulomb and exchange (Coul & x) and with full exchange + correlation functional (Coul & x & c).

These relations remain about the same for small and for large Na clusters. What changes is the level density.

The relative importance of Coulomb versus exchange and correlation terms is much different in dynamical applications. We demonstrate that in Figure 3.11 for the frequency of the Mie plasmon resonance in the same system $Na_{138}$ as in the previous Figure 3.10. At this point in the book, the test case requires some explanations. Optical absorption explores the linear-response dynamics of a system (see Section 4.1.7). It turns out that optical absorption spectra in simple metal clusters are always dominated (often even exhausted) by one large resonance peak, the Mie surface plasmon (see Sections 1.3.3.3 and 4.3.1). Thus one can easily discuss the basic driving forces for TDLDA in terms of one number, the peak position. In Figure 3.10, we discussed the relative contributions of Coulomb force, exchange and correlations to the static binding. Figure 3.11 complements that for dynamic properties in terms of the surface plasmon frequency. The situation is here much different from the static case. The Coulomb force determines almost exclusively the peak position. Exchange and correlations change very little. This is very comforting in view of the fact that we are using instantaneous LDA which is a crude approximation to dynamic density-functional theory. The uncertainty concerns only a very small correction. The dominance of the Coulomb force can be easily understood: dynamics shakes the electron cloud which, in turn, gets displaced from the ionic background. This creates at once a large Coulomb force, for more details see Section 4.3.1.

#### 3.3.3.3 The local-density approximation (LDA)

The theorems of DFT guarantee that there exists a local energy-density functional, in principle. However, this exact functional is not easily known and even if one were able to attain such a functional, it would probably have an extremely involved mathematical structure. To make DFT a practicable scheme, one needs to invoke approximations. Most widely used is the Local Density Approximation (LDA) to the potential energy. Its construction is simple. One computes the ground state of the homogeneous electron gas as exactly as possible. This yields the exchange-correlation energy per particle $E_{\mathrm{xc}}/N \equiv \epsilon_{\mathrm{xc}}$ as function of the (yet homogeneous) density. We rewrite that to the energy per volume, $E_{\rm xc}/V = \rho\epsilon_{\rm xc}$ and integrate that to the total energy. The final, and crucial, step is to extend the validity of this global procedure to the local scale, namely to allow for $\rho \longrightarrow \rho(\mathbf{r})$ in that expression. This amounts to considering the energy as composed piecewise from an infinite electron gas of density $\rho(\mathbf{r})$. The approximation can finally be summarized through the following chain

$$\mathrm{e}^- \text{ gas} \Longrightarrow \frac{E_{\rm xc}^{(\infty)}}{N}(\rho) = \epsilon_{\rm xc}(\rho) \Longrightarrow E_{\rm xc}^{(\infty)} = \int d\mathbf{r}\, \rho\epsilon_{\rm xc}(\rho)$$

$$\Longrightarrow E_{\rm xc}^{(\rm LDA)} = \int d\mathbf{r}\, \rho(\mathbf{r})\epsilon_{\rm xc}(\rho(\mathbf{r}))$$

where the last step is the essence of the LDA. One takes a result which is exact for a homogeneous density $\rho$ and transfers it to the case of an inhomogeneous density distribution $\rho(\mathbf{r})$. This is, at first glance, a bold approximation. It is valid at least for very gently varying densities, as it is typical e.g. for valence electrons in metals. In practice, however, LDA provides a good description for a wide variety of systems. There is an enormous body of literature pondering successes and failures, for a more detailed discussion see e.g. [DG90].

As an example, we discuss briefly the case of the exchange functional in LDA and its contribution to the KS potential. The exchange energy of a homogeneous electron gas has a simple analytical form:

$$\epsilon_{\rm x}(\rho) = -\frac{3e^2}{4}\left(\frac{3}{\pi}\right)^{1/3} \rho^{1/3} \quad , \tag{3.25a}$$

from which one deduces

$$\hat{U}_{\rm x}^{(\rm LDA)}(\mathbf{r}) = -e^2\left(\frac{3}{\pi}\right)^{1/3} \rho^{1/3}(\mathbf{r}) \quad . \tag{3.25b}$$

This is the widely known Slater approximation to exchange [Sla51], which was proposed and employed as a practical scheme long before the theoretical foundation of DFT.

Although the LDA looks straightforward and unique, there are various (slightly) different exchange-correlation functionals available depending on different inputs from the electron gas and on the form in which one parameterizes the final functional. Old functionals started from approximate many-body calculations. For example, the functional of [vBH72] relies on a computation of the electron gas at the level of linear response theory concentrating on polarization diagrams from the collective plasmon mode. The long time widely used form of [GL76] extends that treatment with a diagrammatic approach accounting also for the non-collective modes of the system. Fully exact numerical calculations have become available since the beginning of the 1980s [CA80]. They have been taken as basis for the today widely used functional of [PW92]. Note that the differences between the LDA functionals mainly concern the correlation part. The exchange part in LDA is always the long established Slater approximation.

#### 3.3.3.4 Spin in LDA

A simple and straightforward extension of LDA is to allow for different densities of spin-up $\rho_\uparrow$ and spin-down $\rho_\downarrow$ electrons. One starts from an electron gas with $\rho_\uparrow$ and $\rho_\downarrow$ kept fixed

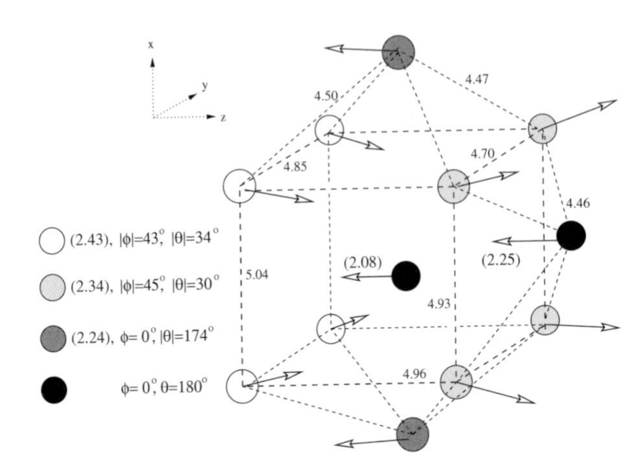


**Figure 3.12:** Geometric and magnetic structure for the energetically lowest non-collinear configuration of $Cr_{12}$. The local magnetic moments including their angles with respect to the x- and z-axis are indicated by arrows. The interatomic distances are given in atomic units and the magnitudes of the local magnetic moments $|\vec{\mu}_{\rm at}|$ (in units of Bohr's magneton $\mu_{\rm B}$) are given in brackets. The gray scale of the dots is related to the detailed angles of the spin directions as specified in the left part of the figure. From [KB99].

separately and carries through the same steps as above to end up with a spin-density functional $\epsilon_{\rm xc}(\rho_\uparrow(\mathbf{r}),\rho_\downarrow(\mathbf{r}))$. This is the so called Local Spin Density Approximation (LSDA). Most practical energy-density functionals are given in that generalized form. The Kohn-Sham equations including these spin-density functionals are outlined in Appendix E.1. The standard LSDA tacitly assumes invariance under spin-rotations and the associated freedom to choose the spin quantization axis in an arbitrary direction. This does not hold any more for magnetic materials, as e.g. Cr, Fe and Ni. These are sensitive to spin orientation. The LSDA can be generalized once more to cope with that situation [KHSW88]. One considers the local one-body density as a 2×2 matrix in spin space, i.e. $\rho_{\sigma\sigma'}(\mathbf{r})$. One transforms that for each space point $\mathbf{r}$ into a local diagonal spin frame with eigenvalues $n_\uparrow(\mathbf{r})$ and $n_\downarrow(\mathbf{r})$. These diagonal spin-densities $n_{\uparrow,\downarrow}(\mathbf{r})$ are then used in place of $\rho_{\uparrow,\downarrow}(\mathbf{r})$ in the LSDA functionals. This then yields the Kohn-Sham potentials first in the local diagonal frame. They need to be rotated back into the laboratory frame. Details for that treatment can be found in [KHSW88, KB99]. Figure 3.12 demonstrates the involved spin structure of the anti-ferromagnetic $Cr_n$ clusters, here for the ground state of $Cr_{12}$. It is immediately obvious from the figure that the spins are not collinear any more. Each ion uses its own spin direction and these directions are fixed by the interplay with the spatial structure. This arises because the finite number of spins and their local relations do not allow the establishment of an anti-ferromagnetic ordering as it appears in bulk or large clusters. This "frustration" leads to escape into non-collinear configurations where some energy is gained by allowing free adjustment of the spin direction for each ion separately. The example also demonstrates that magnetic systems, although accessible to DFT, are an order of magnitude more involved (and more expensive to compute) than simple metals. Their investigation and treatment is a challenging task for future research, see also Section 5.1.5.

#### 3.3.3.5 LDA for the time-dependent Kohn-Sham equation

As already mentioned, the formal foundation of DFT has also been carried forward to time-dependent scenarios [RG84, GK90]. It was shown that one can still write down Kohn-Sham equations with effective one-body potentials which now depend on time. However, a proper treatment is highly non-trivial as it includes all past times in order to incorporate all conceivable memory effects [GDP96].

The simplicity of the local energy-density functionals is regained when invoking additionally an instantaneous approximation. It assumes that the processes to be described by the Kohn-Sham equations are slow as compared to those which generate the correlations in the functional. It is then legitimate to neglect all memory effects and to consider only the mean field at the actual local density $\rho(\mathbf{r}, t)$ yielding the (time-dependent) Kohn-Sham potential in generalization of Eq. (3.24c) as

$$\hat{U}_{\mathrm{xc}}^{(\mathrm{ALDA})}(\mathbf{r}, t) = \epsilon\left(\rho(\mathbf{r}, t)\right) + \rho(\mathbf{r}) \frac{\partial \epsilon_{\mathrm{xc}}^{(\mathrm{LDA})}\left(\rho(\mathbf{r}, t)\right)}{\partial \rho} \tag{3.26}$$

where $\epsilon_{\mathrm{xc}}^{(\mathrm{LDA})}$ is the same energy functional as used in static LDA calculations. This is often called adiabatic LDA (ALDA), not to be mixed up with the adiabatic (Born-Oppenheimer) approximation for ionic motion which is discussed in Section 3.4.2. It is that instantaneous approximation which we have already anticipated for formulating the time-dependent Kohn-Sham equations in Section 3.3.3.2.

A word on nomenclature is appropriate here: The time-dependent application is usually called Time-Dependent LDA (TDLDA). In general, TDLDA poses an initial value problem and can be driven at any rate of excitation. Early TDLDA calculations were nevertheless restricted to small amplitude motion and employed a linearization of the TDLDA equations [Eka84]. This is, in fact, a linearized TDLDA which could also be called a Random-Phase-Approximation (RPA) which is the naming used by cluster physicists with nuclear physics background, e.g., in [YVB93] or [RWGB94] and related publications.

#### 3.3.3.6 DFT beyond LDA

The validity of LDA is a much discussed problem. It depends very much on the system under consideration, on the demands of precision aimed at, and on the observables studied. The mere existence of several variants, generalizations and extensions shows that one is not always fully satisfied with the performance of (TD)LDA. But, even when adding some corrections, one finds that (TD)LDA is a very good starting point and that there are systematic improvements for the more demanding situations which, at the end, deliver sufficiently reliable and precise predictions.

The case of excited electronic states is somewhat similar. Basic DFT theorems concern the ground state of electronic systems and say nothing about excited states. The Kohn-Sham picture allows one to re-inject single orbital electrons in the picture, but there is no guarantee that the latter do indeed truly represent electronic orbitals. They should rather be seen as a calculational, though convenient, ansatz. Still, experience gathered in Kohn-Sham calculations in many systems shows that, after all, Kohn-Sham orbitals are usually much more than convenient calculational tools. And taking them as a first order representation of true electronic orbitals is far from being senseless. The same holds for the corresponding single electron energies, and the reasoning can even be transferred, at least to some extent, to electronic excited states [Goe96], see also the discussion in Section 4.3.2.

In Appendix E, we discuss three directions for improvement of (TD)LDA: the inclusion of inhomogeneities is tackled in Appendix E.2, the problem of the self-interaction error in LDA is addressed in Appendix E.3, and novel ideas to deal better with time-dependent cases are sketched in Appendix E.4.

### 3.3.4 Phenomenological electronic shell models

We have seen above that all KS potentials for ground state metal clusters have a typical profile characterized by a flat bottom and a certain smooth transition to zero (see in particular Figures 3.6, 3.10 and E.1). For neutral clusters, one can approximate that very well by a Woods-Saxon profile

$$U_{\mathrm{WS}}(\mathbf{r}) = -U_0 \left[1 + \exp\left(\frac{|\mathbf{r}| - R(\vartheta,\phi)}{\sigma_{\mathrm{WS}}}\right)\right]^{-1} , \tag{3.27}$$

where, again, a possible deformation can be parameterized by an angular dependence of the radius $R(\vartheta,\phi) = R_0\,(1+\sum_{lm}\alpha_{lm}Y_{lm}(\vartheta,\phi))$, similar to the jellium model of Eq. (3.17) for the density. In Eq. (3.27) the potential depth $U_0$ is taken as the average binding potential in bulk matter. Computations in a fixed potential are somewhat simpler than with the self-consistent KS scheme. The Woods-Saxon model has thus been very often used for describing the electronic shell structure of clusters, particularly in early studies, see e.g. the study of large metal clusters in a spherical Woods-Saxon model [NHM90], a survey over several materials in [Cle91], shell structure and deformation systematics in a deformed model [FP93], or excitations modes [NKdSCI99]. But the computational advantage of fixed external potentials of Woods-Saxon shape shrinks when using properly interlaced iteration algorithms for the comparable self-consistent KS calculations, see Section H for details. They are thus nowadays somewhat out of fashion.

A substantial further simplification is achieved by realizing that shell effects are determined by the states near the Fermi surface and that these states practically see a harmonic potential. This suggests to use for first estimates a simple harmonic oscillator shell model. The harmonic oscillator predicts spherical magic electrons shells at $N_{\mathrm{el}} = 2, 8, 20, 40, 70, 110, \ldots$. It fits observed shells in Na clusters up to $N_{\mathrm{el}} = 40$. Larger Na clusters (and most other alkalines) continue with 58, 92, 138. The problem can be overcome by down-shifting the states with high orbital angular momentum $l$ using a phenomenologically fitted $\hat{l}^2$ term in the model. This yields the Clemenger-Nilsson model for the mean field potential

$$\hat{U}_{\mathrm{CN}} = \frac{m}{2}\left(\omega_x^2 x^2 + \omega_y^2 y^2 + \omega_z^2 z^2\right) - U\hbar\bar{\omega}\left(\hat{l}^2 - n(n+3)/6\right) \tag{3.28}$$

where $n$ is the global shell number ($n = n_x + n_y + n_z$). The three separate curvatures $\omega_i$ allow one to accommodate deformed situations including triaxiality. Volume conservation restricts their choice to $\omega_x\omega_y\omega_z = \omega_0^3$ =constant. The separate values can also be expressed in terms of the quadrupole deformation $\alpha_{lm}$ as introduced above and in Eq. (3.17), e.g. for axially symmetric shapes by $\omega_x = \omega_y = \omega_0 \exp(\alpha_{20}\sqrt{5/4\pi}), \omega_z = \omega_0 \exp(-2\alpha_{20}\sqrt{5/4\pi})$. The parameter $U$ serves to tune the downshift of high $l$ orbits and thus of the shell sequence. The Clemenger-Nilsson model was introduced into clusters physics in [Cle85] taking up a much similar nuclear oscillator model [Nil55]. An extensive review of early applications is given in [dH93]. The computational simplifications in the oscillator model are so dramatic that it

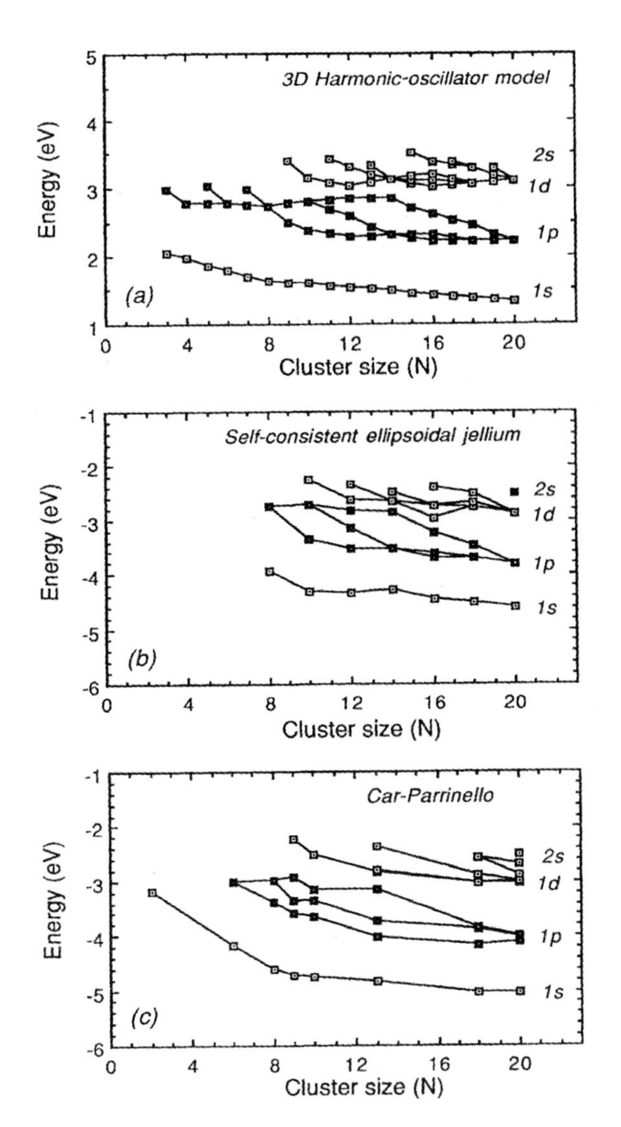


**Figure 3.13:** Single electron spectra versus cluster size $N$ for small Na clusters, computed with three models at different levels of refinement. Panel (a): Clemenger-Nilsson model with potential Eq. (3.28). Panel (b): deformed jellium model Eq. (3.17b), [LRMB91]. Panel (c): LDA with ionic structure [RA91]. From [dH93].

is still quite popular for first explorations of electronic shell structure, see e.g. studies of triaxiality in Na clusters [RFB95, YL95]. For the case of $U = 0$, one can write down a simple analytical solution. The single particle energies are indeed given by $\varepsilon_{\mathbf{n}} = (n_x + \frac{1}{2})\omega_x + (n_y + \frac{1}{2})\omega_y + (n_z + \frac{1}{2})\omega_z$ and the total energy by filling these single-particle levels up to the Fermi energy. This is the form of the model which will be used in the demonstration of the Jahn-Teller effect in Figure 4.10. The model with finite $l$-shift parameter $U$ can still be solved analytically for spherical clusters. But the spectrum of deformed clusters requires already a numerical diagonalization which, however, is still fairly simple. It turns out that the single-particle spectra from that Clemenger-Nilsson model are surprisingly realistic. This is illustrated in Figure 3.13 where one compares the single electron spectra for small Na clusters as obtained from the Clemenger-Nilsson model (3.28) with more elaborate LDA calculations with deformed jellium and with full ionic structure as background. The oscillator energies miss, of course, the binding relative to a continuum (which does not exist for an oscillator). But the energy differences between the occupied single-electron states are very well reproduced by the simple oscillator model with $l$-shift. Quantum mechanical shell effects, as magic electron numbers (see Section 4.2.1.1) or Jahn-Teller deformations (see Section 4.2.2.2) are thus well described by this simple model. However, the model is not so suitable for dynamical studies. These require also the knowledge of the residual interaction between the single-electron states (see e.g. Section 4.3.2 or Appendix G) and their phenomenological parameterization would introduce too many new unknowns to be fixed.

## 3.3.5 Semiclassical approaches

### 3.3.5.1 Classical limits

Classical particles are easier to handle than quantum mechanical wavefunctions. This is already exploited when treating the cluster ions as classical particles, see the basic Hamiltonian Eq. (3.1). One could also simplify the treatment if one could describe the electrons classically as well. Such an approach would become increasingly valid with increasing excitation and/or temperature. The simplest picture emerges if one treats the electrons as classical point particles. That is done regularly in plasma physics and it has been shown there that the fully classical treatment is valid for temperatures $T \gg 4\,\mathrm{Ry}(r_s/\mathrm{a}_0)^{-2}$ [ZTR99]. That is a rather large temperature. It rules out this simplest description for the majority of applications in cluster dynamics. The exceptions are those extremely high-energy events associated with violent Coulomb explosion following irradiation by intense laser beams, where genuine plasma physics models are applied indeed, see [DTS+97a, MBB+03] and Section 5.5.

A larger range of validity applies if one considers a classical limit to a phase space distribution $f(\mathbf{r},\mathbf{p})$. This yields the widely known Thomas-Fermi and Vlasov approximations which we will discuss in the following sections. Both approaches are applicable, in principle, down to temperature and/or excitation zero. They provide a pertinent description of the average cluster properties and miss merely the quantum mechanical shell effects. The latter are known to disappear at around $T \sim \epsilon_\mathrm{F}/20$ for metal clusters, e.g. $T \sim$ 1000–2000 K for Na clusters. Above that point, the phase-space description becomes fully valid. At lower temperature, one nevertheless obtains at least a good picture of gross features [Bra93].

The semi-classical phase-space approaches can be deduced from the quantum mechanical Liouville equation. We sketch that here for the case of a mean-field theory dealing with a mean-field Hamiltonian $\hat{h}$ and one-body density matrix $\hat{\rho}$. Starting point is the quantum mechanical Liouville equation $\imath\partial_t\hat{\rho} = [\hat{h},\hat{\rho}]$. The classical limit can proceed in two slightly different ways, the Wigner transform or the Husimi picture. The Wigner transform of the density operator reads $f(\mathbf{r},\mathbf{p}) = \int d^3y \exp{(\imath\mathbf{y}\cdot\mathbf{p})}\rho(\mathbf{r}-\mathbf{y}/2,\mathbf{r}+\mathbf{y}/2)$ and similarly for the integral kernel $A(\mathbf{r},\mathbf{r}')$ of any operator $\hat{A}$. The Wigner transform of the commutator $[\hat{h},\hat{\rho}]$ yields an involved operation between $f(\mathbf{r},\mathbf{p})$ and $h(\mathbf{r},\mathbf{p})$. Keeping only the lowest order in $\hbar$, one obtains the classical Poisson brackets which are at the root of the Vlasov equation, for details see e.g. [BB97]. However, at second glance, one realizes that this $\hbar$ expansion is at best an asymptotic expansion. It even formally ceases to converge for self-consistent calculations [LSR95] and literally diverges if the Coulomb force is involved [DLRS97]. The Husimi transform is obtained by smoothing the Wigner picture with Gaussian wave packets. The emerging Husimi picture has several interesting advantages: the transformed phase-space function is positive semi-definite, the classical and quantum mechanical Husimi folded distributions agree extremely well even for low quantum numbers, and the Gaussian folding width enters as a free parameter which can be exploited to keep minimum quantum features (the $\hbar$ smoothness) in the classical picture. For detailed discussions on these questions in connection with external field problems see [Pru78, Tak85, Tak86, Tak89] and in connection with self-consistent calculations in clusters see [DLRS97]. At the end, the lowest order approach yields again the Poisson bracket and with it the Vlasov equation, now however on safer grounds.

#### 3.3.5.2 Vlasov-LDA

Whatever way the derivation is done (Wigner or Husimi), in leading $\hbar$ order one obtains from the quantum mechanical Liouville equation the classical Liouville equation for the phase-space distribution $f(\mathbf{r}, \mathbf{p}, t)$ as

$$\frac{\partial f}{\partial t} = -\{f, h\} \quad , \quad \{f, h\} = \nabla_{\mathbf{r}} f \nabla_{\mathbf{p}} h - \nabla_{\mathbf{p}} f \nabla_{\mathbf{r}} h \quad . \tag{3.29a}$$

The classical one-body Hamiltonian $h(\mathbf{r}, \mathbf{p}, t)$ therein is computed self-consistently from the actual local density

$$\rho(\mathbf{r}, t) = \int d^3 p f(\mathbf{r}, \mathbf{p}, t) \tag{3.29b}$$

precisely in the same manner as in full TDLDA. Considering the Coulomb potential Eq. (E.2b) alone yields the much celebrated Vlasov equation which has its stronghold in plasma physics [Vla50]. The classical limit of TDLDA suggests to add in a straightforward manner the exchange-correlation potential (3.24c). This is then called Vlasov-LDA. The ionic background can be either jellium or composed with a local PsP, which both yield a simply local potential. Note that we do not distinguish spin. We use the total phase-space distribution $f$ summed over both spins, and accordingly the local density $\rho$ is the total density. In other words, we deal with spin-saturation $\rho_\uparrow = \rho_\downarrow$. Semi-classical methods including spin explicitely are still in their infancy [PAMB02, KW02]. There nevertheless exist several successful applications of Vlasov-LDA in cluster dynamics, see e.g. [FSCR96, PG99] and the discussion of Figures 3.14 3.15 later on.

#### 3.3.5.3 The Vlasov-Uehling-Uhlenbeck (VUU) approach

The semi-classical limit somewhat simplifies the treatment. This also allows one to deal with extensions beyond LDA. Just in the limit of high excitations where Vlasov becomes valid emerges the need for dynamical correlations. At low energies, electron–electron collisions are efficiently suppressed by Pauli blocking. This fades away at larger excitations and collisions become effective. One has then to complement Vlasov-LDA by a collision term. Strictly classical systems employ the Boltzmann collision term [Bal75]. The quantum statistics of Fermions is taken into account by augmenting the collision term with Pauli blocking. This yields the Uehling-Uhlenbeck collision term [UU32], introduced in the early 1930s. The combination is called the Vlasov-Uehling-Uhlenbeck (VUU) scheme and it has been much used in past decades for nuclear dynamics [BD88, BGM94, AARS96]. It has been successfully transferred to cluster physics and is also named VUU where a LDA mean field is tacitly implied. It allows a pertinent description of electronic thermalization by collisions [DRS98a], its impact on the Mie plasmon resonance [DRS00] and on laser induced cluster dynamics [GRS01]. The VUU equations read

$$\frac{\partial f}{\partial t} = -\{f, h\} + I_{\text{VUU}}(\mathbf{r}, \mathbf{p}) \quad , \tag{3.30a}$$

$$I_{\text{VUU}}(\mathbf{r}, \mathbf{p}_1) = \int d^3p_2 d^3p_3 d^3p_4 W_{1234} \left[ f_1 f_2 \bar{f}_3 \bar{f}_4 - \bar{f}_1 \bar{f}_2 f_3 f_4 \right] \tag{3.30b}$$

with the collision rate $W_{1234} = W(\mathbf{p}_1, \mathbf{p}_2, \mathbf{p}_3, \mathbf{p}_4)$ given by

$$W_{1234} = \frac{d\sigma}{d\Omega}(\{\mathbf{p}_n\})\, \delta\left(\mathbf{p}_1 + \mathbf{p}_2 - \mathbf{p}_3 - \mathbf{p}_4\right) \delta\left(\frac{\mathbf{p}_1^2}{2m} + \frac{\mathbf{p}_2^2}{2m} - \frac{\mathbf{p}_3^2}{2m} - \frac{\mathbf{p}_4^2}{2m}\right) \tag{3.31}$$

where again $h$ is computed according to the standard LDA recipes using the local density Eq. (3.29b) and where we have used the short hand notation $f_i = f(\mathbf{r}, \mathbf{p}_i, t)$ and $\bar{f}_i = 1 - \frac{f_i}{2}$. The term $\frac{d\sigma}{d\Omega}(\{\mathbf{p}_n\})$ represents the differential cross section for in-medium electron–electron scattering. It has to be computed with the properly screened Coulomb interaction for given Wigner-Seitz radius $r_s$ [DRS98a, DRS00, GRS02]. The phase space weight contains the typical terms $(1 - f_i/2)$. These are the Pauli blocking factors. Note that the total phase-space density for spin 1/2 particles is thus limited correctly to $f \leq 2$.

The success of VUU crucially depends on an efficient numerical implementation. The Vlasov equation as such describes dynamics in six-dimensional phase space. A direct solution on a mesh is still at the edge of, or beyond, present day computing facilities; for a few attempts in reduced symmetries see [RMM$^+$99, HZ02, HZT02]. One cannot even dream of treating the collision term that way because this involves additionally a nine-fold integration. The most widely used and most efficient technique to circumvent these problems is the test particle method, see e.g. [BD88] and Appendix H.4 for details.

Figure 3.14 demonstrates the impact of the electron–electron collisions in the case of laser irradiations of medium size sodium clusters. It shows the yield for photo-ionization as a function of laser photon frequency. This yield is related to the photo-absorption strength. Calculations have been performed with full ionic motion included (Section 3.4) and considering various models for electron dynamics (TDLDA, Vlasov-LDA, VUU). Comparing Vlasov-LDA and TDLDA, one sees that the semi-classical mean-field description agrees nicely with the fully quantum mechanical one, both showing a still well concentrated surface plasmon peak. The peak is broadened as compared to optical response in the linear regime. This is due to the fact that we are considering here a rather large laser intensity which drives the system into the non-linear regime. There is some broadening already at the mean-field level merely due to non-linear effects [CRS97]. Further broadening comes into play when stepping from Vlasov-LDA to VUU. This is due to the collisional damping which is included only in VUU. The effect of these collisions strongly depends on the strength of the excitation, see also Figure 1.15 in Chapter 1.

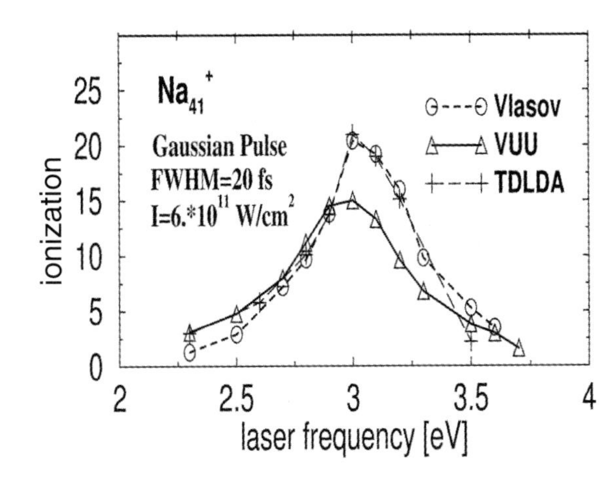


**Figure 3.14:** The total ionization as a function of laser frequency for fixed laser intensity $I = 6\times10^{11}\ \mathrm{W/cm}^2$ and a Gaussian pulse profile with FWHM= 20 fs. We compare TDLDA-MD with Vlasov-LDA-MD and VUU-MD. The test case is the cluster $\mathrm{Na}_{41}{}^+$ for which the Mie plasmon lies around 3 eV. After [GRS01].

#### 3.3.5.4 Thomas-Fermi approach

Vlasov-LDA and VUU are schemes for dynamical propagation. The initial condition requires an appropriate semi-classical ground state distribution $f_0$. It is found by the Thomas-Fermi approach. It consists in applying LDA also for the kinetic energy thus delivering a model which is formulated completely in terms of the local density $\rho(\mathbf{r})$. The Thomas-Fermi functional for the kinetic energy reads

$$E_{\mathrm{kin}}^{(\mathrm{TF})} = \beta_k \frac{3}{5}\frac{\hbar^2}{2m}\int d\mathbf{r}\,\rho(\mathbf{r})^{5/3} \quad , \quad \beta_k = (3\pi^2)^{2/3} \quad . \tag{3.32}$$

This, together with the LDA functional for the potential energy in the ansatz Eq. (3.23), provides the starting point. Variation with respect to the local density yields a self-consistent equation directly for the local density

$$3\pi^2\hbar^3\rho_0(\mathbf{r}) = \Big[2m\max\Big((\varepsilon_{\mathrm{F}} - U_{\mathrm{KS}}^{(\mathrm{LDA})}(\mathbf{r})),0\Big)\Big]^{3/2} \tag{3.33}$$

where $U_{\mathrm{KS}}^{(\mathrm{LDA})}$ depends on $\rho_0$ as given in Eq. (3.29b) and where the Fermi energy $\varepsilon_{\mathrm{F}}$ is to be adjusted such that the normalization condition $\int d\mathbf{r}\,\rho_0 = N_{\mathrm{el}}$ is fulfilled. This is the much celebrated Thomas-Fermi equation which provides a reliable estimate of gross trends for energies and densities in many-Fermion systems. A classical phase-space interpretation of that equation is to assume that at each local point $\mathbf{r}$ a Fermi sphere in momentum space is filled up to the given global Fermi energy $\varepsilon_{\mathrm{F}}$, i.e. up to the local Fermi momentum $p_{\mathrm{F}} = \left[3\pi^2\hbar^3\rho_0(\mathbf{r})\right]^{1/3}$ which, in turn, is related to the local density in an obvious manner.

The stationary Thomas-Fermi equation is related to the local Fermi sphere in momentum space with radius $p_{\mathrm{F}}(\mathbf{r})$. In a similar spirit, one can develop a time-dependent Thomas-Fermi approach (TDTF). An analysis of full Vlasov-LDA for the electron dynamics in metal clusters has shown that the local momentum distribution at all times stays close to a sphere [DRS98b]. The center of this momentum sphere is shifted according to the local irrotational flow. This suggests a collective description of the dynamics in terms of local density $\rho(\mathbf{r})$ and local velocity distribution $\mathbf{u}(\mathbf{r}) = \nabla\chi(\mathbf{r})$. Note that the irrotational veloc-

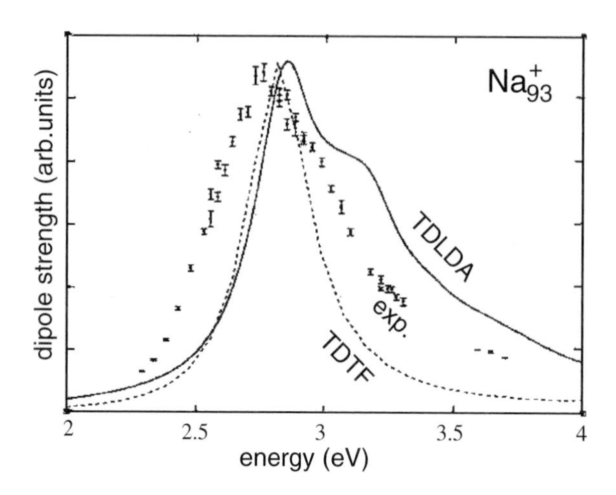


**Figure 3.15:** Dipole strength for $\mathrm{Na_{93}}^+$ in soft jellium approximation comparing results from TDLDA (full line) with those from TDTF (dashed line). TDLDA means here linearized TDLDA (=RPA) and the results are taken from [MGJ95]. Experimental results from [Re95] are shown for comparison (error bars).

ity field can be described by a scalar generating field $\chi$. The static kinetic energy functional $E_{\mathrm{kin}}^{(\mathrm{TF})} = (3/10m)(3\pi^2)^{(2/3)}\hbar^2 \int d\mathbf{r}\,\rho^{5/3}$ has then to be complemented by a flow term, $E_{\mathrm{kin}}^{(\mathrm{TDTF})} = E_{\mathrm{kin}}^{(\mathrm{TF})} + (m/2)\int d\mathbf{r}\,\rho\,\mathbf{u}^2$. Variation with respect to $\rho(\mathbf{r},t)$ and $\chi(\mathbf{r},t)$ then yields the TDTF equations

$$\frac{\partial\rho}{\partial t} + \nabla\cdot(\rho\nabla\chi) = 0 \quad , \tag{3.34a}$$

$$m\frac{\partial\chi}{\partial t} + \frac{m}{2}(\nabla\chi)^2 = -\frac{\delta T_0}{\delta\rho} - U_{\mathrm{KS}}^{(\mathrm{LDA})} \quad . \tag{3.34b}$$

These look like hydrodynamical equations, and indeed, one can view TDTF as a fluid dynamical picture of fermionic dynamics in a similar manner as hydrodynamics is obtained in a classical fluid by describing the phase-space distribution $f(\mathbf{r},\mathbf{p},t)$ through its lowest moments [Bal75]. The simplification due to the TDTF treatment is obvious: we have to propagate only one complex field $(\rho,\chi)$ instead of $N_{\mathrm{el}}$ different wavefunctions in full TDLDA. This becomes particularly interesting for very large systems. The simplification with respect to Vlasov-LDA consists in treating $\rho = \int d^3p\,f$ and $m\mathbf{u} = \int d^3p\,\mathbf{p}f$ instead of the full $f(\mathbf{r},\mathbf{p},t)$. However, TDTF underestimates damping because it misses some damping mechanisms which rely on coupling to single-electron motion (Landau damping in Vlasov-LDA, collisional relaxation in VUU). It is possible, in principle, to extend TDTF by terms describing viscosity. To our knowledge, there do not yet exist such considerations in connection with cluster dynamics.

Figure 3.15 compares the performance of TDTF with full TDLDA and with experiment for the optical response of $\mathrm{Na_{93}}^+$. For technical reasons a Husimi type folding has been applied to the density in the TDTF propagation, see Appendix H.5. TDTF visibly provides a reliable description of the average plasmon peak position. It misses, of course, the fragmentation of the TDLDA spectrum which stems from interplay of the Mie plasmon resonance with the detailed $1ph$ states in the energetic vicinity (Landau fragmentation).

There is a close relation between linearized TDTF and other models employing the picture of collective flow. There thus exists the local or collective-current RPA which can be derived from variation of sum rule moments [Bra89, RBG90]. For more recent detailed discussions see [KB01]. In fact, simple sum rule models [LS91] are the ancestors of local RPA and TDTF and they can be recovered as limiting cases. For an extensive comparison of all levels of approaches in the linear regime see [RGB96a].

The straightforward Thomas-Fermi approximation Eq. (3.32) provides, so to say, a zeroth order estimate of the smooth trends in energy and density. It is possible to enhance the descriptive power while still remaining at the level of a pure density functional. This is achieved by adding gradient corrections. The result is called the Extended Thomas-Fermi (ETF) method. Starting point is the extended kinetic-energy functional

$$E_{\rm kin}^{\rm (ETF)} = E_{\rm kin}^{\rm (TF)} + \frac{\hbar^2}{2m}\int d\mathbf{r}\,\frac{(\nabla\rho(\mathbf{r}))^2}{36\rho(\mathbf{r})} \quad . \tag{3.35}$$

This approach is still missing quantum mechanical shell effects. But it provides an excellent description of the averages, for details and applications in cluster physics see [BB97]. A dynamical ETF scheme for clusters was also presented in [BBG96].

## 3.4 Putting things together

### 3.4.1 Coupled ionic and electronic dynamics

In the previous sections, we have discussed at length the propagation of electronic dynamics and the coupling between electrons and ionic background. We finally add the ionic dynamics to complete the picture. Ions are described by classical Molecular Dynamics (MD), i.e. classical equations of motion, under the influence of their mutual Coulomb force, the forces experienced from the electrons, and possibly external forces. We start from the final expression for the total energy

$$E_{\rm total} = E_{\rm total,el} + E_{\rm kin.ion}(\mathrm{P_I}) + \mathbf{E}_{\rm pot,ion}(\mathrm{R_I}) \quad , \tag{3.36a}$$

$$E_{\rm kin,ion} = \sum_I \frac{\mathrm{P}_\mathbf{I}^\mathbf{2}}{2M_I} \quad , \tag{3.36b}$$

$$E_{\rm pot,ion}(\mathrm{R}_I) = \frac{1}{2}\sum_{J\neq I}\frac{e^2}{|\mathbf{R}_I-\mathbf{R}_J|} + U_{\rm ext}(\mathbf{R}_I,t) \quad , \tag{3.36c}$$

where $E_{\rm total,el}$ is the expectation value of $\hat{H}_{\rm el} + \hat{H}_{\rm coupl}$ in the basic Hamiltonian Eq. (3.1). In the case of TDLDA, it can be taken directly from Eq. (3.23) with details given in Eq. (E.1). The $U_{\rm ext}$ describes the action of an external field on the ions (which is usually negligible as compared to the effect on the electrons). Let us now take a closer look at the ionic parts of the total energy. This energy is considered as a classical Hamiltonian (with respect to the ions) to be fed into a classical action. By standard variation we obtain the classical equations-of-motion for the ions of positions $\mathbf{R}_I$ and momenta $\mathbf{P}_I$

$$\partial_t\mathbf{P}_I = -\nabla_{\mathbf{R}_I}\left[E_{\rm pot,ion}(\mathrm{R}_I) + \sum_{\alpha=1}^{N_{\rm el}}(\varphi_\alpha|V_{\rm PsP}(\mathbf{r}-\mathbf{R}_I)|\varphi_\alpha)\right] \quad , \tag{3.37a}$$

$$\partial_t\mathbf{R}_I = \frac{\mathbf{P}_I}{M_I} \quad . \tag{3.37b}$$

They are to be propagated simultaneously with the electrons, here represented by wavefunctions $\varphi_\alpha$.

When electron dynamics is described at the LDA level (Section 3.3.3.3) the simultaneous propagation scheme is called TDLDA-MD. It applies to all dynamical situations including those that are far from the adiabatic limit and embraces truly diabatic scenarios. The practical realization adds the (simple) classical propagation to the evolution of the electronic states (for some more numerical details see Appendix H). Although technically straightforward, the actual propagation with TDLDA-MD can become cumbersome due to the dramatic span of time scales: the full resolution of electronic dynamics requires a very small time-step of a fraction of a fs while it takes ps to explore the global ionic dynamics. Less expensive approaches which are applicable for adiabatic motion will be discussed in Section 3.4.2.

The ionic propagation Eq. (3.37) cooperates equally well with semi-classical approximations to electron dynamics. It can be coupled to VUU yielding VUU-MD. A slight modification is required in the second term of Eq. (3.37a). We assume local PsP and use that term in the form

$$(\varphi_\alpha|V_{\mathrm{PsP}}(\mathbf{r}-\mathbf{R}_I)|\varphi_\alpha) \longrightarrow \int d\mathbf{r}\,\rho(\mathbf{r})\,V_{\mathrm{PsP}}(\mathbf{r}-\mathbf{R}_I) \tag{3.38}$$

where the local density is now computed from the classical phase-space distribution $f(\mathbf{r},\mathbf{p},t)$ according to Eq. (3.29b). The semi-classical propagation cooperates nicely with PsP by virtue of the smoothing through Husimi folding implied in the test-particle method of resolution of VUU or Vlasov equations. The collision term in VUU guarantees sufficient long-time stability for propagation at the much longer ionic time scales [DGRS00, GRS02]. A similar coupling with TDTF to TDTF-MD, again in connection with local pseudo-potentials, has been explored in [BBG96]. The technique described there allows, in principle, fully coupled non-adiabatic dynamics. However, this first application employs a Car-Parinello technique and thus explores basically the dynamical Born-Oppenheimer surface at the level of a BO-MD, see Section 3.4.2.

At the level of that BO-MD, there also exist CI calculations with molecular dynamics (MD), although the details of the calculations are somewhat involved [MM97, KET98]. There are also recent calculations which do not refer at all to the BOA and which treat even the ions fully quantum mechanically [WIHS02]. All these advanced approaches are worked out up to now for small molecules. A fully coupled method beyond the adiabatic approach does not, to the best of our knowledge, exist for clusters.

## 3.4.2 Born-Oppenheimer MD

### 3.4.2.1 The Born-Oppenheimer surface

Ionic motion in molecules and clusters is usually much slower than electronic dynamics. There are many dynamical situations where the electrons stay close to an electronic ground state configuration. One can then make the approximation that the electronic ground state readjusts almost instantaneously to the actual ionic background. The ionic motion, in turn, is then accompanied by a comoving electron cloud in state $\Phi_{\mathbf{R}}$ where $\mathbf{R} = (\mathbf{R}_1 \ldots \mathbf{R}_{N_{\mathrm{ion}}})$ stands here for the actual ionic positions. The adiabatic separation Eq. (3.2) still allows a fully

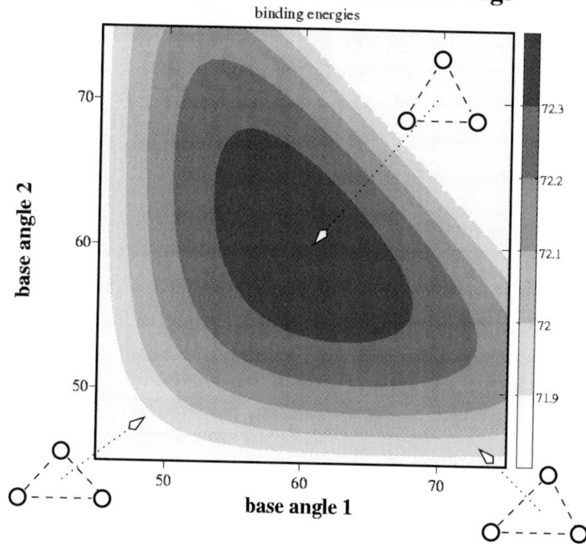


**Figure 3.16:** 2D cut of the 3D Born-Oppenheimer (energy) surface (BOS) for $Mg_3$ computed with LDA using the energy functional of [PW92] and the non-local pseudopotential of [GTH96]. The energy is drawn as a gray-scale plot with scale as indicated. The coordinates are two base angles of the triangle built by the three Mg ions. The radius is kept fixed at $3.2056\,a_0$, the value of the minimal configuration. The typical shapes are illustrated for three cases where the associated points in the angular plane are indicated by an arrow.

quantum mechanical treatment of ionic motion [Wei78]. This aspect was addressed briefly at the beginning of Section 3.1.3 and a few more details can be found in Appendix C.1. We consider here adiabatic motion in connection with classical molecular dynamics (MD) for ionic dynamics. This is used, e.g., in the Car-Parinello method [CP85]. A transparent way to look at the problem is to consider it as the adiabatic limit of TDLDA-MD. In that case, the classical approximations for the ions have already been made. The actual ionic positions are then identified with the instantaneous ionic configuration $\mathbf{R} = \mathbf{R}(t)$. The essence of the Born-Oppenheimer, or adiabatic, approximation is here that the electrons follow the ions along instantaneous ground states $\phi_{\mathbf{R}}$ which are computed for fixed ionic configurations $\mathbf{R}$ with one of the above discussed methods, e.g. with CI, or with LDA, or with Thomas-Fermi. Having solved the electronic problem for each shape $\{\mathbf{R}\}$, one can concentrate in the second step exclusively on the ionic dynamics. The effective potential for the ions is then a complicated multi-dimensional energy surface $V_{\mathrm{BO}}(\{\mathbf{R}\})$, often called the Born-Oppenheimer surface (BOS). It contains the coupling from electrons to ions through the instantaneous energy of the electron cloud. It reads, e.g., in connection with the DFT functional (3.23)

$$V_{\mathrm{BO}}(\mathbf{R}) = E_{\mathrm{total}}(\mathbf{R}, \{\varphi_{\alpha,\mathbf{R}}\}) - E_{\mathrm{kin,ion}} \tag{3.39}$$

where $\varphi_{\alpha,\mathbf{R}}$ stands for the set of single-electron wavefunctions in the BO state $\Phi_{\mathbf{R}}$ at instantaneous ionic positions $\mathbf{R}$. Figure 3.16 shows a cut through the three-dimensional BOS of the the $Mg_3$ cluster computed from LDA at each instant of ionic configurations. The minimum configuration is clearly the symmetric triangle with $\alpha = \beta = 60^o$. Its binding energy is $|E_B| = 72.377\,\mathrm{eV}$. The area of deepest black here covers the space of angles most likely occupied in an ensemble of configurations at about 900 K. Note that $Mg_3$ is a very small cluster, and yet, one can show only a part of the energy surface. The chances for visualization are much worse for larger clusters. A way out is to confine the plots to the most relevant degrees of freedom. These are usually the global multipole moments, headed by the quadrupole and octupole moments with sometimes crucial contributions from the hexadecapole. From that viewing angle one can also, consider,for example, BO surfaces for the jellium model (3.17) where the multipole deformations $\beta_{lm}$ serve as coordinates characterizing the ionic background, see e.g. [MHM+95].

#### 3.4.2.2 Dynamics on the Born-Oppenheimer surface

There are several ways to deal with the BO approximation subsequently. The standard procedure in molecular physics is to solve the Schrödinger equation for ionic motion yielding eventually the rotational and vibrational spectrum of molecules [Wei78]. Femtosecond spectroscopy gives access to the quantum dynamics of wavepackets in the space of ionic coordinates related to the solutions of the time-dependent Schrödinger equation within the given BOS [GS95]. A fully quantum mechanical treatment has not yet been undertaken in the much more complex case of cluster dynamics. But BOS have been used as a basis for a semi-classical propagation of wavepackets in Wigner space (containing at that level some quantum aspects) [HPBK+98].

Most often, one uses BOS for subsequent classical propagation of the ionic coordinates. This procedure is particularly efficient because one does not need to map out the unsurmountable huge BOS in all conceivable ionic configurations. One effectively performs a Born-Oppenheimer molecular dynamics BO-MD which steps along one path of ionic configurations. This proceeds as follows. For a given ionic configuration $(\mathbf{R}(t), \mathcal{P}(t))$, one first computes the instantaneous electronic ground state $\Phi_\mathbf{R}$. This then allows one to exploit the adiabatic theorem $\nabla_\mathbf{R}\langle\Phi_\mathbf{R}|\hat{H}|\Phi_\mathbf{R}\rangle = \langle\Phi_\mathbf{R}|(\nabla_\mathbf{R}\hat{H})|\Phi_\mathbf{R}\rangle$ for computing the effective forces on the ions as

$$-\nabla_{\mathbf{R}_I} V_\mathrm{BO} = -\nabla_{\mathbf{R}_I} E_\mathrm{pot,ion} - \sum_\alpha (\varphi_{\alpha,\mathbf{R}}|\nabla_{\mathbf{R}_I} U_\mathrm{back}|\varphi_{\alpha,\mathbf{R}}) \tag{3.40}$$

where the $\varphi_{\alpha,\mathbf{R}}$ are again the single-electron states of the adiabatic state $\phi_\mathbf{R}$ and the derivative is taken with respect to the ionic coordinates embodied in the background potential $U_\mathrm{back}$. One then propagates the ionic equations of motion (3.37) for a short time step $\delta t$. This yields finally a new $(\mathbf{R}(t+\delta t), \mathbf{P}(t+\delta t))$ as starting point for the next step. The validity of the BO-MD relies on two conditions: The first criteria for the BO approach is that the ionic motion is slow and the electrons follow ions instantaneously along ground state configurations. The second criteria for TDLDA-MD is that the ionic motion can be handled classically (large mass, no ionic interference effects). The savings as compared to the full TDLDA-MD are that one can use a larger time step for ionic propagation than for TDLDA. An example for BO-MD are the simulations of fission of small Na clusters in [BLNR91, BLR91] and for small K clusters in [BCC+94]. Figure 3.17 shows results for the example $K_{12}^{++} \longrightarrow K_9^+ + K_3^+$. It is, of course, hopeless to visualize the detailed trajectory in the multi-dimensional phase space. Results are then visualized in terms of global features as fragment distance and energies. The potential energy curve (panel (a)) shows barriers and minima relevant for the fission process. Note that the minimum configuration (denoted A in the plot) looks already like a preformed fission configuration. And yet, there is still a large barrier towards the scission point. The panel (b) shows the actual time evolution. The process is very slow initially (very small slope

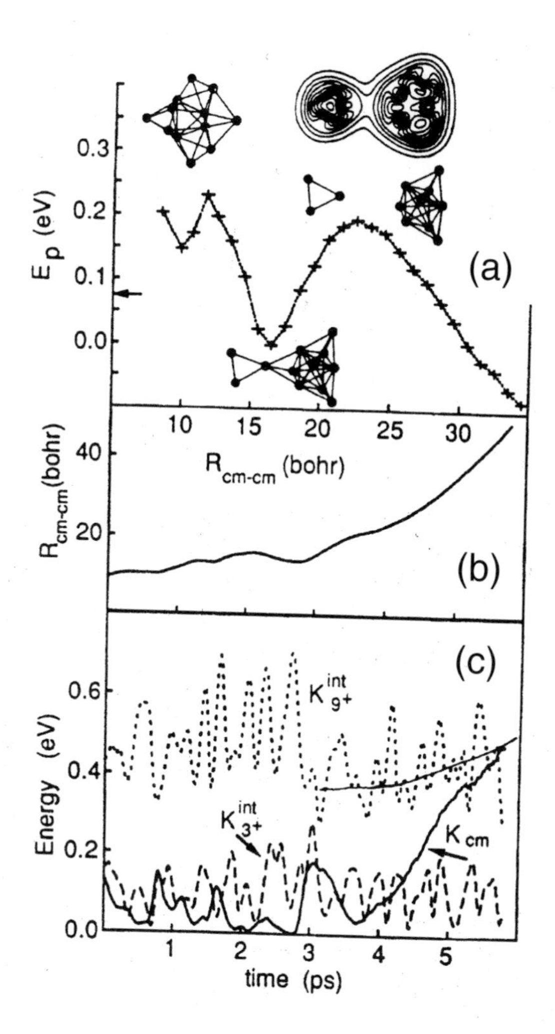


**Figure 3.17:** BO dynamics of the fission process $K_{12}^{++} \longrightarrow K_9^+ + K_3^+$. Panel (a): potential energy along the fission path as a function of the distance $R_{\rm cm-cm} = R_{\rm cm}(K_9^+) - R_{\rm cm}(K_3^+)$. The minimization has been computed with constraint on $R_{\rm cm-cm}$. The insets show typical configurations at various stages along the fission path. Panel (b): time evolution of fragment distance $R_{\rm cm-cm}$. Panel (c): time evolution of the various pieces of the kinetic energy as indicated. After [BCC+94].

of $R_{\rm cm-cm}$). A fast Coulomb separation sets on in the late stages, once the system has drifted over the barrier. The kinetic energies behave accordingly. The intrinsic parts for each fragment fluctuate almost unaffected by the ongoing fission process. The kinetic energy of the relative center-of-mass motion grows at the time when the Coulomb force drives the separation of the fragments, see panel (c).

The coding of the BO-MD is more involved than full TDLDA-MD because ions and electrons are treated at different levels. The TDLDA-MD is here conceptually much simpler as both species are propagated in time. There is nevertheless a way to maintain the structure of TDLDA-MD while employing a larger time step. The Born-Oppenheimer method relies on the fact that the electrons adjust very quickly to the present ionic configuration. When the Born-Oppenheimer MD is applicable, the full TDLDA-MD yield results which are identical to BO-MD. That is due to the enormous ratio of masses between electrons and ions, e.g. the factor is about 50000 for Na. One can thus introduce a pseudo-mass $m_{\rm CP}$ for the electrons which is much larger than the physical mass $m$ and still stay in a regime where the electronic pseudo-dynamics follows almost immediately the ionic configuration. The TDLDA-MD with the pseudo mass then reproduces again BO-MD. But the time step can be chosen larger by the factor $\delta t_{\rm CP} = \delta t\sqrt{m_{\rm CP}/m}$. This is the basic idea of the Car-Parinello method which was initiated in [CP85] and has found widespread applications since then. From the dynamical point of view pursued in this section, the Car-Parinello method is an efficient technique to compute a BO dynamics at an ionic scale. A variant of the method is also used for structure optimization, as will be discussed in the next section.

The BO-MD propagation becomes invalid in the vicinity of level crossings, i.e. if two BO surfaces come close to each other during the dynamics. The coupling term $\propto \partial_t \mathbf{R}_I \langle \Phi^{(n)} |\nabla_{\mathbf{R}_I}|\Phi^{(n')}\rangle$ between two different electronic states $n$ and $n'$ (see appendix C.1) becomes non-negligible and drives transitions $n \longleftrightarrow n'$. To cope with that situation, the BO-MD can be complemented by a hopping dynamics between electronic states which are energetically close, for details see e.g. [Tul90]. This generates an ensemble of BO trajectories from which one can deduce reaction rates, averages and variances. For details of the hopping elements see Appendix C.1. A more general sketch of stochastic mean-field theories is found in [RS92] where also the relation to the Fokker-Planck and Langevin description of ensembles is discussed.

### 3.4.3 Structure optimization

#### 3.4.3.1 Minimum energy configurations by dynamical methods

Before starting any dynamics, one has first to establish the electronic and ionic ground-state configurations. The stationary solutions for the electrons are found by standard techniques, see e.g. Appendix H. The simultaneous optimization of the ionic configuration is more tedious because there are often many isomers around which have to be carefully discriminated. We will address that aspect briefly in this section.

A conceptually simple and direct optimization scheme can for example be deduced from TDLDA-MD. One propagates electrons and ions simultaneously in the standard manner. Once in a while one removes all kinetic energy from the ions by simply setting $\mathbf{P}_I \longrightarrow 0$. This extracts steadily energy until the system comes to a halt in an energetic minimum configuration. To find appropriate times for resetting $\mathbf{P}_I$, one can take a protocol of the ionic kinetic energy $E_{\rm kin,ion}$. One best resets the ionic momenta if $E_{\rm kin,ion}$ has reached its maximum increase or a maximum. This method is straightforward and drives usually directly into a minimum.

However, the huge mass difference between ions and electrons makes it rather slow because many electronic cycles are computed until one explores a significant change of ionic motion. One way to accelerate that straightforward minimization scheme is to use Born-Oppenheimer MD (see Section 3.4.2). One propagates ionic motion with that until a sufficient amount of potential energy has been converted into kinetic energy. Then one removes the kinetic energy by resetting ionic velocities to zero. Repetition of that cycle finally ends up in a minimum-energy configuration.

Separate coding of a Born-Oppenheimer MD can be somewhat cumbersome. One can exploit the adiabatic approximation while reusing a given TDLDA-MD with minimal modifi-

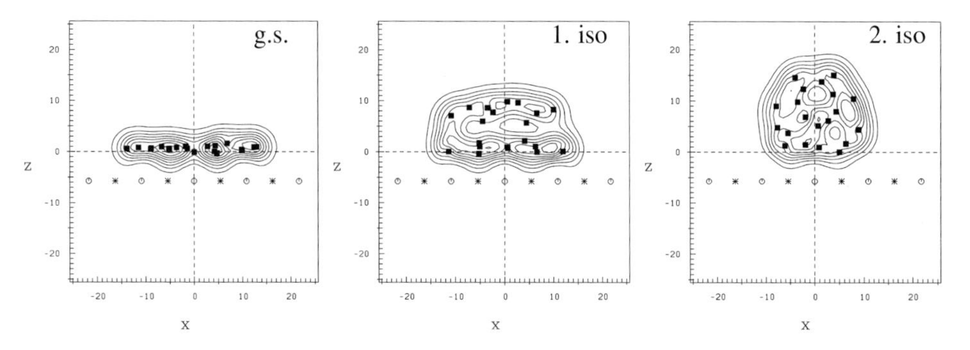


**Figure 3.18:** Ground state (g.s.) and isomers (1. and 2. iso) of $Na_{20}$@NaCl as indicated. Ions are denoted by squares, the electron density by equi-contour lines. The NaCl surface is indicated by a horizontal row of stars (Na) and circles (Cl). After [Koh97].

cations. Adiabaticity means that the electron cloud readjusts very quickly to the instantaneous ionic configuration. The process for the physical ratios is so fast that it still works if one reduces the mass ratio by one to two orders of magnitude. While cooling the system, we are not interested in the dynamical details but only on the end point of the relaxation which is independent of the mass ratio. Thus we are free to reduce the mass ratio for the purpose of a cooling TDLDA-MD. This is the essence of the widely used Car-Parinello method [CP85]. Two strategies can here be distinguished, either one reduces the ion mass or one enhances the electron mass. Both ways can be used for pure structure optimization although reduced ion mass is preferable here because one then obtains immediately the correct electronic energy. The other choice to work allows one to go even further. TDLDA-MD with enhanced electron pseudo-mass and without resetting of the ionic velocities is a simple and efficient technique to map a Born-Oppenheimer MD.

#### 3.4.3.2 Minimum energy configurations by stochastic methods

All these direct cooling methods have the problem that they can easily get stacked in a local minimum, i.e. in an isomeric configuration. In order to map the variety of isomers and to find the true ground state amongst them, one has to restart the cooling from a large variety of initial configurations. A more systematic way to encircle the minimum and/or to map the whole landscape of isomers is provided by simulated annealing with Monte-Carlo (MC) techniques. It consists in a random walk through the the possible stationary states of the system. In our case, a state is characterized by a point $\mathbf{R}$ in the $3 * N_{\rm ion}$-dimensional coordinate space of the ions together with the wavefunction $\phi$ of the $N_{\rm el}$ electrons. The key quantity for the walk is the total energy $E(\mathbf{R}, \phi)$ which is a functional of the state. Only this energy is needed, no forces and no explicit functional derivatives. Random changes of the state are checked for their change in energy. A lowering is always accepted. A step with increasing energy is occasionally followed to avoid being trapped in a secondary minimum. The method is explained in more detail in Appendix H.3.

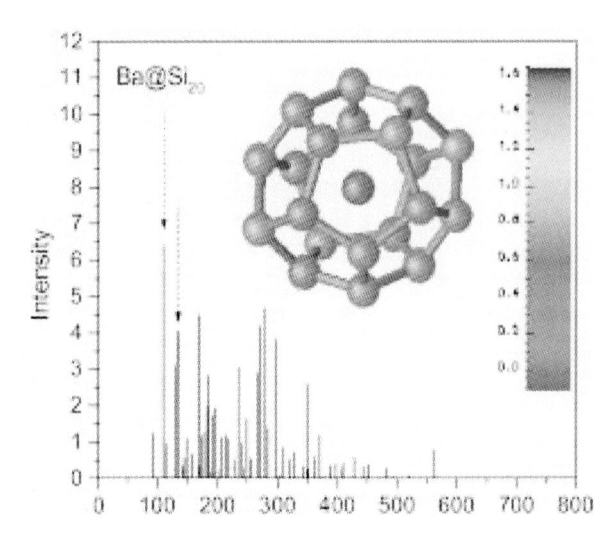


**Figure 3.19:** Configuration of the cluster $BaSi_{20}$ and the corresponding vibrational spectrum, computed with DFT in a basis of atomic orbitals. From [SWB$^+$02].

An example for ground state and isomers explored with simulated annealing is shown in Figure 3.18. The test case is the cluster $Na_{20}$ attached to a NaCl surface. The modeling of the effective cluster–surface interaction is described in [KR97]. It is strongly attractive such that the ground state becomes a planar cluster at the NaCl surface, see lower left panel. An alternative is the free $Na_{20}$ cluster placed gently on the surface, see upper left panel. The surface interaction distorts somewhat the initially spherical shape but it leaves a stable isomer. One can spot three layers of Na ions in that isomer. In between is an isomer with two layers (upper right panel). All these configurations have been found by several runs of simulated annealing. The isomers are spotted by careful tracking of the walk. There appear regions where the random walk resides for a while before jumping over to the final minimum. These "rest areas" have to be retraced with direct minimization schemes converging to the nearest local minimum.

Another case is displayed in Figure 3.19 which corresponds to an interesting situation in which one tries to stabilize a silicon cage, similar to the cages obtained with carbon. These silicon cages are interesting for possible technological applications, in view of the wide range of accessible band gaps. It turns out that the optimization of such structures is by no means simple and stabilization has been achieved here by implementation of a metal inside the cage. Figure 3.19 illustrates the case of a barium atom encapsulated inside a $Si_{20}$ cage [SWB$^+$02]. Also indicated is the corresponding vibrational spectrum of the structure with frequencies lying in the IR domain, as expected for vibrational modes. The study of vibrations has constituted here a means to evaluate the actual stability of the formed structures. The figure also provides some indications, in terms of grey scales, hardly visible here, on the repartition of electrons. It turns out that this repartition does significantly depend on the encapsulated metal atom. In the displayed case of barium the electrons are relatively homogeneously shared between all silicon atoms, but this is by no means a rule.

#### 3.4.3.3 Thermal ensembles

In Sections 3.4.3.1 and 3.4.3.2, we have discussed the methods to find the combined electronic and ionic ground state configuration at zero temperature. Experiments are done at finite temperature and one often wants to simulate a thermal ensemble. In fact, the dynamical scheme (TDLDA-MD, Car-Parinello) as well as simulated annealing offer the option to generate a thermal ensemble. There are two ways for TDLDA-MD. One can add a Nosé-Hoover thermostat to the classical ionic motion [Hoo91]. This generates a canonical ensemble. Or one can propagate for sufficiently long time a straightforward TDLDA-MD at a properly tuned (small in the ionic range) excitation energy. This generates a micro-canonical ensemble. The difference between canonical and micro-canonical treatment is small in large clusters. In simulated annealing, one simply stops the reduction of temperature $T$ and runs the loops for a long time at fixed final $T$. This generates a canonical ensemble. This technique is more efficient in exploring a landscape of many different isomers because the stochastic jumps are faster in stepping over barriers.

### 3.4.4 Modeling interfaces

The example of Figure 3.18 has shown a cluster deposited on a substrate. This situation requires particular modeling which we are going to discuss in this section. The conceptually simplest approach is to describe the atoms of the substrate at the same level as those of the cluster, e.g. DFT or CI, and represent the substrate by a few layers (typically about 5). This was done so, e.g., in the very detailed calculations of $Na_n$ clusters on NaCl [HM96]. That dealt with a simple metal in combination with a comparatively simple substrate, and yet, calculations are very expensive, thus limited to rather small cluster sizes. One may reduce the expense by a diligent choice of the substrate atoms entering the calculation which allows one to minimize their number [JFA+01, JFS+02]. And yet, the calculations remain elaborate. One needs to develop simple models for the substrate if one aims at large-scale calculations (large clusters, long-time dynamics, systematics). This can be done for the many situations where the surface remains rather inert. An example for a totally inert substrate model with microscopic details is discussed in Section 3.4.4.1. The polarizability of the substrate often plays a role. The simplest way to include that is a jellium model for the substrate augmented with a continuous dielectric description, as done e.g. in [RS93]. A more microscopic approach considers the polarization potentials for each of the substrate ions. An example for such a model is discussed in Section 3.4.4.2. Note that these cases are to be considered as examples which need to be re-tailored in new situations. They serve to give an idea about what to do in the modeling.

#### 3.4.4.1 An inert interface model

The fully microscopic calculations of $Na_n$ clusters on NaCl substrate from [HBL94, HM96] have shown that NaCl is very inert and remains basically unchanged when the cluster resides on it. This allows one to develop an inert-interface model where the substrate is acting simply like a local pseudo-potential. The electrons and ions of the $Na_n$ cluster see the substrate as an effective interface potential [KR97]

$$U_{eff}^{\text{ion,el}}(\mathbf{r}) = \sum_{\alpha\in\{Na,Cl\}} \frac{Z_\alpha}{|\mathbf{r}-\mathbf{R}_\alpha|} + U_\perp^{\text{ion,el}}(z)U_\parallel^{\text{ion,el}}(x,y) \quad , \tag{3.41a}$$

$$U_\perp^{\text{ion,el}}(z) = C^i\left(\zeta^6-\zeta^4\right) \quad , \quad \zeta = \frac{z_0\sqrt{\frac{2}{3}}}{z+z_0} \tag{3.41b}$$

$$U_\parallel^{\text{ion,el}}(x,y) = D^{\text{ion,el}} + E^{\text{ion,el}}\sin(k_x x)\sin(k_y y) \tag{3.41c}$$

where the model parameters are

$$\begin{aligned} z_0 &= 5.76\,a_0, & V_{\text{cutoff}} &= 0.5\,eV, & & \\ C^{\text{el}} &= 3.0\,eV, & D^e &= 1.157, & E^{\text{el}} &= 0.629 \\ C^{\text{ion}} &= 2.56\,eV, & D^i &= 0.98, & E^{\text{ion}} &= -0.218 \end{aligned} \tag{3.41d}$$

The first term in $U_{eff}$ accounts for the Coulomb forces from the Na and Cl sites. The second term consists in a Lennard-Jones like potential well $U_\perp(z)$ orthogonal to the surface and a corrugation potential $U_\parallel(x,y)$ parallel to it. The parameters are adjusted such that the properties of $Na_n$ clusters in the calculations of [HM96] are well reproduced. These detailed calculations are limited to very small clusters. The effective interface model (3.41) allows one to compute much larger clusters or detailed cluster dynamics, see e.g. [Koh97, KCRS98, KRS00]. This is a strategy similar to the development of pseudo-potentials (see Section 3.2.2.2): one relies on very detailed calculations of a small system, uses this as benchmark for fitting en effective potential, and transfers that to large-scale calculations.

#### 3.4.4.2 Dynamical cluster–substrate interaction

The inert-interface model of Section 3.4.4.1 neglects any response from the substrate. This is rather extreme even for an insulator as substrate. We thus continue with insulating materials where we can assume well localized electrons and only some small response from the substrate. The first effect beyond the frozen approximation is the polarization of the substrate ions. In fact, it is confirmed empirically that the interface of the cluster with the surrounding material and the polarization interaction considerably modify its dynamical response. For a recent review see, e.g., [Bin01]. Effects of the environment had been studied fairly early. It was shown, e.g., in [WH93] that a surrounding rare gas matrix modifies the optical response of small Ag clusters. Similarly came early theoretical explanations which used a dielectric description of the environment [RS93]. Nonetheless, the case is not yet settled and new data are coming up. There is, for example, a recent study in which the rare gas surroundings have been systematically varied, see [DTMB02] and Section 5.2.1.1.

In order to illustrate such investigations from the theory side, we show in Figure 3.20 an example of the, most important in this context, Ar–electron pseudo-potential [LBA86]. The asymptotics is the polarization potential induced by the electronic charge. It falls off as $-\alpha_0/r^4$ where $\alpha_0$ is the atomic polarizability as given in Table 3.1.

The inner part is fitted to electron-atom scattering data and parametrized as a high-order polynomial. The pattern look similar for the other rare gases. There are, of course, quantitative changes in asymptotics (see again Table 3.1) and in the core part. The polarizability and thus

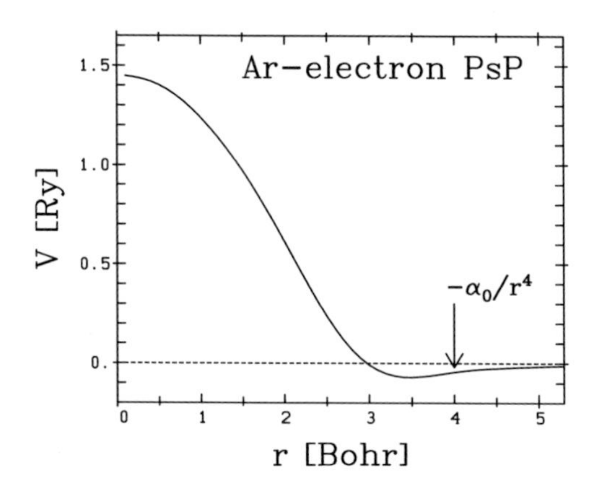


**Figure 3.20:** The local Ar–electron pseudopotential from [LBA86]. The asymptotic attraction is dominated by the polarization potential $-\alpha_0/r^4$ where $\alpha_0$ is the polarizability of the Ar atom.

**Table 3.1:** Polarizability $\alpha_0$ for rare gas atoms.

| | He | Ne | Ar | Kr | Xe |
|---|---|---|---|---|---|
| $\alpha_0$ [$a_0^3$] | 1.3845 | 2.672 | 11.080 | 16.782 | 27.318 |

the asymptotic attraction increases with atom size. What is more important for dynamical applications is that the polarizability depends on the applied frequency in the same way as a dielectric constant does, see Section 1.3.3.2. A dynamical modeling can be achieved by taking care of the dynamical polarizability $\alpha(\omega)$, e.g. in a single-pole approximation. We have here an explicit example on how dynamical effects are going to modify the mean-field picture.

## 3.4.5 Approaches eliminating the electrons

### 3.4.5.1 The tight binding approach

There are many situations where one is interested only in changes at the length scale of ionic positions and not so much in the finer details of the electronic distributions. Only their contribution to the global binding and to the electron transport between the ionic sites is of interest. This suggests employing a minimalistic treatment of the electronic degrees of freedom. This idea leads in particular to the widely used class of Hückel or Tight-Binding (TB) approaches [Wei78]. An example for the case of a binary molecule was discussed in Section 1.1.2.3. The idea is to work with a set of given atomic orbitals and to deduce the binding properties of the more complex systems through overlap integrals of the orbitals at different sites. The ansatz for the electronic single particle wavefunctions is thus a Linear Combination of Atomic Orbitals (LCAO)

$$\varphi_\alpha(\mathbf{r}) = \sum_{nI} \phi_n(\mathbf{r} - \mathbf{R}_I) c_{nI} \tag{3.42}$$

where $\phi_n$ are atomic single electron wavefunctions obtained, for example, from a DFT calculation. The label $n$ counts the atomic single particle levels and $I$ the various ionic centers. This ansatz is a valid and very general method if one considers a large basis of atomic levels $n$. It is

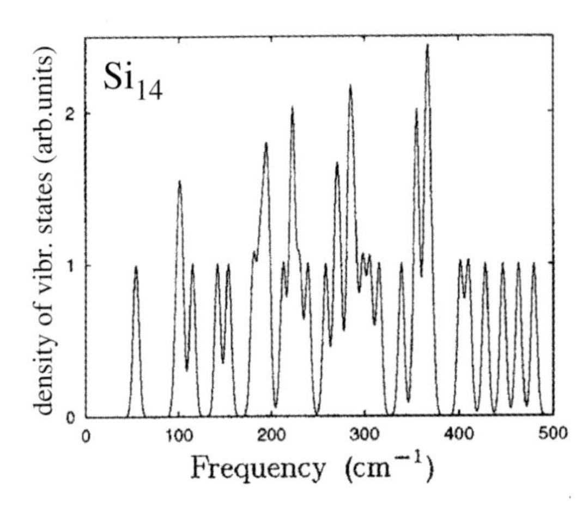


**Figure 3.21:** Vibrational spectrum of the $Si_{14}$ cluster computed with a tight-binding approach. After [SPF+97].

used as such, e.g., in the extensive TDLDA-MD calculations of [SS96, SS98]. And it becomes a very simple and efficient tight-binding (TB) approach, if one considers very few levels, often only one or two per ionic site. In the framework of LCAO the electronic wavefunctions are now represented by the coefficients $c_{nI}$. These are determined by matrix equations of the type Eq. (3.22b) with the Hamiltonian matrix (see e.g. [PFK+95])

$$H_{nI,mJ} = \delta_{IJ}\delta_{nm}\varepsilon_n + (1-\delta_{IJ})t_{nI,mJ}(\mathbf{R}_I, \mathbf{R}_J) \tag{3.43}$$

where $\varepsilon_n$ is the atomic single-electron energy and $t_{nI,mJ} = \int d\mathbf{r}\phi_n(\mathbf{r}-\mathbf{R}_I)H\phi_m(\mathbf{r}-\mathbf{R}_J)$ stands for the transfer integrals between wavefunctions at different sites (taking into account corrections for ortho-normalization of the LCAO basis). These transfer integrals are similar to the bond integrals introduced in Section 1.1.2.3 but are built from the whole Hamiltonian $H$ rather than only the potential part. This Hamiltonian matrix is, in principle, computed from the overlaps of the basis functions $\phi_{nI}$. In practice, one re-tunes it somewhat to compensate for incompleteness of the basis and other aspects. The norm kernel from non-orthogonal overlap of the $\phi_{nI}$ is sometimes considered and sometimes effectively parameterized into the Hamiltonian matrix.

The simplifications of the TB approaches open the door to demanding applications. For example, large scale BO calculations of ionic vibrations over long time spans now become feasible. Figure 3.21 shows results for vibrational spectra of the $Si_{14}$ cluster computed with a TB approach [SPF+97] whose actual details are explained in [PFK+95]. This variant takes into account all electron states of the valence shell of the Si atom. Particular care is taken to reconstruct an appropriate core–core repulsion. Note the enormous spectral resolution of better than $10\,\mathrm{cm}^{-1} \approx 0.2\,\mathrm{meV}$ which requires simulations over about 100 ps. That is beyond present day capacities of TDLDA-MD.

#### 3.4.5.2 The distance-dependent tight binding approach

The TB ansatz provides a flexible framework for a wide span of approaches. And many variants thereof can be found in the literature. The straightforward TB model is mostly used for covalent binding, in which electrons are known to remain localized, at least along bonds (Section 1.1.2.2). An extension which allows also a pertinent description of metallic binding is achieved by the the Distance Dependent Tight-Binding (DDTB) approach. It has indeed been very often used in connection with (metallic) clusters [SP95] up to truly dynamical simulations [ZSS$^+$94]. The DDTB is designed to take care especially of the polarization effects from $s$–$p$ state mixing, for example in Na. It starts with taking into account, from the atomic states $n$ in ansatz Eq. (3.42), the valence $s$ states plus the first available $p$ states per atom. Coherent admixture of $p$ states builds the polarization interaction. The $p$ states are eliminated in second order perturbation theory. This finally yields the Hamiltonian matrix (for Na)

$$H_{II} = \sum_{K\neq I} \left\{ \rho_{ss}(R_{IK}) - \frac{t_{sp}^2(R_{IK})}{\epsilon_p - \epsilon_s} \right\} \quad , \tag{3.44a}$$

$$H_{IJ} = t_{ss}(R_{IJ}) - \sum_{K\neq I} \frac{t_{sp}^2(R_{IK})t_{sp}^2(R_{KJ})}{\epsilon_p - \epsilon_s} \frac{\mathbf{R}_{IK}\cdot\mathbf{R}_{KJ}}{R_{IK}R_{KJ}} \tag{3.44b}$$

where $\rho_{ss}$ summarizes the effective ion–ion repulsion and the $t_{sp}$ are the inter-atomic transfer integrals. By virtue of the perturbative elimination, one has achieved a simple Hamiltonian matrix which only depends on ionic sites, and yet, the essence of the polarization potentials is incorporated. Electrons are still treated as being localized. Nonetheless, proper tuning of the $s$–$p$ polarization reproduces metallic systems pretty well. It is, of course, also a substantial improvement for other materials, e.g. rare gases. The polarization effects are in any case an important ingredient, see also the modeling of a substrate in Section 3.4.4.2 and the variant of TB in the following paragraph.

#### 3.4.5.3 The extended Hubbard model

The TB approach has a strong relationship with the Hubbard model which is widely used for describing electrons in solids [Cal91, Mah93] but also in quantum liquids [Vol84]. The Hubbard model, as a tight binding approach, is particularly well suited for dealing with localized electrons as in the case, for example, of noble metals. In the case of clusters, an extended Hubbard model, properly complemented by polarization effects, has proven to be a versatile tool also for structure problems [GPB91, ODDP99]. Electronic transport is modeled through the basic Hubbard Hamiltonian as sketched in Figure 3.22. Similarly to DDTB, one considers for each atom $I \in \{1, \ldots, N\}$ the two electronic states $\alpha = \mathrm{v}$ (valence) and $\alpha = \mathrm{c}$ (conductance) and the spin degree-of-freedom $\sigma \in \{-1, +1\}$ which produces a two-fold degeneracy. The model Hamiltonian in Figure 3.22 is composed of three terms: first, an intra site Hamiltonian $\hat{H}_0^{(\mathrm{Hubb})}$ which describes the energetic separation $\varepsilon$ of valence and conductance band; second, an electronic Coulomb contribution $\hat{H}_{\mathrm{Coul}}^{(\mathrm{Hubb})}$ which acts as an on-site repulsion punishing the occurrence of two electrons on the same site; and third, a hopping term $\hat{H}_{\mathrm{hopp}}^{(\mathrm{Hubb})}$ connecting two nearby ionic sites. The ingredients are the energetic separation $\varepsilon$, the parameter for Coulomb repulsion, and the hopping matrix element from site $I$ to site $J$. The hopping and Coulomb matrix elements sensitively depend on the ionic distance $\mathrm{r}_{IJ}$. This Hubbard Hamiltonian provides a pertinent picture of electronic transport properties once the ionic positions are given. This is typically the case in studies of solids where the crystal structure is taken

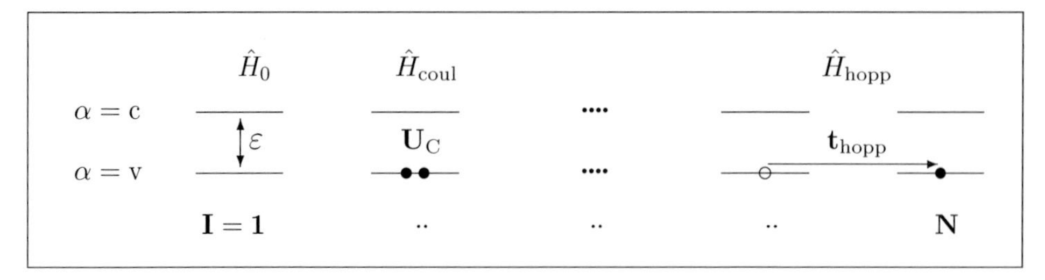


**Figure 3.22:** Schematic picture of the Hamiltonian of the Hubbard model.

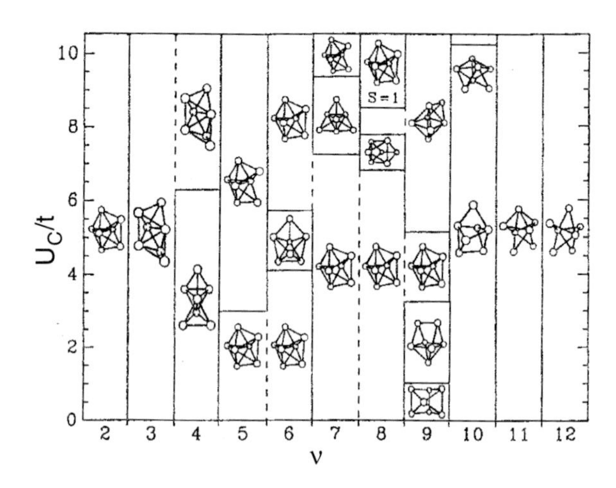


**Figure 3.23:** Structures of a cluster with $N = 7$ atoms as obtained from the Hubbard model for varying $U_C/t_{\mathrm{hopp}}$ and number $\nu$ of active electrons. The ground state spin is minimal, i.e. $S = 0$ or $S = 1/2$. The one exception with $S = 1$ is indicated. From [PB99].

from empirical information. Cluster physics, however, calls for a self-consistent determination of the ionic configuration. This requires extensions of the above Hubbard model. A crucial ingredient for determining the appropriate equilibrium positions is the Coulomb force between ions (atoms). One takes into account here the two leading terms, the monopole and the dipole interactions between the various ionic sites. At the level of dipole–dipole interactions, there also exists a correlation effect from simultaneous $s \rightarrow p$ excitation at the two sites, the van-der-Waals interaction (see also Section 1.1.2.2). Although the composition of this model from these various influences may look a bit involved, it deals with only very few electronic degrees of freedom and thus constitutes an enormous simplification of full CI or DFT calculations. Accordingly there are several applications which were not easily possible with the more involved methods, as e.g. a study of the systematics of bond change in $\mathrm{Hg}_n$ clusters [GPB91] or the impact of laser excitation on ionic configurations in a fully time-dependent simulation [ZGB99]. On the other hand, the extended Hubbard model can be viewed as being nothing else than an extremely simplified electronic Hamiltonian for which all approaches discussed in Sections 3.3 (CI, TDLDA) and 3.4 (TDLDA-MD, BO) apply. For example it has been used in connection with trajectory hopping (see Section 3.4.2 and [Tul90]) in an application to charge transport in $\mathrm{Xe}_n^+$ clusters [AS98].

An example for the landscape of parameters and associated cluster structures, obtained with the extended Hubbard model, is shown in Figure 3.23. It is obvious that the model with its few basic parameters (on-site Coulomb repulsion $U_C$, hopping matrix element $t_{\text{hopp}}$, and number of active electrons $\nu$) covers a very broad range of possible bonds and materials. Not visible in the figure is that the model can also be used to describe magnetic properties of clusters [PB99] where it becomes particularly efficient as compared to the very cumbersome full LDA calculations (see e.g. Figure 3.12). Anyway, almost every material can be accommodated by a proper choice of the model parameters. But this careful fitting is a necessary step at the beginning, as it is the typical price to be paid for any effective phenomenological approach. The reward is then this very efficient scheme.

#### 3.4.5.4 Effective atom-atom potentials

There are many situations where one can take for the TB Hamiltonian Eq. (3.43) only one reference state per atom site. This reduces the problem to the simple treatment of atoms taken as classical point particles interacting by effective atom–atom potentials. Such models are well established in molecular physics. Take, for example, the effective potential between Ar atoms which can very well be parameterized in the Lennard-Jones form (see e.g. [AM76])

$$V_{\text{Ar-Ar}}(r) = V_0 * \left[ \left( \frac{\sigma_{\text{vdW}}}{r} \right)^{12} - \left( \frac{\sigma_{\text{vdW}}}{r} \right)^{6} \right] , \tag{3.45a}$$

$$V_0 = 4 * 7.6 * 10^{-4}\,\text{Ry} = 480\,\text{K} , \tag{3.45b}$$

$$\sigma_{\text{vdW}} = 6.43\,\text{a}_0 , \tag{3.45c}$$

$$m_{\text{Ar}} = 7.35 * 10^5 m_{\text{el}} . \tag{3.45d}$$

What remains is a simple classical MD for the atoms following Eqs. (3.37), now of course without the ion–electron coupling. This allows large scale studies of cluster dynamics e.g. the computation of thermodynamic properties and phase transitions, for an example with molecular clusters see e.g. [PNB02]. The success of effective two-body potentials is not surprising for rare gases. But the approach is also applicable to other materials. Effective atom–atom potentials have even been developed for metals and were used for thermodynamic studies of small clusters [JB93].

Polarization effects are a crucial ingredient for the long range molecular attraction. They are embodied in these effective atom–atom potentials. However, there are situations where one needs to describe them explicitely. To that end, one can treat polarization explicitely by associating to each atom (or ion) a dipole polarizability. The result is polarization potentials of various kinds, see e.g. [DS96].

The case of van der Waals clusters with simple atom–atom potentials has been extensively studied for many years. It has in particular served as test case for systematic studies of temperature effects in clusters. The latter can be then attacked by direct MD simulations over very long times, which is relatively easily accessible with such simple Hamiltonians, in moderate size systems. An example of such calculations is shown in Figure 3.24 in the case of an $Ar_{13}$ cluster. In this figure is plotted the (short-time averaged) kinetic energy of the system as a function of time. Note the huge time scales involved here (500 ps), and the time step of 10 fs, well above most electronic time steps. The calculation shows a bimodal behavior: the cluster spontaneously oscillates between two phases, the transitions themselves being quite brief. The system then remains very long (as compared for example to typical ionic vibrations

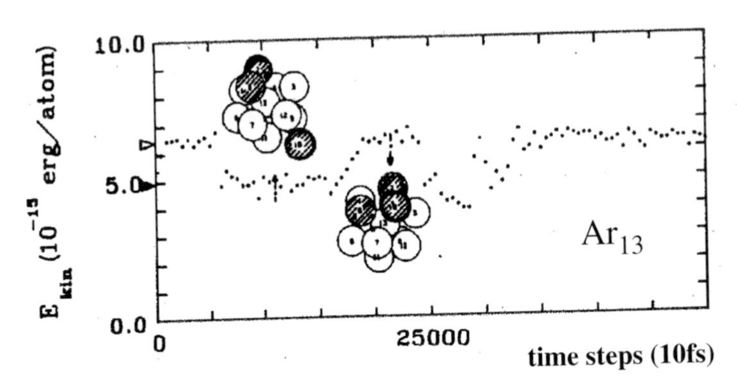


**Figure 3.24:** Time dependence of the kinetic energy (short-time averaged) for a MD simulation of $Ar_{13}$ near the phase transition between “solid” and “liquid” phase (total energy -4.16*$10^{-14}$ erg/atom $\simeq$ -26 meV/atom). After [JBB86].

time scales) in one or the other phase, depending on the total energy initially stored. The two phases can be characterized a bit more: the high kinetic energy phase is manifestly a vibrating icosahedron, while the low kinetic energy phase exhibits a more floppy behavior typical of a non-rigid structure.

A particular case is He clusters. This material never freezes out to a solid state. Due to quantum smoothing, it remains a liquid, to be more precise a quantum liquid, down to the lowest temperatures. There exist reliable atom–atom potentials for He. But the large repulsive core makes He a highly correlated quantum liquid whose fully microscopic description is extremely involved, but very successful [PPW86, Vol84, Kro00]. There even exist variational calculations for finite clusters [KZ01]. For simpler handling and large scale studies, one has however developed effective energy functionals for liquid He and clusters thereof [SGB+91, WR92, WR93].

## 3.5 Conclusions

Because clusters exhibit specific characteristics in between molecules and bulk, their theoretical description requires dedicated methods. Of course these methods are linked to somewhat similar methods developed for molecular or bulk systems. But transfers to clusters is not direct and one has to consider the cluster many-body problem as a specific and original question deserving a proper theoretical handling. In this chapter we have given a rapid survey of the basic methods used for attacking this question.

After having identified and briefly analyzed the properties of components of clusters (ions and electrons) we have focused a sizable part of our discussions onto the methods used to treat the electronic problem, by far the most demanding part of the whole program. Indeed while ions can, in general, be treated as point like classical particles, electrons have to be described at a quantal level. A first simplification, though, lies in the possibility to restrict the electronic problem to the valence electrons only. Although the separation between core and valence electrons requires due caution, the simplification it provides is at the heart of the possibility to treat scalable systems like clusters. The electron–ion interaction is then treated in terms of, possibly involved, pseudo potentials.

Electron–electron interactions, and in particular exchange, constitute, by far, the toughest part of the cluster many-body problem. There have been several theoretical works dedicated to this question over the years. We have in particular presented the so called ab initio methods of quantum chemistry and applications of density functional theory. In ab initio methods the interaction Hamiltonian is treated exactly and approximation schemes are developed to restrict the Hilbert space in which electrons are allowed to stay. This usually sets severe restrictions to the size of systems possibly described in such a way. In density functional methods, on the contrary, the Hamiltonian is simplified and the theory can be formulated in a simple Hilbert space. As a result, such an effective mean field method is usually applicable to systems of various sizes, a welcome feature for scalable systems like clusters.

Having at hand theoretical tools for describing both ions, electrons and their interactions, we have put all pieces together to attain a full description of clusters, a key aspect for addressing dynamical questions, as is the aim of this book. This was also the occasion of discussing simplified methods in which ionic and electronic degrees of freedom are packed together in an effective Hamiltonian, like in the case of the tight binding approximation. We have also illustrated the merits of these various (microscopic as well as macroscopic) approaches on several physical examples.

# 4 Gross properties and trends

As mentioned at several places in this book, one of the specifics of clusters is their scalability, a property which differentiates them from both molecules and bulk. A key issue of cluster physics is thus the study of physical properties, in particular as a function of size. And indeed many cluster properties exhibit interesting, and often quite telling, features as a function of size. Of course size is not the only relevant physical characteristic property of a cluster and we shall see that many other quantities can be identified as key parameters. But, in any case it remains an important task to summarize cluster properties in a systematic way. This is the aim of the present chapter. The two previous chapters presented the experimental and theoretical methods: we are going now to collect in a combined view the basic features of clusters as they have been reported over the past two decades. This concerns mostly structure and excitation spectra, the latter being associated with dynamics in the linear regime. Dynamics beyond the linear regime is mostly a recent topic, which will be addressed more specifically in Chapter 5. We shall thus not discuss such questions in the present chapter. Although we shall try to cover all sorts of materials, metal clusters will appear more often than others. This is due to the fact that metal clusters display pronounced electronic shell effects and resonance excitations (Mie plasmon resonance). But this also reflects their predominance in the past studies on cluster structure and dynamics. As a first step we shall discuss how to compute, from the theoretical side, physical observables, and especially measurable ones (even if theoretical and experimental means of measurements differ). The latter observables provide the link between theoretical models and measurements, which is essential for a proper progress in a field of physics. This aspect is discussed in Section 4.1.1. The next Section 4.2 is devoted to static properties of clusters. This means that we focus our discussions onto structure characteristics of clusters. Especially important here are the effects due to electronic or ionic shell closures which lead to the existence of particularly stable cluster sizes. Dynamic properties at the level of linear response are addressed in Section 4.3. This is the realm of the optical response, a question particularly relevant and documented in the case of metal clusters, which provide here most of the examples. Properties of metal clusters often show similarities to atomic nuclei. These similarities have actually already been pointed out at many places, in particular in Chapter 3. The aim here is thus to try to quantify this question by considering well defined physical quantities. We shall discuss this question in Section 4.4 in terms of both structural and dynamical quantities.

*Introduction to cluster dynamics.* Paul-Gerhard Reinhard, Eric Suraud

ISBN: 3-527-40345-0

# 4.1 Observables

## 4.1.1 Excitation mechanisms

A typical experiment consists in three steps: preparation, excitation, and observation. The theoretical modeling can be divided into the same steps. We have explained in Chapter 3 how to prepare the ground state of a cluster (stationary solution of the electronic problem and optimization of ionic structure). And we have outlined in Chapter 3 all the tools to describe the dynamic evolution. But before we can extract observables from the dynamic states, we need to specify the modeling of excitation mechanisms. The notion of an external excitation field $U_{\rm ext}(\mathbf{r},t)$ allows one to cover nearly all relevant excitation mechanisms. In this section, we will sketch briefly a few typical choices. Before doing that, we ought to mention that there is often useful information extracted already from the ground states, as e.g. energies, ionization potentials (IP), or radii. Excitation or reaction mechanisms are then not needed for the modeling because experimental analysis has already dealt with that and delivered these key data as such.

### 4.1.1.1 Laser excitation

Lasers are the most important and very flexible means for a dedicated, well tuned excitation of electronic systems. They produce a strong coherent electromagnetic field which can be well approximated, in the majority of cases, by classical time-dependent electromagnetic fields $\propto \exp(\imath\mathbf{k}\cdot\mathbf{r} - \imath\omega t)$ times a temporal profile function $f(t)$. Typical wavelengths are in the range of several hundreds of nm. This is a huge distance as compared to atoms, molecules, and (most) clusters. To give an example, blue light (a short wavelength in the visible) at a frequency of about 3 eV has a wavelength of about 300 nm. Na clusters have a Wigner-Seitz radius of $r_s = 4\,{\rm a}_0 = 0.2\,{\rm nm}$. It requires a clusters as large as $N \approx 1.7\times10^9$ until the diameter equals the wavelength. Actual cluster sizes stay usually far below that value. One can thus treat the laser field in the limit of long wavelengths ($k \longrightarrow 0$), i.e. we deal with a spatially homogeneous electrical field $\mathbf{E}$

$$\mathbf{E}_0 \exp(\imath\mathbf{k}\cdot\mathbf{r} - \imath\omega_{\rm las}t)f(t) \longrightarrow \mathbf{E}_0 \exp(-\imath\omega_{\rm las}t)f(t) \tag{4.1}$$

and we can neglect the effect of the magnetic field. The coupling Hamiltonian is formulated in terms of the associated potentials. These are not unique due to the freedom of gauge transformation. One way is to express the electrical field through a vector potential $\mathbf{A}(\mathbf{r},t)$ such that $\mathbf{E} \propto \partial_t\mathbf{A}$. The vector potential is also spatially homogeneous in the limit of long wavelengths. The coupling $e(\mathbf{A}\hat{p} + \hat{p}(\mathbf{A})$ becomes in detail

$$U_{\rm ext} = e\mathbf{E}_0 F(t)\hat{\mathbf{p}} \quad , \quad F(t) = \int dt'\, f(t')\exp(-\imath\omega_{\rm las}t') \quad . \tag{4.2}$$

One often prefers the same electrical field expressed in terms of the Coulomb potential such that $\mathbf{E} \propto -\nabla\Phi$. The coupling Hamiltonian then reads

$$U_{\rm ext} = e\mathbf{E}_0 f(t)\cdot\hat{\mathbf{r}}\exp(-\imath\omega_{\rm las}t) \tag{4.3}$$

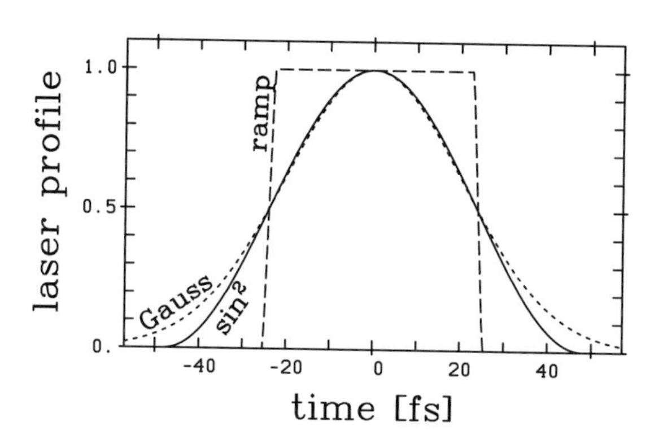


**Figure 4.1:** Time profiles for three types of laser pulses: full line = $\sin^2$ according to Eq. (4.5), dotted line = Gaussian $\exp\left(-(t/T_{\text{pulse}})^2\right)$, dashed line = ramp. All pulses have the same FWHM = 48 fs.

i.e. the laser field acts here simply as a time-dependent spatial dipole operator. Both formulations are fully equivalent. It is a matter of practical considerations which formulation one prefers. The form Eq. (4.2) is less convenient because it involves the momentum operator. However, it provides much more stable conditions in critical applications (e.g. looking selectively for high orders in multi-photon processes). The local form Eq. (4.3) is much simpler to implement. It is the form of choice for most applications. But the growth $\propto \mathbf{r}$ can cause trouble in highly excited situations where electronic density is flowing to large distances. For this case, one better switches to (4.2).

In any case, the laser field is characterized by the frequency $\omega_{\text{las}}$, by the peak field strength $E_0 = |\mathbf{E}_0|$, the polarization $\mathbf{E}_0/E_0$, and by the time profile $f(t)$ of the pulse. The field strength is usually given in terms of the laser intensity $I$ as

$$E_0 = c_{EI} I^{1/2} \quad , \quad c_{EI} = 1.07 * 10^{-8} \frac{\text{eV}}{\text{a}_0} \left(\frac{\text{W}}{\text{cm}^2}\right)^{-1/2} \quad . \tag{4.4}$$

The profile is a matter of debate. The experimental pulse profiles are not precisely known either. It is assumed to be a well peaked function with a certain full-width at half maximum (FWHM), in practice well approximated by a Gaussian. Theoretical applications often approximate that by a simple ramp, switching on linearly in the interval $t \in [0, \tau_{\text{sw}}]$, staying constant at 1 over almost all pulse length $t \in [\tau_{\text{sw}}, T_{\text{pulse}} - \tau_{\text{sw}}]$, and disappearing again linearly for $t \in [T_{\text{pulse}} - \tau_{\text{sw}}, T_{\text{pulse}}]$. This is a rough prescription and it is not too well peaked in frequency space due to discontinuous first derivatives at $\tau_{\text{sw}}$ and $T_{\text{pulse}} - \tau_{\text{sw}}$. A widely used smoother alternative is the Gaussian pulse $f(t) = \exp -(t/T_{\text{pulse}})^2$ which has optimal concentration in frequency space and which is probably close to the experimental situation. The practical difficulty in computations is that the signal spreads over all time. A good compromise, still close to experimental shape, is the $\sin^2$ pulse

$$f(t) = \begin{cases} \sin^2\left(\pi \frac{t}{2T_{\text{pulse}}}\right) & \text{for} \quad t \in \{0, 2T_{\text{pulse}}\} \\ 0 & \text{else} \end{cases} \quad . \tag{4.5}$$

It combines high spectral selectivity with finite bounds. Note that the form (4.5) is scaled such that the pulse parameter $T_{\text{pulse}}$ is identical with the full-width at half-maximum (FWHM).

The three types of pulse are sketched in Figure 4.1. They all have the same FWHM of 48 fs in the example shown on this figure. From experience with Fourier transformation one knows that the Gaussian pulse has the best spectral resolution for given FWHM. The $\sin^2$ pulse is similar and has almost comparable resolution. The ramp is visibly less smooth and this leads to less resolution due to long tails in frequency space. A further practical aspect to be considered is the temporal extension of the pulse. A calculation has to start with the pulse. The Gaussian profile now extends, in principle, from $-\infty$ to $+\infty$. Starting at a finite time means a sudden switching which inevitably induces a long tail in the frequency distribution. The effect can be reduced by taking a sufficiently long time interval before the peak. For example, starting at a time four FWHM before the peak produces the switch-on step with only $10^{-7}$. The $\sin^2$ pulse, on the other hand starts precisely at zero for the time of one FWHM before the peak and it has still a good spectral resolution. That is the best compromise for practical simulations. But all three pulse profiles miss one feature which plagues experiments with short pulses. There is a background of a "pre-pulse" with much longer time extension. Carefully tuned laser setups can suppress this by several orders of magnitude. But, e.g., four orders below an intensity of $10^{13}\,\mathrm{W/cm}^2$ can still mean something like $10^9\,\mathrm{W/cm}^2$ which is not faint a pulse either. One needs to be aware of such background effects when interpreting data with high-intensity pulses.

Another track to complications deals with wanted effects. Even though, the profile Eq. (4.5) represents a simple single pulse. Modern laser setups are much more flexible than that. Typical pump-and-probe experiments employ a succession of two such pulses with a well defined time difference [Zew94]. There are even lasers in which one can tailor the pulse with a dedicated chirp (small variation of frequency during the pulse) in order to conduct wanted chemical reactions [DDK+02]. The variability of lasers is so huge that one cannot discuss all conceivable profiles on general grounds. Each case has its own very specific features and needs to be discussed separately.

An interesting extreme is to think in terms of very short pulses. Profile and frequency become irrelevant features if $2\pi/T_{\rm pulse}$ is larger than any relevant frequency in the studied system. This regime is reached for most clusters at $T_{\rm pulse} < 0.5\,\mathrm{fs}$. Analysis with short pulses is particularly useful for theoretical spectral analysis. All frequencies are contained with equal weight. The response of the system thus contains a pertinent picture of the strength distribution, see Section 4.1.7 for more details. Short pulses are universal in that they all become simply $f(t) \propto \delta(t)$ in the limit $T_{\rm pulse} \longrightarrow 0$. The numerical realization of this limit skips the explicit use of $U_{\rm ext}$ because the effect can be incorporated into the initial condition. The potential Eq. (4.3) with $f \propto \delta(t)$ then modifies the initial wavefunction by a phase profile

$$\varphi_{\mathrm{gs},\alpha} \longrightarrow \varphi_\alpha(\mathbf{r}, t{=}0) = \exp\left(\imath \mathbf{p}\cdot\mathbf{r}\right)\varphi_{\alpha,\mathrm{gs}} \tag{4.6}$$

where $\mathbf{p}$ has the dimension of a boost momentum and tunes the amplitude of excitation. The phase profile means that a homogeneous velocity field is generated. The short pulse thus acts like an instantaneous boost.

#### 4.1.1.2 Charged projectiles

Probing electronic systems by beams of charged particles is a standard tool in atomic and molecular physics, see e.g. [Bra83, BM92]. It is also used in cluster physics, see for example Section 5.4 and citations therein. Highly charged ions, protons and also electrons can be considered as being structureless in such situations. What counts is only their Coulomb field.

Charged ions are heavy and can be treated with classical trajectories $\mathrm{R}_{\mathrm{ext}}(t)$. One can approximate these trajectories by straight lines for fast and heavy ions which are not deflected at all by the Coulomb field of the cluster. One has to take into account curved trajectories for very slow or light ions. In any case, the effect of the ion (charge $Z_{\mathrm{ext}}$) on the cluster can again be described by a time dependent external field

$$U_{\mathrm{ext}}(\mathbf{r},t) = \frac{Z_{\mathrm{ext}}e^2}{|\mathbf{r} - \mathrm{R}_{\mathrm{ext}}(t)|} \quad . \tag{4.7}$$

Magnetic effects are neglected. They may play a role only for very fast ions and/or heavy elements but are neglected here for simplicity. (A fully relativistic description in terms of a self-consistent time-dependent Dirac equation exists already for molecular physics problems [ASG+00, FAS+97] and can be imported in cluster physics if needed.) The ionic trajectories are characterized by the ion velocity and the impact parameter $b$ (which is the distance of closest approach when considering a straight continuation of the incoming trajectory). The velocity is more or less well defined by the experimental setup. But the impinging ion beam will cover a broad range of impact parameters. From the theoretical side, one has then to run several calculations with systematically varied impact parameters. Reaction cross sections are then computed by integration of the reaction probability over impact parameters, for examples see [RSU98].

Electron collisions require a full quantum mechanical treatment with standard methods of scattering theory [GW64]. The simplest description is provided by the Born approximation using plane waves for incoming and outgoing electrons. The differential cross section for elastic scattering is then just the Fourier transform of the potential from the cluster seen by the scattering electron (in DFT it would be the Kohn-Sham potential of the cluster). However, this simple approach is acceptable only for very fast electrons where the potential applies only to a small perturbation. The more typical situation is a strong perturbation. That requires the distorted-wave Born approximation (DWBA) where the projectile electron explores the scattering potential in all detail (by techniques of phase analysis). An example for a DWBA analysis of electron scattering on a water molecule is given in Section 5.4.4. Applications in cluster physics can be found e.g. in [CHH02]. DWBA treats the cluster still handled as an inert scattering potential. That is not applicable for low-energy electrons and for inelastic scattering. A more involved coupled channel calculation is then required, see e.g. [CGIS00].

#### 4.1.1.3 Atomic and cluster–cluster collisions

Related to ion impact, but more involved, is the case of collisions with atoms, molecules or other clusters. The Coulomb field still plays a crucial role. But both collision partners have now intrinsic electronic structure which is relevant for the reaction and both systems need to be treated on equal footing. An external excitation field is inappropriate. One should then "simply" take the whole machinery of electronic and ionic propagation without external field. The actual reaction is defined by setting appropriate initial conditions. This proceeds as follows. First, the ground state configurations for cluster target and actual projectile (atom or cluster) are determined separately in a standard manner. Second, the initial state is prepared by placing cluster and projectile at sufficient distance and giving them the wanted relative velocity vector (which defines velocity and impact parameter). Third, the dynamical evolution is followed with a standard method of choice. Fourth, observables are recorded either at all time or at final time. Similarly to the case of ion impact, reaction cross sections have to be obtained by finally integrating over a series of calculations with scanned impact parameter. For an example of such a calculation see e.g. [KJSS00] for the case of a Cs on $\mathrm{Na}_n$ collision (experimental data are presented in Section 5.4.3).

It is to be noted that this procedure involves a classical concept in that the initial state has well defined relative velocity and distance of the reaction partners. This is appropriate for mean field models of any sort. Fully quantum mechanical reaction theories would require CI calculations with continuum states in the basis. For a discussion of the state of the art in atomic physics see, e.g., [KLKD99] and for a practical code [SNBB02]. We are not aware of any fully quantum mechanical reaction calculations in cluster physics.

### 4.1.2 Energies

The total binding energy is a most prominent observable. Almost any computational scheme provides a value. However, it is not easily accessible in experiments. Measurements can provide differences of energies. There is for example the monomer separation energy as the adiabatic energy difference $E_{\mathrm{mon}} = E(N_{\mathrm{el}}, \mathbf{R}^{(N_{\mathrm{ion}})}) - E(N_{\mathrm{el}} - 1, \mathbf{R}^{(N_{\mathrm{ion}}-1)})$ where the reference energies are to be taken from fully relaxed ionic configurations. And there is the vertical ionization potential (IP) $\mathrm{IP} = E(N_{\mathrm{el}}, \mathbf{R}^{(N_{\mathrm{ion}})}) - E(N_{\mathrm{el}} - 1, \mathbf{R}^{(N_{\mathrm{ion}})})$ where the new electronic state in the $N_{\mathrm{el}} - 1$ system has relaxed but the ions are kept in their original configuration, see also the distinction between vertical and adiabatic IP's in Section 2.3.2.

As a byproduct of mean field calculations, one obtains also the series of single-electron energies $\varepsilon_\alpha$ for the occupied states. DFT does not guarantee that these are correctly described. And indeed, one finds, for example, that mere LDA underestimates the IP which should be identified with the energy of the HOMO level, see Figure E.1 and its discussion. But there are corrections (SIC and variants) which make single-particle energies more reliable. They can be observed experimentally by photoelectron spectroscopy, see Sections 2.2.3, 2.3.8 and 5.2.3.

Calculations allow one to observe processes in more detail than experiments. For example, it is useful to disentangle the various contributions to the energy, see e.g. Figure 3.17 where the potential energy was displayed as an analyzing tool. The ionic kinetic energy can be used as a "thermometer" in dynamical simulations. An electronic temperature can also be deduced from the electronic kinetic energy by

$$E_{\mathrm{therm,el}} = E_{\mathrm{kin,el}} - m \int d\mathbf{r}\, \mathbf{j}^2/(2\rho) - E_{\mathrm{ETF}}(\rho) \tag{4.8}$$

where $\mathbf{j}$ is the electron current and where the last term is the minimal quantum mechanical energy in Extended TF approximation and the second last the collective flow energy, for details see [CRSU00]. This quantity proved to be useful in analyzing the heating of clusters by laser impact.

### 4.1.3 Shapes

The detailed shape of a cluster is given by the detailed ionic coordinates. These are often characterized in terms of symmetries, molecular ones for small and crystal ones for large clusters. A useful analyzing tool is provided by global shape parameters in terms of multipole moments. The leading quantity is the monopole moment, i.e. the r.m.s. radius $r$. The further multipoles are best quantified in terms of dimensionless moments

$$\alpha_{lm} = \frac{4\pi}{5} \frac{\langle r_{\rm rms}^l Y_{lm} \rangle}{N r^l} \quad , \tag{4.9a}$$

$$r_{\rm rms} = \sqrt{\frac{\langle r^2 \rangle}{N}} \quad , \tag{4.9b}$$

which can be read in two ways depending on the meaning of the brackets $\langle ... \rangle$. For electrons, $\langle ... \rangle$ is the moment of the electron density ($\langle ... \rangle = (1/N) \int d\mathbf{r} ... \rho(\mathbf{r})$) and $N \equiv N_{\rm el}$, while for ions, one considers the classical moment $\langle f(\mathbf{r}) \rangle = \sum_I f(\mathbf{R}_I)$ and identifies $N \equiv N_{\rm ion}$. The moments are, in principle, different for ions and electrons. But the Coulomb force always tries to keep the two values aligned, at least for the low moments. There is of course a similarity between the moments $\alpha_{lm}$, which are observables, of the ionic or electronic structure and the $\alpha_{lm}$ in the soft jellium model Eq. (3.17b) which serve there as coordinates generating the jellium shape. The values of both these moments are almost identical for small deformations and start to deviate for larger ones. We thus do not introduce separate notations. There is a close relation between the low moments and the Mie plasmon response in metal clusters. The radius $r$ determines the average peak position, the quadrupole $\alpha_{2m}$ is responsible for the collective splitting, and the octupole deformation plays a role for the fragmentation width in small clusters [MR95b], see Sections 4.2.2.2 and 4.3. The optical response thus provides a direct connection to the global shape of clusters.

Axially symmetric systems are distinguished by having non-zero deformation only for $\alpha_{l0}$. Practically speaking, in a calculation, the appearance of $\alpha_{lm \neq 0} \neq 0$ can have two causes: first, the system is not aligned along its principal axes, and second, there is a true breaking of axial symmetry. The first action is then to rotate the system such that the $z$-axis is identical with one principal axis of the cluster. Axial symmetry is broken if we then still find some $\alpha_{lm \neq 0} \neq 0$. As far as the quadrupole is concerned, the only remaining case is $\alpha_{22} = \alpha_{2-2} \neq 0$. One thus often regroups the quadrupole deformation parameters into total deformation $\beta_2$ and triaxiality $\gamma$ as

$$\beta_2 = \sqrt{\sum_m \alpha_{2m}^2} \quad , \quad \gamma = \mathrm{atan}\left( \frac{\sqrt{2}\alpha_{22}}{\alpha_{20}} \right) \quad . \tag{4.10}$$

This convention has been originally introduced to characterize shapes of nuclei [HW53] and has been taken over for clusters at several places, e.g. [LRMB91, RFB95, YL95]. It is to be noted that $\gamma = 0^o$ as well as $\gamma = 60^o$ represent axially symmetric shapes. The case $\gamma = 0^o$ corresponds to prolate shapes and $\gamma = 60^o$ to oblate ones.

The quadrupole shape is often alternatively characterized by the moments of inertia

$$I_{ii} = M \langle r^2 - r_i^2 \rangle \quad \text{for} \quad i \in \{x, y, z\} \quad , \tag{4.11}$$

to be evaluated in the frame of principal axes of the cluster. The mass $M$ is here the ion mass because ions dominate that moment of inertia in practice. The expectation value means then a classical average. One can, of course, also apply this definition to compute an electronic moment of inertia. For this one replaces $M \longrightarrow m$ and employs quantum mechanical expectation values. In any case, these moments of inertia can be related to the above quadrupole moments as

$$r_{\rm rms} = \sqrt{\frac{I_{xx}+I_{yy}+I_{zz}}{2MN}} \quad , \tag{4.12a}$$

$$\alpha_{20} = \frac{4\pi}{5}\sqrt{\frac{5}{16\pi}}\frac{\langle I_{xx}+I_{yy}-2I_{zz}\rangle}{NMr^2} \quad , \tag{4.12b}$$

$$\alpha_{22}+\alpha_{2-2} = \frac{4\pi}{5}\sqrt{\frac{15}{8\pi}}\frac{\langle I_{yy}-I_{xx}\rangle}{NMr^2} \quad . \tag{4.12c}$$

### 4.1.4 Emission

Electron emission delivers a series of interesting observables: the number of emitted electrons (average ionization), their angular distribution, their kinetic energy spectrum, and the detailed ionization probabilities for each final charge state. There exist two different mechanisms for electron emission: thermal evaporation and direct emission. The thermal emission mechanism is conceptually simple. A heated electron cloud collects once in a while enough kinetic energy to one of its electrons such that it can escape. There exist several estimates for evaporation rates. There is, e.g., the RRK(M) theory which has been developed and refined in several steps, see [RR27, Kas28, MR51, Mar66, Eng86, PL65, Klo71, CB77]. We present here a simple estimate with the Weisskopf formula [Wei37]

$$W_{\rm evap} = \frac{2\sigma_{\rm capt}}{\hbar^3\pi^2}mT^2\exp\left(-E_{\rm IP}/T\right) \tag{4.13}$$

where $T$ is the actual temperature, $E_{\rm IP}$ the ionization potential, and $\sigma_{\rm capt}$ the cross section for electron capture. The latter may be approximated by the geometric cross section in the case of slow electrons. The associated relaxation times $W_{\rm evap}^{-1}$ are very long at room temperature (ns to $\mu$s). But they can shrink into the ps and fs range for high excitations reaching temperatures of 1000 K or more, see Figure 1.15 in Chapter 1. Thermal emission is distinguished by isotropic angular distributions and yields smooth exponentially decreasing photonelectron spectra [SKvIH01].

The direct emission is related to an electron following immediately the force pulse of an external electrical field (laser or bypassing ion), possibly amplified by response fields of the cluster [RS98]. Very strong fields lower the barrier so much that the electron can flow out over the barrier. Less strong fields still can trigger emission by tunneling. The typical characteristics of these processes are strong anisotropy of the angular distributions and possibly signatures in the photonelectron spectra. Very little can be said here in general. The actual distributions depend much on the details of the excitation and on the subsequent dynamical evolution. One needs numerical simulations at a microscopic level.

Such calculations of direct emission require two crucial ingredients: first, it ought to be a time-dependent method to track the dynamical process in detail, and second, the numerical representation should deal properly with continuum states. The various time-dependent methods have been explained in Chapter 3. The second condition is fulfilled by those numerical methods which represent the wavefunctions in plane waves or on a coordinate space grid. The grid has however to be augmented with absorbing boundary conditions, see e.g. Eqs. (46,47) in [CRSU00]. More elaborate techniques work with outgoing-wave boundary conditions [BSK97]. In any case, the average number of emitted electrons is then computed by the loss in normalization of all wavefunctions together. The remaining norm of each separate single-particle wavefunction provides sufficient information to deduce the detailed ionization properties by combinatorial analysis [CRSU00, Hor91, Ull00]. In grid based techniques the angular distribution has to be sampled from the outgoing flow one grid layer before the absorbing bounds start. The computation of the kinetic energy spectrum requires one to keep a protocol of the wavefunction one grid point before the absorbing bounds. One can then assume that only outgoing waves contribute at that point and disentangle the kinetic energy components by Fourier analyzing the time-dependent wavefunction [PRS00, PRS01].

The evaluation and analysis of these observables from emission is much simpler in semi-classical VUU with test-particle propagation. One defines an analyzing volume which is typically a sphere with radius much larger than the cluster. The emitted electrons are sampled at the instant where a test-particle leaves the analyzing volume. One can easily disentangle angular and energy distributions while sampling. However, one cannot extract the detailed ionization probabilities because one does not have single-particle states any more.

### 4.1.5 Polarizability

The static polarizability is a key observable of clusters as discussed in Section 2.3.3. The computation is as straightforward as its definition. We discuss it here for the (most important) dipole polarizability $\alpha_D$. One applies a static external dipole field $U_{\rm ext}(\mathbf{r}) = e\mathcal{E}_0\cdot\mathbf{r}$ and one performs static calculations for various values of $\mathcal{E}_0$ using the method of choice, see Chapter 3 for the possible options. As a result, one obtains a dipole momentum $\bar{\mathbf{D}} = \langle e\mathbf{r}\rangle$ as a function of $\mathcal{E}_0$. The tensor of static dipole polarizability is then simply given by

$$(\alpha_D)_{ij} = \left.\frac{\partial \bar{D}_i}{\partial \mathcal{E}_{0,j}}\right|_{\mathcal{E}_0=0} \quad . \tag{4.14}$$

This form is advantageous if the system has low symmetry anyway. It causes extra workload for systems with high symmetry because the external dipole field breaks some symmetries. In cases of highly symmetric systems, one often uses an alternative which goes through the excitation spectrum of the system as obtained, e.g., by linear response (see Appendix G). The polarizability is then computed as

$$(\alpha_D)_{ij} = 2e^2 \sum_n \frac{\langle\Psi_0|r_i|\Psi_n\rangle\langle\Psi_n|r_j|\Psi_0\rangle}{E_n - E_0} \tag{4.15}$$

where $n = 0$ is the ground states and $n$ labels the excited states with associated energies $E_n$.

### 4.1.6 Conductivity

There exist several attempts for a theoretical description of the conductivity of clusters, see e.g. [Lan97, YBL98, LA00a, LA00b, GGKS01]. The techniques and boundary conditions are the same as for any mesoscopic system [Dat95]. A recent review with emphasis on very small systems can be found in [Nit01]. The basic step is the same as in macroscopic physics, namely to consider the response to an applied voltage. Conductivity is obtained in the linear regime of arbitrarily small voltage. This amounts to computing the electron transmission function $T(\varepsilon_{\mathrm{F}})$ from one contact to another. This can be done by Green's function techniques of RPA in the linear regime and by time-dependent methods in general. The methods are standard. They reside in between stationary and dynamical problems. The state of the system is dynamical in the sense that time-reversal invariance is broken. However, it is stationary as nothing changes in time. We have a typical steady-state scenario. The particular problem with steady-flow at the nano-scale is that the contact is an integral part of the setup and influences sensitively the resulting transport properties. Careful modeling of the contacts is required, see e.g. the detailed discussion of conductivity of small Na clusters in [GGKS01]. We skip these details here because the topic lies somewhat outside the scope of this book which concentrates on cluster dynamics.

### 4.1.7 Spectral analysis

Optical response is a key observable in cluster physics, see Sections 2.2.2, 2.3.4 and 5.2. All methods discussed in Section 3.3 deal explicitly with electrons and can be used to compute the spectrum of dipole excitations. The conceptually most straightforward scheme is provided by CI. It computes the excitation spectrum directly. One has then to evaluate the dipole transition elements between the ground state and the excited eigenstates of CI to compose the dipole strength. The thus emerging sum of Dirac delta-distributions is smeared to a sum of Lorentzians to account for broadening mechanisms not contained in CI (e.g. incoherent thermal broadening) and thus to make the spectra comparable to experiment. However, this way to proceed is limited to small clusters. A reliable computation of excitations requires an even larger basis than would be sufficient for mere ground state calculations.

There are two ways to compute a dipole strength from TDLDA, either by solving the secular equation for a linearized TDLDA or by considering full TDLDA with an instantaneous dipole excitation and subsequent spectral analysis. The linearized TDLDA is a most basic scheme in many-body physics. There are various names for it, e.g. linear response theory, or random-phase approximation (RPA) motivated by a diagrammatic derivation. There are equally many ways for a formal representation of it, Green's function techniques, Liouville equations, operator algebra, etc. Almost any textbook on many-body quantum mechanics contains a representation of RPA. An extensive discussion of small amplitude oscillations for finite Fermion systems in general can be found in [BB94]. Linearized TDLDA has been much used in cluster physics, most often in cases with additional symmetries (jellium, spherical, axial), e.g. [Eka84, EP91, BG93]. The linearization maps TDLDA into a stationary eigenvalue problem. The solution consists in discrete spectra with corresponding transition strengths. As in CI, one smoothes the spectra by folding them with an appropriate Lorentzian. For a more detailed discussion of the method and notations see Appendix G. The selection rules in problems with symmetries reduce dramatically the expense of the eigenvalue problem. Linearized TDLDA (as well as CI) in symmetry-unrestricted cases can become very cumbersome dealing with huge matrices. The derivation of optical response from truly dynamical calculations is often more efficient here [YB96, CRS97]. It will be discussed in the following paragraph.

The computation of dipole spectra from full TDLDA explores the response of the system to laser pulses. The most efficient approach in the linear regime is to use the concept of an extremely short pulse, see Section 4.1.1.1. One starts from a well relaxed ground state and applies the instantaneous initial excitation by a small boost $\mathbf{p}$ of the center of mass of the electrons (dipole mode), see Eq. (4.6). One then propagates electrons with TDLDA and samples a protocol of the dipole moment with respect to ionic background

$$\mathbf{D}(t) = \int d\mathbf{r}\, \mathbf{r}\rho(\mathbf{r},t) \quad .$$

After a sufficient time ($T_{\max}$), one Fourier transforms the dipole signal and finally obtains the spectral strength in $x$, $y$, or $z$ directions as

$$S_{D_i}(\omega) = \Im\{\tilde{D}_i(\omega)\} \quad , \quad \tilde{D}_i(\omega) = \int dt e^{\imath\omega t} D_i(t) \quad . \tag{4.16}$$

The maximum possible spectral resolution is given by $\delta\omega = 2\pi/T_{\max}$. In practice, however, one should perform spectral filtering to avoid artefacts due to the fact that the dipole signal has usually not relaxed to zero at the end of the analyzing time [PTVF92]. This costs a factor of two or so in spectral resolution. It is interesting to note that this treatment in connection with absorbing boundary conditions allows one to compute correctly the escape width of spectral states lying in the electron continuum. For details of spectral analysis and variants thereof see [CRS97]. We exemplify it in Figure 4.2 with a result from [YB96] for a rather large Li cluster. The left panel shows the dipole signal in the time domain and the right panel the corresponding dipole strength (full line).

The same spectral analysis can be performed with other observables, as e.g. higher multipoles, spin modes etc. The full TDLDA furthermore allows one to go beyond the linear regime. The more appropriate observable is then the power spectrum. It is computed with the same steps as in the linear domain and the spectral power is obtained as the absolute squared Fourier transform, $\mathcal{P}_{D_i}(\omega) = |\tilde{D}_i(\omega)|^2$. The evaluation of the dipole moment has to be taken with care in the case of violent excitations to properly suppress the unwanted contributions from emitted electrons, for details see e.g. [CRSU00].

By far most calculations of optical absorption are done with fixed ionic background. Mie plasmon oscillations are much faster than ionic motion such that one can consider them as an instantaneous snapshots for a given ionic configuration. There is, however, an important effect from ionic motion. Clusters at finite temperature proceed slowly through their whole deformation energy landscape (for an example see Figure 3.16). This samples a thermal ensemble of cluster shapes. Experiments thus see an incoherent mixture of spectra from all these different shapes in the ensemble, see Section 4.3.5. This is the basic mechanism for the width of the plasmon peak in small clusters [BT89, YPB90] whereby the octupole deformations play a crucial role [MR95b, MLY01]. Fully fledged TDLDA-MD driven over ionic time scales allows one to resolve a dynamical plasmon–phonon coupling as well, see Section 4.3.5.3.

The same principles of spectral analysis apply, of course, also for computing the vibrational spectra of clusters. The analyzing times are then to be taken much longer to supply sufficient spectral resolution for the excitations in the meV range, characteristics of such ionic vibrational states.

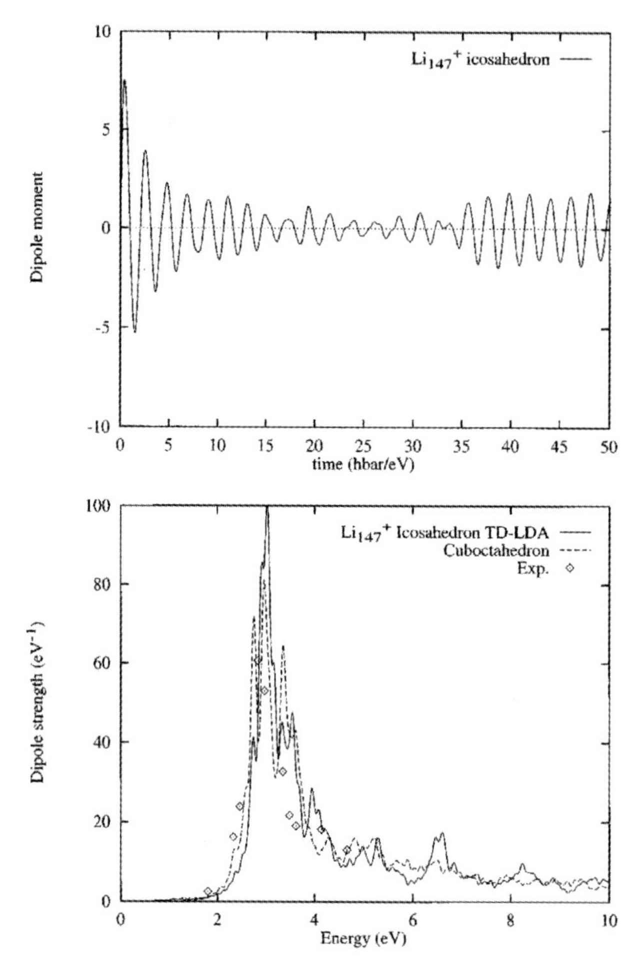


**Figure 4.2:** Spectral analysis in $Li_{147}^+$. Upper panel: time dependent dipole signal. Lower panel: corresponding dipole strength (full line) together with results from an alternative geometry and from experiment. From [YB96].

## 4.2 Structure

The most prominent observable for clusters and the first one to be looked at is the spectrum of mass abundances, i.e. the distribution of cluster sizes in a given production setup. As discussed in Section 2.3.1, mass abundances contain and reflect a bunch of important structural information. Furthermore, a proper identification of masses is already an involved task requiring proper mass spectroscopy (Section 2.2.1). Figure 4.3 shows experimental abundances for Na (upper panel) and for Xe clusters (lower panel). These are two very different materials, and yet, they both show pronounced spikes of particularly preferred cluster sizes $N$, often called magic numbers or shell closures. The phenomenon occurs for almost all materials, see also the case of carbon clusters in Figure 2.10 in Section 2.3.1. But its origin depends on the material, or more precisely on the type of binding. There are two different mechanisms which produce these magic numbers, cluster geometry and electronic quantum effects. The latter are only possible in metals where the electronic mean free path is larger than the cluster

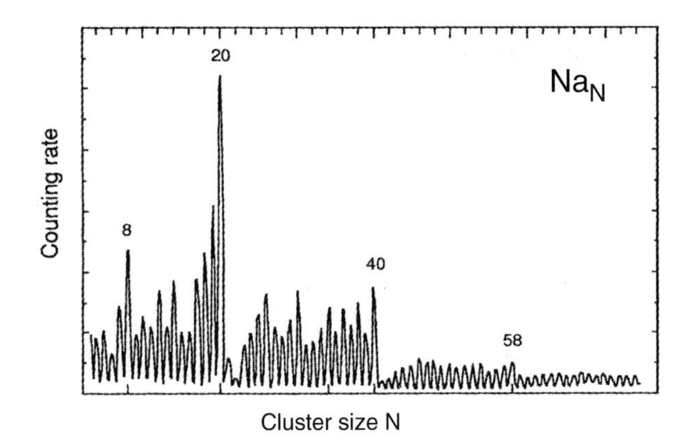


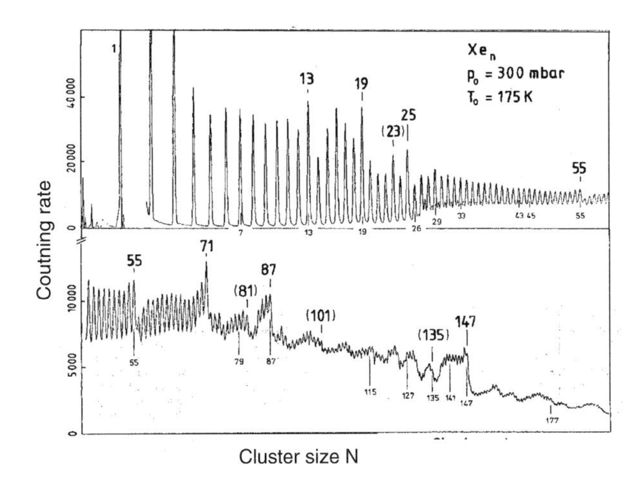


**Figure 4.3:** Experimental abundances of Na clusters (upper, after [dH93]), and Xe clusters (lower, after [ESR81]).

diameter, while the former has been observed in both metals and non-metals. Both types have been observed already in the earliest stages of clusters physics (for metals with electronic shell effects see [KCdH$^+$84], for rare gas clusters with atomic shells see [ESR81]) and the search for magic numbers was a prevailing task at that time. In the following, we will discuss shell effects in more detail, considering both mechanisms, electronic and atomic, separately.

## 4.2.1 Shells

### 4.2.1.1 Electronic shell effects

The case of Na in the left panel of Figure 4.3 shows abundances for neutral Na clusters which had been measured first in [KCdH$^+$84]. It serves as an example for electronic shells. We see the pronounced peaks at mass number $N$ = 8, 20, 40, and 58. These can be associated to electronic shell closures for spherical shapes containing the same number of valence electrons (2, 8, 20 ...) in this case of sodium in which Na atoms each provide 1 valence electron in

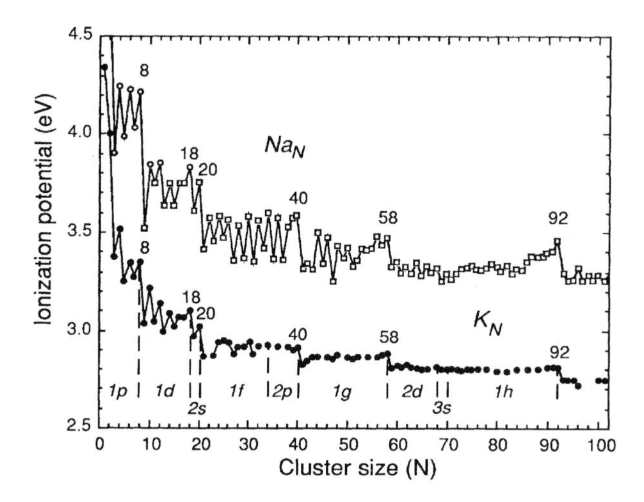


**Figure 4.4:** Ionization potentials (IP) of Na clusters. Taken from [dH93].

the cluster (Section 3.2.1). For an illustration of this shell closure effect have a look back at Figure 3.10 which shows the level sequence for a large cluster where one can see the shell closures $N = 58$ ($1g \to 2d$), 92 ($1h \to 2f$), and 138 ($1i \to 2g$). A calculation for mid-size clusters confirms the magic $N = 20$ and 40 as well. The situation is the same as with rare gases amongst the atoms. The binding is strongest if the number of electrons matches a full electronic shell. For clusters, the strong binding makes the magic systems particularly stable in an evaporative ensemble and this yields then the high peaks in the abundance spectra.

Corroborating information on shell closures can be drawn from the systematics of ionization potentials (IP). They are shown in Figure 4.4. Steps in the IP signal electronic shell closures. The results from the IP agree with the shells seen in the abundances, and, not surprisingly, the simple shell model reproduces correctly the position of the steps. On the other hand, it requires more elaborate modeling such as for example SIC (see appendix E.3) to describe correctly the detailed values of the IP.

A similar reasoning applies for the dissociation energy $D(N) = E(N) - E(N-1)$, i.e. the energy required to separate one monomer from a cluster of $N$ atoms. In fact, this is close to the quantity which determines the relative abundance in a thermal ensemble, namely the difference of free energies $F(N) - F(N-1)$ (where $F = E - TS$ is the free energy), which directly enters the Boltzmann factors in the equilibrium distribution. A better visualization is given by the difference of dissociation energies, i.e. by the second difference of free energies $\Delta_2 F = F(N+1) - 2F(N) + F(N-1)$. Large values of $\Delta_2 F$ signal large abundances. Theoretical results for $\Delta_2 F$ show peaks in nice agreement with data [Bra93]. Note that the above reasoning assumes naively equilibrium conditions in the evaporative ensemble. This suffices for the qualitative consideration here. A more careful analysis of the ensemble conditions and possible transient effects can be found in [HM94].

The pioneering measurements of [KCdH$^+$84, KCdHS85], as reproduced in Figures 4.3 (left panel) and 4.4, discovered the magic shells up to $N = 92$. This has triggered a long search for ever larger clusters and their magic shells in a combined effort, for a theoretical exploration see e.g. [NHM90] and for experiments [PBB$^+$91]. A summary is shown in Fig-

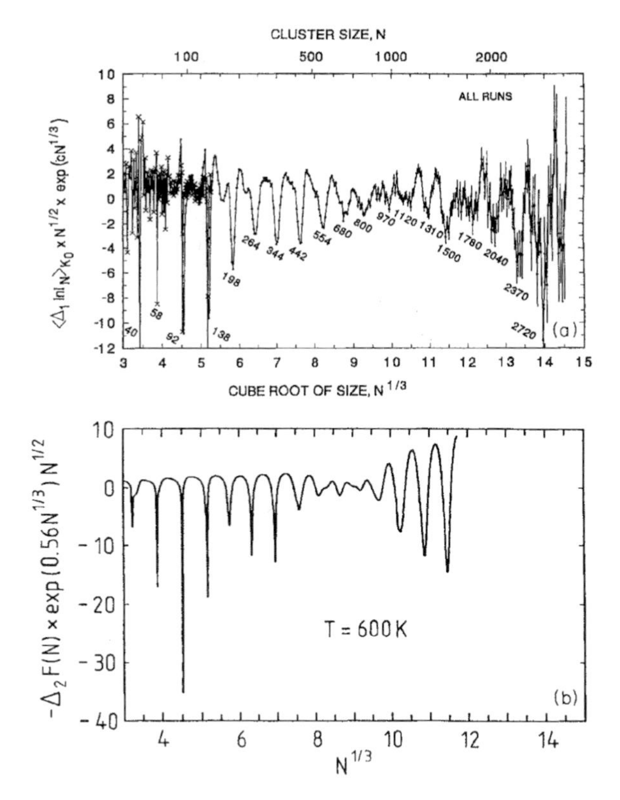


**Figure 4.5:** Upper part: first difference of logarithmic experimental yields $I_N$ augmented with a scale factor to equilibrate the average trend; magic numbers as associated with each negative spike are indicated, from [PBB+91]). Lower part: negative second difference of free energies from KS calculations with jellium background. From [Bra93].

ure 4.5. Note that abundances drop over several orders of magnitude when going so far up in particle number. A scaling factor has thus been applied which counterweights the average trend and allows one to draw all results on the same scale. But one should not forget that the yield for the largest clusters was very small and it required patient collecting of data to establish this systematics. The shell closures are distinguished here by deep negative spikes. The associated magic electron numbers are provided in the upper panel showing the experimental results. The lower panel shows results from KS calculations with a jellium model. The quantity to be compared with is $-\Delta_2 F$ as discussed above. It is drawn with the same rescaling as the experimental data. There is a nice agreement between theory and experiments. A particularly interesting feature, well reproduced by theory, is the trend one can observe in the shell effects. They shrink for increasing cluster size, go through a minimum (for around $N \sim$ 600–800) and grow again when stepping further up ($N \gtrsim$ 1000). This beating of the shell gap overlaying the sequence of magic numbers, is called super-shells. The effect was predicted for cavity modes long ago [BB74] and metal clusters have provided the first quantum system to display this effect experimentally. The mechanism for producing super-shells is understood more deeply in a semi-classical picture of shell quantization (periodic orbit theory). It emerges from a competition of the two dominant periodic orbits, those forming a triangle versus those forming rectangles. A thorough discussion of these interesting but subtle points is outside the scope of this book. For more details see [BB97, Bra97].

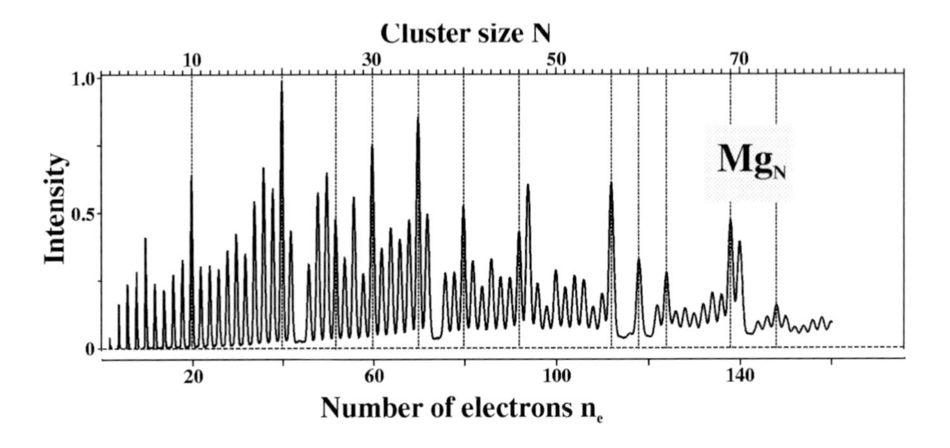


**Figure 4.6:** Abundances of Mg clusters, from [DDTMB01].

Electronic shell effects can be found in all metal clusters. But the actual sequence of magic electron numbers depends somewhat on the material. From the theory side, the sensitivity to details of the KS potential grows with growing size due to the higher level density. This was discussed in detail in [LBP+93, LPC+95] where it was shown that the large magic electron numbers depend on the surface softness. Also the position of the super-shells (the shell-beating point where shell correction goes through a minimum, see Figure 4.5) is found to be extremely sensitive to the surface profile.

A particularly interesting situation exists in metals with valence 2 as, e.g. Be, Mg, Ca, Sr, or Ba. The two electrons of the atom constitute a closed sub-shell which tends to act against metallic binding in small clusters. Even more interesting is the divalent metal Hg where (similar as in noble metals) some core electrons are close to the valence electrons and strongly contribute to the binding. The case has been extensively discussed and a transition from covalent to metallic character is expected for medium large clusters ($N \sim 50$) [GPB91]. However, it is not easy to pin down corresponding experimental signatures [HvIY+93]. More recently, the case of the simple divalent metal Mg has been studied. Figure 4.6 shows mass abundances for Mg clusters as obtained from evaporating a He cluster with a large content of Mg [DDTMB01]. One reads off peaks at $N_{\rm el} = 10, 20, 40, 60, 70, 80, 94$ etc. The magic numbers 20, 40 and 70 can be well understood as spherical shell closures. The other numbers are unconventional and still wait for explanation. It is conceivable that we see here a competition between electronic and atomic shells. The latter will be discussed in the following.

#### 4.2.1.2 Atomic shell effects

Up to now, we have only pondered electronic shells. These dominate in warm clusters where the ionic background is smoothed by thermal fluctuations and where the situation resembles more a liquid drop rather than a piece of a crystal. The picture changes at very low temperatures. The ions freeze out into well defined positions. Large clusters arrange themselves in the crystal symmetry of the bulk material. It happens only for certain numbers of atoms that

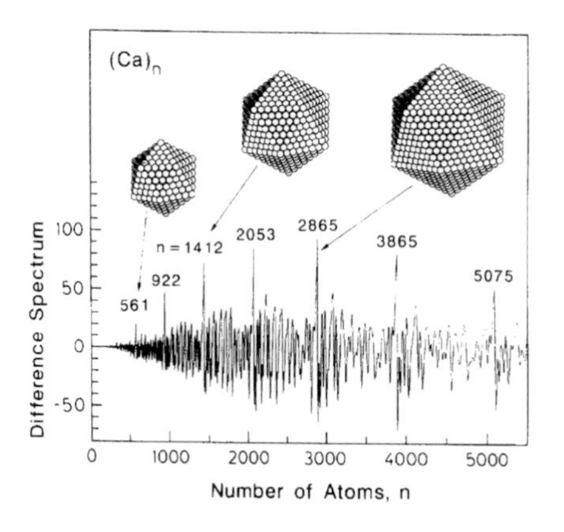


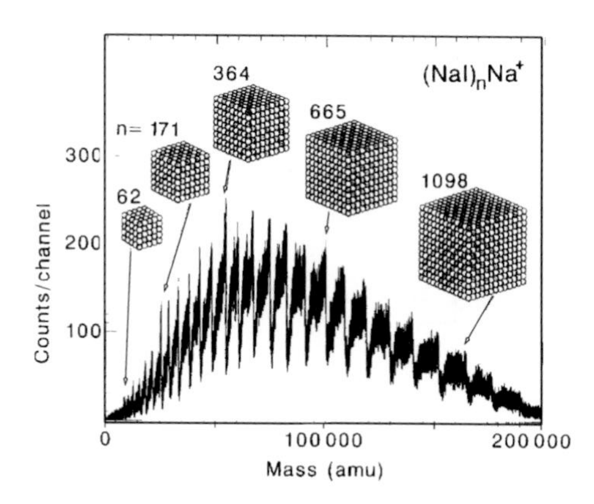


**Figure 4.7:** Difference of mass abundances for Ca clusters (left) and NaI clusters (right). The shapes associated with pronounced atomic shell closures are shown and the link is indicated by an arrow. Taken from [Mar93].

all surfaces of the cluster form regular planes. Such systems have a large surface energy and are significantly better bound. This produces a step in the monomer separation energies similar as for the electronic shell closures. The effect is called atomic shells and a new series of magic numbers is associated with them. It is obvious that the sequence of magic numbers for atomic shells crucially depends on the crystal symmetry of the material. The case of atomic shells has been discussed extensively and in a well readable manner in the review [Mar93]. Figure 4.7 shows two examples, Ca clusters (left) with icosahedral symmetry and NaI (right) with cubic symmetry. The large peaks in the abundance spectra correspond to atomic shell closures due to completion of a full icosahedral or cubic shape. Each (although small) oscillation corresponds to the completion of sub-faces. The figure also demonstrates that atomic shells sensitively depend on the material, more than electronic shells do. There are at least as many shapes and shell sequences as there are crystal symmetries.

One has thus to be aware of two different sorts of shell closures, atomic and electronic ones. Electronic shells become apparent in metal clusters at large temperatures (around above the melting point) while atomic shells are seen best at very low temperatures. Due to their different nature, the shells scale very differently. Figure 4.8 shows a summary of both sorts of shell closures in Na clusters (and one sequence for Li). The magic electron numbers follow in a much denser sequence than the atomic ones. The larger steps in atomic magic numbers hint that it is a much more demanding condition to arrange a nice piece of a crystal with clean flat surfaces all around. The figure indicates also the theoretical methods which can be used to check or predict the sequences of shell closures. Simple geometric symmetry consideration are completely sufficient for producing the magic atom numbers. Quantum mechanical calculations with single-electron degrees-of-freedom are required to describe the electronic magic numbers. The simplest ones at that level are a spherical shell model or spherical Kohn-Sham calculations assuming a smooth jellium background (which is, in fact, an appropriate model for a liquid state ionic background). And these are completely sufficient for determining the magic shells as indicated in the plot.

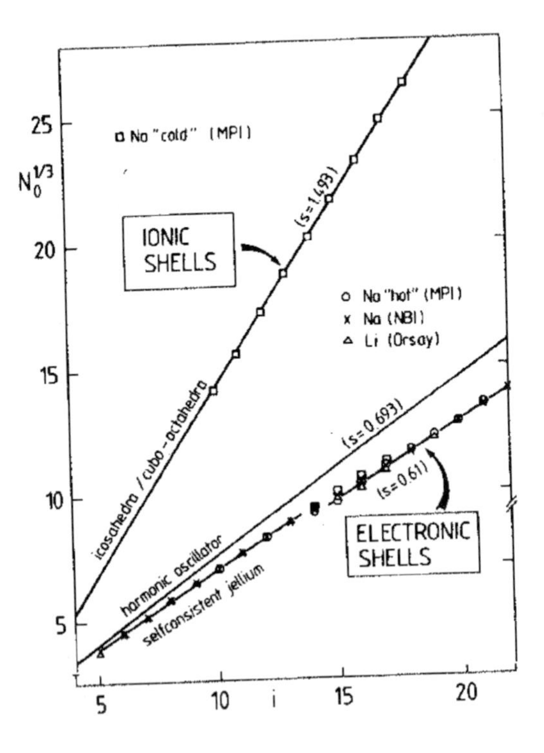


**Figure 4.8:** "Shell radius" $N_0^{1/3}$ versus shell index $i$ for spherical metal clusters. $N_0$ denotes magic numbers and $i$ counts the number of major shells. "Ionic shells" correspond to icosahedral or cubo-octahedral ionic configurations. "Electronic shells" correspond to spherical electron shell closures. The symbols are associated with different measurements as indicated: open rectangle = Na "cold" from [MBGL90, MBGL91], open circles = Na "hot" from [MBB$^+$91], crosses = Na from [PBB$^+$91], open triangles = Li from [BCdF$^+$92]. The solid lines are theoretical estimates with their slopes indicated in parentheses. After [Bra93].

### 4.2.2 Shapes

The shapes of clusters depend, of course, on the details of binding. A remarkable example are carbon clusters which are as they are only due to the particular chemistry of carbon, see e.g. $C_{60}$ shown in Figure 1.1. The binding, on the other hand, emerges from a subtle interplay between ionic and electronic effects. Thus we have the typical situation of molecular physics: little can be said in general, most depends on the actual situation. Nonetheless, it is useful to discuss as a guideline and as examples two limiting cases, first the shapes emerging from atomic arrangement and, second, shapes dictated by electronic shell effects.

#### 4.2.2.1 Ionic shapes

The cleanest cases for atom–atom potentials are rare gases. The interaction is purely van-der-Waals like and spherically symmetric. A sequence of Ar clusters are shown in Figure 4.9. For clusters with more than 7 atoms one obtains a pentagonal (five fold) symmetry, actually a bipyramid with a ring of 5 atoms and 2 caps. Pentagonal symmetry is not allowed in standard infinite lattices, as it cannot lead to complete space filling without distortions. But in cluster physics this symmetry becomes possible. And it shows up again, to some extent for clusters with 13 atoms, the first one with icosahedral symmetry. The cluster consists indeed of an interior atom with two pentagonal caps. Larger clusters grow by addition of atoms on their sides, until the next larger icosahedron has been built. This yield the sequence of sizes, 55, 147, 309, 561, ..., as shown in the lower part of Figure 4.9. The number of nearest neighbors

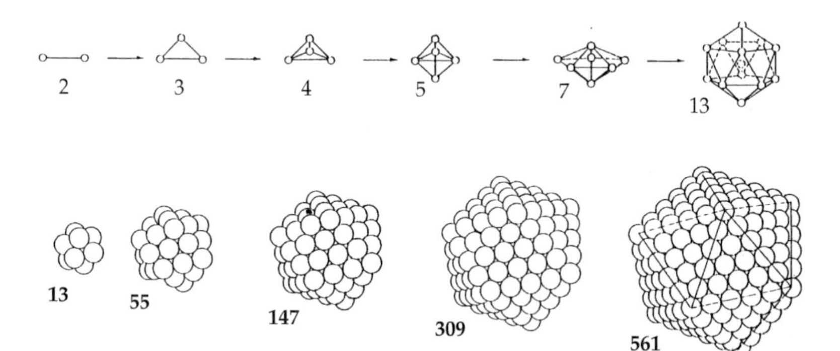


**Figure 4.9:** Structure of Ar clusters for various sizes. The larger clusters in the lower part represent atomic shell closures for icosahedral symmetry. After [Hab94].

in an icosahedral structure turns out to be larger than in a piece if crystalline structure, and this is the reason for the special stability of this structure. Surface energy is minimized by the flat triangular surfaces which are indicated by guiding lines for the largest sample, $Ar_{561}$.

#### 4.2.2.2 Electronically induced shapes

Electronic shell effects play a crucial role for the shape of metal clusters. They constitute a typical realization of the Jahn-Teller effect [Eng72]. The guiding idea here is that the electronic ground state does not like to be degenerate. Atoms escape a degenerate ground state configuration by spin alignment leading to Hund's rule [Wei78]. Molecules (and clusters) can change their shape and that turns out to be an even more efficient mechanism to reach an unambiguous ground state. This is demonstrated in Figure 4.10 for small metal clusters where a deformed harmonic oscillator provides a pertinent shell model. The figure shows the single electron levels as a function of deformation. Each level can carry two electrons (spin up and spin down). The configuration for a given electron number is found by filling the levels from below. The labels in the plot indicate the electron number reached at a given energy and deformation. At spherical shape ($\alpha_{20} = 0$), we recognize the magic electron numbers 2, 8, and 20 which we discussed in Section 4.2.1. It is obviously impossible to obtain another electron number at $\alpha_{20} = 0$ without running into degeneracies. But different gaps open up at different deformations and the cluster moves to that deformation where it may find a gap at given $N_{\rm el}$. We can read off from Figure 4.10 that $N_{\rm el} = 4$ and 10 are prolate with $\alpha_{20} \approx 0.3$ and $N_{\rm el} = 6$, 14 and 18 are oblate with $\alpha_{20} \sim -0.4$. One misses the cases $N_{\rm el} = 12$ and 16. These do not fit into the present scheme (the point marked 16' turns out to be only an isomeric state). Axial symmetry produces a degeneracy of the states with angular momentum $\pm m$ for $m \neq 0$. One needs to break that degeneracy and to deal with fully triaxial shapes to create space for a gap at $N_{\rm el} = 12$ and 16. However, this symmetry breaking is not as efficient as the first step from spherical to axial symmetry. As a consequence, there is a competition between Hund's rules (spontaneous spin alignment) and the Jahn-Teller effect for clusters with $N_{\rm el} = 12$ and 16 [KFR97]. But these are details which go beyond the scope of this book.

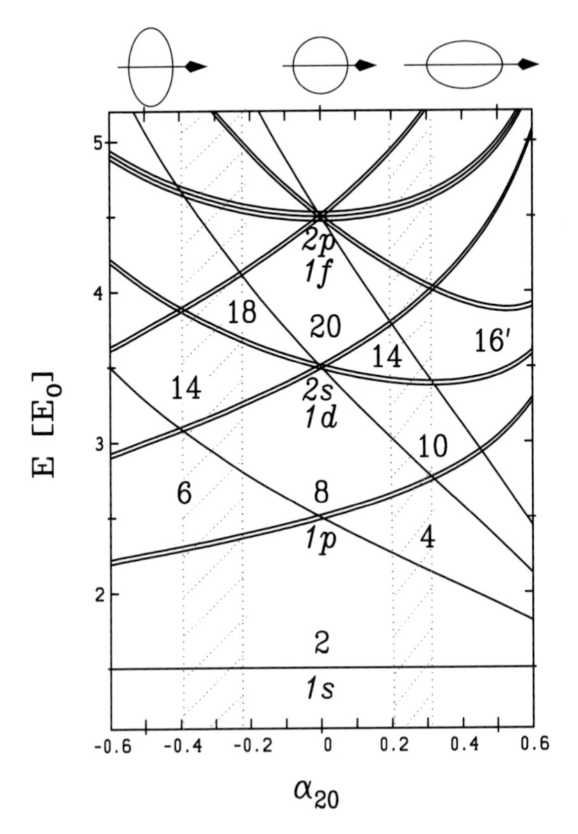


**Figure 4.10:** Single electron energies versus deformation $\alpha_{20}$ for the axially symmetric deformed harmonic oscillator. The numbers indicate (deformed) shell closures, up to $N = 20$. The shaded areas indicate bands of deformed magic shells. Typical shapes associated with given deformation $\alpha_{20}$ are indicated on top.

Figure 4.10 illustrates the principel mechanism generating deformed shapes. The actual determination of the optimal configuration in microscopic calculations uses simulated annealing for detailed ionic models (see Section 3.4.3) or dedicated search in the landscape of deformations $\alpha_{lm}$ for the jellium model [MHM$^+$95]. Figure 4.11 shows the systematics of ground state quadrupole deformations for small Na clusters. The lower panel gathers results from KS calculations with deformed soft jellium background (see Section 3.2.3.2). The dimensionless quadrupole deformation $\alpha_{20}$ is used throughout (Eq. (4.9)). One sees the spherical shape ($\alpha_{20} = 0$) for the magic electron numbers (2, 8, 20, 40). Substantial deformation develops mid-shell between the magic numbers. Prolate shapes ($\alpha_{20} > 0$) prevail just above magic numbers and oblate ones ($\alpha_{20} < 0$) just below, as indicated. The transition from the prolate to the oblate regime is accompanied by truly triaxial shapes, indicated by stars in the open circles in the lower panel of the plot.

The shape of small metal clusters is experimentally assessed through the splitting of the Mie plasmon resonance. This will be explained in Section 4.3. The Mie resonance is constituted by one sharp peak for spherical clusters. This peak splits into two pieces for axially symmetric deformed clusters. The mode oscillating along the longest axis has the lowest frequency while the mode along the shortest axis the highest one. Three peaks appear for triaxial clusters and they are sorted inversely to the length of the axes. In anticipation of that, the

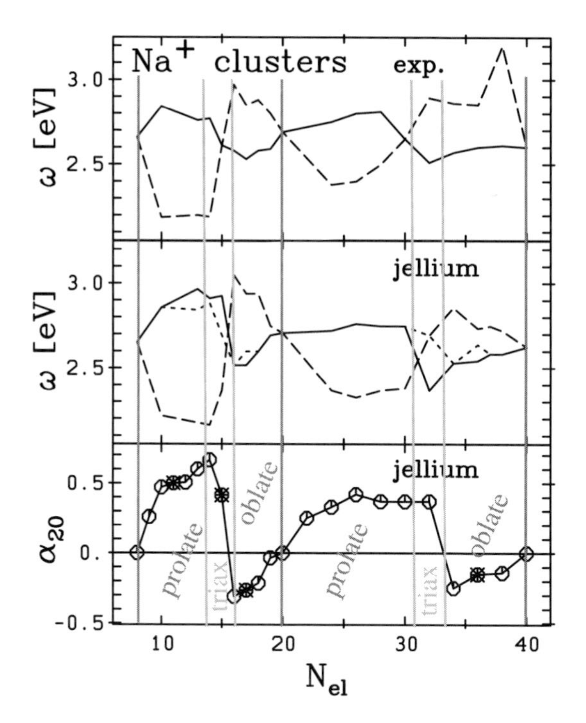


**Figure 4.11:** Deformation and collective splitting of the Mie plasmon resonance. Test case are $Na_n{}^+$ clusters. Theoretical calculations are done with (TD)LDA using soft jellium for the ionic background (see Section 3.2.3.2). Theoretical results are taken from [RS99]. Lower panel: dimensionless quadrupole deformation $\alpha_{20}$ for the ground state configurations. Middle panel: theoretical results for the frequencies of the Mie resonance peaks for the three modes along $x$, $y$ and $z$ axis computed with the triaxial deformed jellium model. Upper panel: the frequencies of the two dominant peaks (major peak = full, minor peak = dashed) in the experimental spectra, from [BCK+93] and [HS99].

upper panels of Figure 4.11 summarize the systematics of Mie plasmon peaks. The middle panel shows theoretical results produced with the deformations as shown in the lower panel. The three modes in $x$-, $y$- and $z$-direction are drawn. Two of them are degenerate for axially symmetric clusters (which are the majority). Three separate peaks indicate a triaxial shape. The upper panel shows experimental results where the positions of the dominant peaks of photoabsorption cross sections have been deduced from a fit with Gaussian profiles. The full line shows the strongest peaks. These are to be associated with the theoretical peaks where two modes coincide. Axially symmetric, prolate clusters have the strongest mode (=two modes) at higher frequency than the weaker mode (=1 mode) while the relation is reversed for oblate clusters. This is confirmed by the figure. And we also see that there is a good agreement between theory and experiment which confirms the theoretical deformation systematics and the underlying interpretation in terms of the Jahn-Teller effect. However, the occurrence of triaxial shapes is not yet clearly discriminated by the experimental data. It is extremely hard to disentangle the triple splitting in the rather broad absorption spectra.

## 4.3 Optical response

The optical absorption strength as a function of frequency is a key tool to investigate structure and dynamics of a quantum system. The detection of discrete absorption lines in atomic hydrogen was one of the cornerstones of quantum mechanics. The pattern of optical absorption gives crucial clues on the structure of molecules; corresponding selection rules often allow one to pin down the underlying point symmetry by means of group theory. It is not surprising

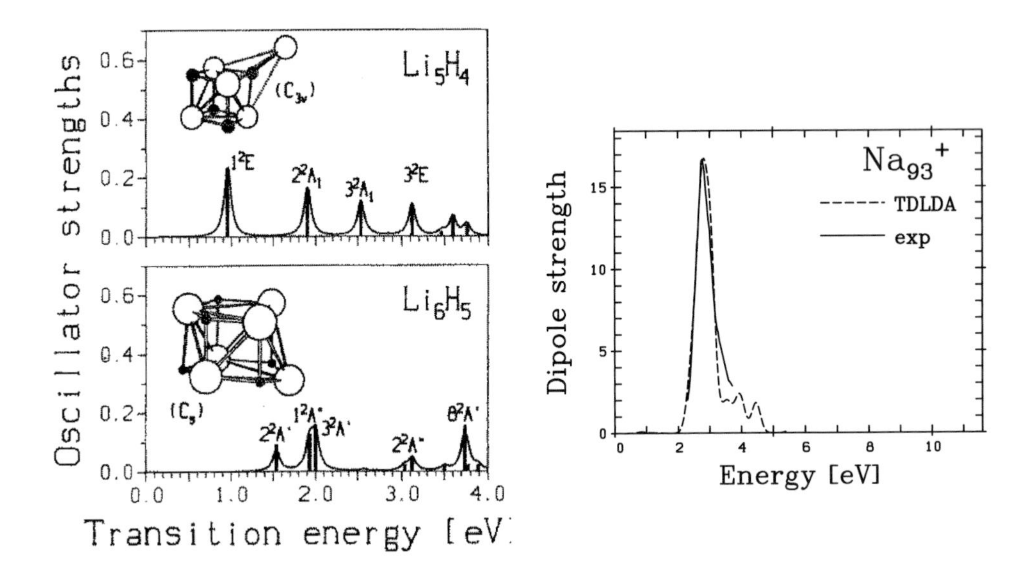


**Figure 4.12:** Photo-absorption spectra for two $Li_nH_{n-1}$ clusters (left, after [BKPF+96]) and for the metal cluster $Na_{93}{}^{+}$ (right, experimental strength from [RESH97]).

then that optical measurement also plays a key role in cluster physics. We will discuss in this section a few typical examples. There is an emphasis on the case of simple metal clusters because these follow very general patterns due to the dominance of the Mie plasmon resonance (Section 1.3.3.1). This allows a prototype discussion which can be transferred with few changes to any other metal. Non-metallic materials have more specific spectra which can only be discussed case by case. This is much harder to cover in the limited space here.

Figure 4.12 shows the optical absorption strength for clusters from two very different materials (for the measurement of optical absorption see Section 2.3.4 and for theoretical calculations Chapter 3). The left panels exemplify the ionic crystal and insulator LiH. The spectrum is distributed over several small peaks and extends over a broad range of energies. This is a typical situation in many molecules (and clusters) with covalent, ionic or van-der-Waals binding. The situation is the same as with the shape for this sort of system: little can be said in general because the actual spectra very much depend on the details of ionic and electronic structure. This is different for metals. The right panel of Figure 4.12 shows the spectrum of a typical Na cluster. It is dominated by one well concentrated peak, the Mie plasmon resonance. The picture looks much the same for all other Na clusters. The frequency resides always in the same range and there are small (but informative) variations in the shape and width of the resonance. All features can be explained and predicted on very general grounds. This will be discussed in the following.

In metals the optical response can be characterized by one prominent peak, the Mie plasmon resonance. This allows one to display the trends as a function of cluster size. Figure 4.13

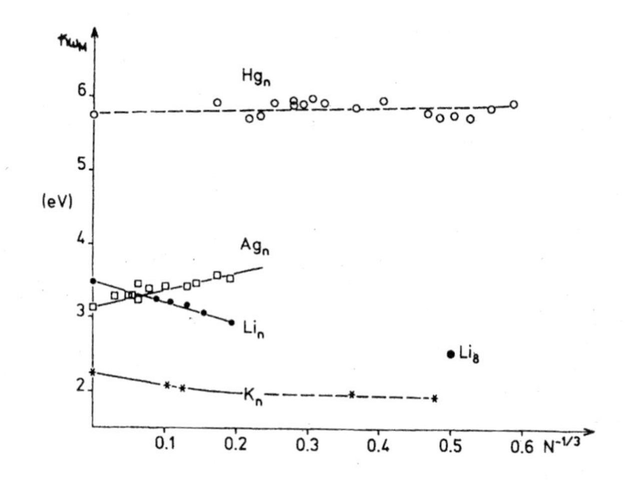


**Figure 4.13:** Mie plasmon frequency versus inverse cluster size $N^{-1/3}$ for selected metals. From [BC95].

shows the resonance frequencies for a selection of different metal clusters. The Mie estimate, discussed in Section 1.3.3.1, is a guideline for the observed frequencies, particularly at very large $N(=$ small $N^{-1/3})$. There are indeed weak trends with cluster size. The frequencies depend linearly on $N^{-1/3}$ for large clusters because the electron oscillations are caused by effects at the surface of the cluster (and the ratio of surface to volume is just $\propto N^{-1/3}$, see Figure 1.16). The slopes differ for the various materials, hinting that metals differ in details of their optical response. K and Li are simple metals (see the discussion in Section 3.2.1) and their optical responses exhibit negative slopes as a function of $N^{-1/3}$. This arises because the surface weakens the plasmon oscillations as we will discuss below. The noble metal Ag (and other noble metals) have positive slopes. This is due to the interplay of the plasmon resonance with the loosely bound $d$ shell core states (see the discussion of figure 3.2). The surface effects weaken the polarization of the $d$ shell and that causes the positive slope [TKMBL93]. The case of Hg shows almost no size dependence. It seems that the surface effects on the Mie plasmon and on the core polarization just counterbalance each other for that material. We will mainly discuss in the rest of this section the structure of the Mie plasmon resonance which shows up in a most clean manner in simple metals. The test case will be Na clusters. We will briefly address the modifications in noble metals and in the non-standard geometry of carbon chains.

### 4.3.1 Mie plasmon, basic trends

The simplest picture for the surface plasmon resonance is provided by the Mie model [Mie08] which was already introduced in its simplest fashion in Section 1.3.3.1. We provide here in the second round a slightly more detailed version of the model, based on a steep jellium picture of the ionic background. We consider again a small metal drop. The basic mode is an oscillation of the center-of-mass of the electron cloud against the positive ionic (jellium) background. Its frequency is $\omega = \sqrt{C/M}$ where $M = N_{\rm el} m_{\rm el}$ is the mass of the electron cloud and $C$ the spring constant of the restoring force. It can be computed from the energy of the shifted electron cloud in the Coulomb field of the jellium background $E(\mathbf{r}) = -\int d\mathbf{r}'\,\rho_{\rm el}(\mathbf{r}')U_{\rm jel}(\mathbf{r}-\mathbf{r}')$ where $\rho_{\rm el}$ is the electron density. The spring constant for oscillation along $x$ is then $C = \partial_x^2 E$ and similarly for $y$ and $z$. All three are identical for a spherical drop. We combine to $\partial_x^2 + \partial_y^2 + \partial_z^2 = \Delta$ and exploit the Poisson equation $\Delta U_{\rm jel} = -4\pi\rho_{\rm jel}$ relating the jellium potential $U_{\rm jel}$ to the jellium density $\rho_{\rm jel}$ (Section 3.2.3). This yields altogether

$$\omega_{\rm spl}^2 = \frac{1}{3}\frac{\Delta E\big|_{\mathbf{r}=0}}{N_{\rm el}m_{\rm el}} = \frac{4\pi e^2}{3}\frac{\int d\mathbf{r}\,\rho_{\rm el}(\mathbf{r})\rho_{\rm jel}(\mathbf{r})}{N_{\rm el}m_{\rm el}} \tag{4.17}$$

where the index "spl" stands for the Mie surface plasmon. The simplest estimate can be obtained by assuming homogeneous spheres for the densities, i.e. $\rho_{\rm el} = \rho_{\rm jel} = \rho_0\theta(R - |\mathbf{r}|)$ where $R = r_s N^{1/3}$ and $\rho_0 = 3/(4\pi r_s^3)$. This is exactly the basic Mie model (Section 1.3.3.1) and this yields $\hbar\omega_{\rm Mie} = \sqrt{\hbar^2 e^2/(m_{\rm el} r_s^3)}$. In the particular case of Na ($r_s = 4\,\mathrm{a}_0$) one obtains $\hbar\omega_{\rm Mie} = 3.4\,\mathrm{eV}$. The actual peak for ${\rm Na_{93}}^+$ in Figure 4.12 is found at around 2.8 eV. That is pretty close in view of the extreme simplicity of the Mie estimate with steep surface. A big step towards a more realistic picture is achieved by realizing that the electron density $\rho_{\rm el}$ can never produce a truly steep surface due to quantum smoothing. A certain amount of electron density will spill out beyond the bounds of the steep jellium. This modifies the Mie estimate (1.18) for spherical steep jellium to [Bra93]

$$\omega_{\rm spl}^2 = \omega_{\rm Mie}^2\left(1 - \frac{\Delta N_{\rm el}}{N_{\rm el}}\right) \quad \text{with} \quad \Delta N_{\rm el} = 4\pi\int_R^\infty dr\,r^2\,\rho_{\rm el}(r) \quad . \tag{4.18}$$

The spill out $\Delta N_{\rm el}$ is proportional to the surface area. Thus the correction reads $-\Delta N_{\rm el}/N_{\rm el} \propto -N_{\rm el}^{-1/3}$ which explains the negative slope for the trend of Mie plasmon frequencies for simple metals as observed, e.g., in Figure 4.13. The steep jellium is still rather rough an approach. An improvement employs Eq. (4.17) with the soft jellium background and the correct electron density from a static KS calculation. This yields $\hbar\omega_{\rm Mie}({\rm Na_{93}}^+) = 3.1\,\mathrm{eV}$ which is already a much better estimate. It is still about 10% above the experimental frequency. The reason is that the Mie model packs all dipole strength into one collective mode (like in a sum rule, see [RGB96b] and the next paragraph). In practice, small pieces of dipole strength are distributed over higher lying modes (second surface plasmon, volume plasmon). A summarized view then tends naturally to somewhat larger frequencies.

The above formulae deal with a simple picture of one valence electron cloud and ignores other possible dynamical degrees of freedom. This is a valid approach in simple metals. Other materials, however, have core electrons taking part in the polarization of the medium, see e.g. the forthcoming Section 4.3.4.1. The Mie model can still cope with these more involved situations. One merely has to take into account the effective dielectric function $\epsilon(\omega)$ of the whole medium, see also Section 1.3.3.2. In that formulation, the Mie model leads to the dipole strength distribution as

$$S_{\rm Mie}(\omega) = \frac{4\pi\omega R^3}{c}\Im\left\{\frac{\epsilon(\omega)-1}{\epsilon(\omega)+2}\right\} \tag{4.19}$$

where $R$ is the radius of the sphere and $\epsilon(\omega)$ the bulk dielectric constant. This form is applicable for practically all metals. In any case, the Mie model and variants thereof are extremely

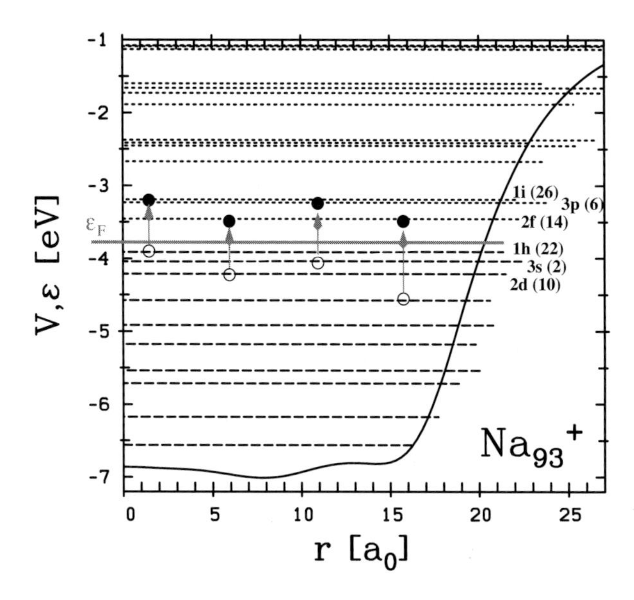


**Figure 4.14:** Single electron states of $Na_{93}^+$ with soft jellium background ($r_s = 3.96\,a_0$, $\sigma_{jel} = 0.9\,a_0$). The Kohn-Sham potential is shown by a solid line. The dashed lines show occupied states and the dotted lines empty ones. The Fermi energy $\varepsilon_F$ is indicated by an horizontal line. For some states the spectral assignment is given together with their multiplicity in brackets. The low lying $1ph$ excitations with most dipole strength are shown by arrows and symbols (open circles for emptying a state and closed symbols for filling it).

powerful tools for a collective description of the plasmon resonance in metal clusters. It has been discussed and exemplified extensively in [KV93].

The estimate Eq. (4.17) can be derived from linearized TDLDA by invoking a sum rule approximation. This is outlined in detail in Appendix G.2. The reasoning relies on the assumption that one collective dipole, the Mie resonance, exhausts the dipole sum rule, i.e. $\sum_n \omega_n |\langle\Psi_n|\hat{D}|\Psi_0\rangle|^2 \approx \omega_{\rm spl}|\langle\Psi_{\rm spl}|\hat{D}|\Psi_0\rangle|^2$. Such sum rule estimates for resonance frequencies have been developed as a versatile and efficient tool in nuclear physics [BLM79] and successfully transplanted to metal clusters in [LS91]. They are the starting point for more detailed and yet simple models in terms of collective flow (fluid dynamics) [Bra89, RGB96a, KB01]. For details see Appendix G.2.

## 4.3.2 Basic features of the plasmon resonance

The microscopic structure of the plasmon resonance is, of course, more involved than in the simple Mie model. The stationary Kohn-Sham solution provides a large choice of single-electron states and associated $1ph$ transitions. This is exemplified in Figure 4.14 for a medium large Na cluster. It is impossible to draw all $1ph$ transitions with dipole strength. Only those with the largest strength (and accordingly lowest excitation energy) are indicated. One may then wonder how a collective mode near the Mie plasmon energy could emerge from that starting point.

Figure 4.15 tries to answer that question. It serves to illustrate the composition of the Mie surface plasmon resonance for the chosen test case. The static KS ground state provides a sequence of single electron levels. Excitations within the static KS potential generate one-particle-one-hole ($1ph$) states, i.e. transitions from an occupied to an empty electron level. Panel (d) shows the energies of $1ph$ states accessible by dipole transitions, see also the detailed illustration in Figure 4.14. This defines the configuration space for the computation of excited

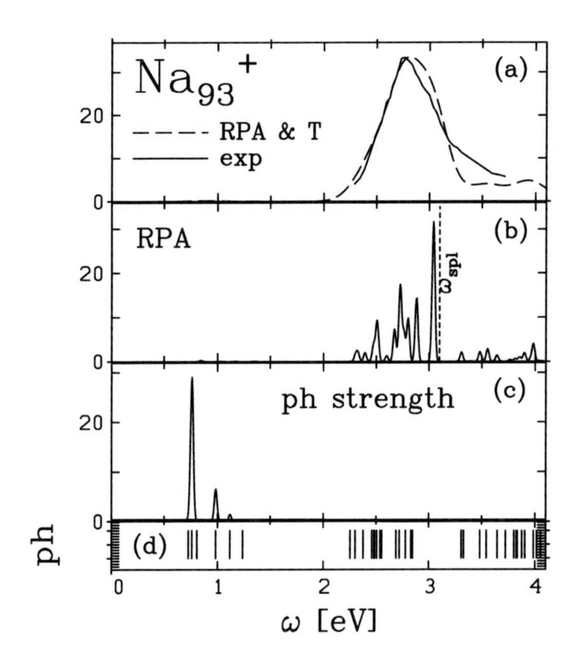


**Figure 4.15:** Composition of Mie plasmon resonance. Test case is $Na_{93}{}^{+}$ with soft jellium background ($r_s = 3.96\,a_0$, $\sigma_{\rm jel} = 0.9\,a_0$). The dipole strength distributions are shown at various stages of calculation. Panel (d): the $1ph$ energies as such. Panel (c): dipole strength from pure $1ph$ states. Panel (b): strength distribution from TDLDA at small amplitude (=RPA), the position of the Mie surface plasmon frequency Eq. (4.17) is indicated by a vertical dashed line ($\omega_{\rm spl}$). Panel (a): experimental strength (solid line) from [RESH97] and TDLDA results folded with a Gaussian of width 0.15 eV to simulate thermal line broadening.

states in linearized TDLDA (=RPA). Panel (c) shows the dipole strength distribution from pure $1ph$ states, $S_{ph}(\omega) = \sum_{ph} \langle \hat{D} \hat{a}_p^+ \hat{a}_h \rangle \delta(\omega - \varepsilon_p - \varepsilon_h)$. The $\delta$ distributions have been smoothed by a small width to provide a smooth shape appropriate for plotting. The pure $1ph$ distribution gathers at rather low excitation energies ($\sim 1\ eV$) corresponding to transitions over one shell (see Figure 4.14), having by far the largest dipole elements. It is far from the experimentally observed spectrum. Unlike the case of atoms, the pure $1ph$ states are not at all an approximation to the excitation spectrum of a metal cluster. They merely serve as expansion bases for the true excited states which can be represented, in the linear regime, as coherent superpositions over all the $1ph$ states, for details see Appendix G.

The key ingredient to tailor the true, physical, response is the Coulomb force which acts between the various $1ph$ states and recouples them substantially. The strong Coulomb repulsion moves the peak of dipole strength to dramatically higher energies than the ones of the pure $1ph$ strength. The essence of that is correctly described in the Mie model where the spring constant for the oscillating cloud is determined by the Coulomb force. The position of the Mie frequency Eq. (4.17) is indicated in panel (b) of Figure 4.15 by a vertical dashed line. The dramatic shift from the $1ph$ peak strength is obvious. As mentioned above, the Mie model produces an average peak position. The Mie peak is indeed an appropriate average through the dipole distribution. It is to be remarked that both collective modes show up as one isolated peak each. Note that the collective peaks reside in a region of high density of $1ph$ states, see panel (d) of the figure.

An approach of choice here is TDLDA which correctly incorporates all microscopic details of time-dependent single particle structure, i.e. all features of a coherent superposition of $1ph$ states. The result of linearized TDLDA is shown in panel (b) of Figure 4.15. The resonance peak remains where it was in the collective picture. But it is obviously distributed over the

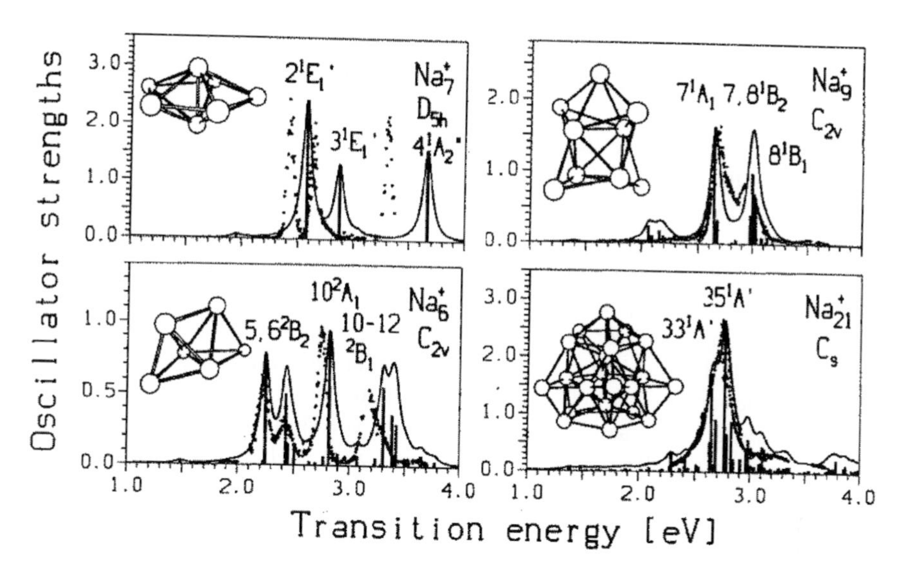


**Figure 4.16:** Photo-absorption spectra for small Na clusters computed with CI, after [BKPF+96]. Experimental results are indicated by dots.

nearby $1ph$ states. This is called Landau fragmentation in analogy to Landau damping in the case of a continuous electron gas [LP88].

The surface plasmon resonance in large Na clusters thus already gathers some width due to Landau fragmentation. This is obviously not enough to explain the full experimentally observed width, see the uppermost panel. The point is that this measurement was performed at a rather high temperature of 400 K. This causes large fluctuations of the cluster shape. The resonance changes with shape and a broad peak is obtained from incoherent superposition of all these different spectra. We will discuss that mechanism in more detail later. Here the thermal broadening has been summarized by folding the TDLDA result with a Gaussian of full width at half maximum (FWHM) of 0.14 eV [BR97]. The result agrees very nicely with the experimental spectrum. Figure 4.15 thus demonstrates the various contributions from which the optical absorption strength is composed. TDLDA embraces the dominant (collective) effect of the Coulomb repulsion and all details of $1ph$ structure. Thermal effects have to be added in terms of an incoherent superposition of TDLDA results at various shapes.

There remains the question of what the impact of higher particle-hole excitations may be. Small clusters are still accessible to full CI calculations which can go rather far in these higher excitation structures, at least for small clusters. Figure 4.16 shows optical absorption spectra for Na clusters as obtained from CI calculations [BKPF+96]. The vertical bars mark the discrete eigenenergies of the CI spectrum and the height of the bar the associated dipole strength. The discrete distribution is smoothed by folding with a certain width and compared with experimental data. There is generally a good agreement. Some cases seem to perform better than others. But that should not be taken too seriously. It is extremely hard to reproduce optical spectra in every detail including the relative weights of the peaks. The variation of the agreement demonstrates the predictive value of present day theories for excitations. The situation is similar for full TDLDA in connection with a careful optimization of ground state structure.

A most interesting aspect in the CI results of Figure 4.16 is revealed when looking at the discrete states (vertical bars) and their strengths. There are only few states for ${Na_7}^+$ much similar as in TDLDA results. A substantially higher density of states is found in the larger clusters, most of them having small strength. And there are much more states than available $1ph$ excitations in that energy range. These are obviously $2ph$ states or possibly even higher configurations. The feature persists for the larger samples in Figure 4.16: there are a few states with large strength and very many states with very small strength. The strong states are similarly found in TDLDA calculations and can be associated with $1ph$ excitations. The weak states are $2ph$ states and beyond. The CI results thus contain the broadening due to coupling to $2ph$ states. We see very many $2ph$ states but with very little strength. The folded spectra would not look much different if these $2ph$ states were excluded. Figure 4.16 thus hints that the coupling to $2ph$ states has only a small effect on the spectra. TDLDA calculations which account only for $1ph$ excitations are thus not missing much in that respect. (The question remains open, of course, whether they treat dynamical exchange and correlations well enough.)

### 4.3.3 Effects of deformation

Another interesting aspect in Figure 4.16 is that ${Na_7}^+$ shows a strong splitting of the spectrum with clearly separated peaks. This is not so pronounced in the three other clusters. Looking at the ground state configurations, we realize that ${Na_7}^+$ is by far the most (oblate) deformed cluster in the sample. The strong splitting of the excitation spectrum is indeed due to the large quadrupole moment of the cluster. The connection is illustrated in Figure 4.17. The lower part shows the example of the prolate cluster ${Na_{15}}^+$. The right part sketches the deformed shape and the corresponding basic dipole modes as imagined in the Mie model. In the case of $Na_{15}^+$ (lower panel of the figure) the oscillations along the $z$-axis (horizontal) are more extended and thus have lower frequency while the orthogonal axes are squeezed and thus yield higher oscillation frequencies. Note that there are two modes orthogonal to the symmetry ($z$) axis. The left lower part of the figure shows the optical absorption strength from TDLDA. One clearly sees the splitting into two peaks where the upper peak has twice as much strength. The TDLDA results agree with experimental data which is taken as a clear hint that ${Na_{15}}^+$ is a prolate cluster with ground state deformation $\alpha_{20} \approx 0.57$ (see also Figure 4.11 and associated discussion). The middle part of Figure 4.17 shows the example of the oblate ${Na_{19}}^+$. The $z$-axis is now the shorter axis and produces the upper peak while the other two axes feed the lower peak. The deformation splitting is again seen clearly in the spectrum and the relative weights (now more strength at lower frequency) indicate an oblate system. Finally, the upper part of the figure exemplifies a spherical ${Na_{21}}^+$ where all three modes are degenerate and produce one unique, large plasmon peak. This direct relation between spectral splitting and ground state deformation has been exploited to deduce the systematics of quadrupole deformations and related surface plasmon peaks as shown in Figure 4.11. One determines ground state deformations in the soft jellium model, computes the associated dipole spectra in TDLDA and compares with experiment. Good agreement with data allows one to conclude that the

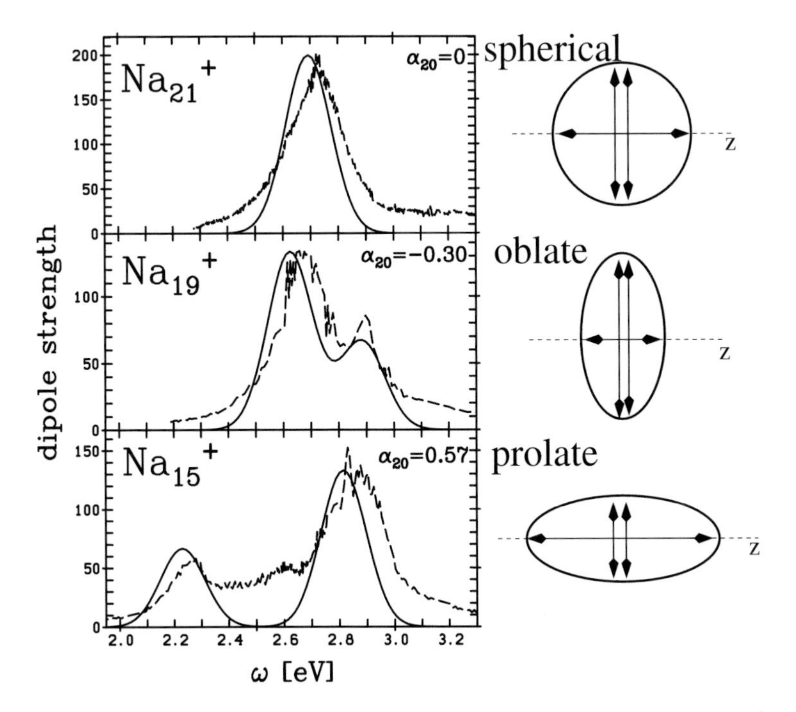


**Figure 4.17:** Illustration of deformation splitting of the Mie resonance. Test cases are $Na_{21}{}^{+}$ (spherical), $Na_{19}{}^{+}$ (oblate) and $Na_{15}{}^{+}$ (prolate). The left panels show the optical absorption strengths, experimental from [HS99] and theoretical results from linearized TDLDA on soft jellium background [NKR02]. The right part sketches the shapes (the deformations are exaggerated to make the point). The lines with arrows indicate the directions of the basic plasmon oscillations. The symmetry axis ($z$ axis) is indicated by a fine dashed line.

underlying ground state deformation was correct. This analysis works reliably well for small clusters up to $N \approx 40$. Strong Landau fragmentation sets on for larger clusters. This blurs the information about deformation splitting for medium large free clusters [NKR02]. One has to be aware of that interference and better checks for each new case the relation between the damping width and the size of the splitting. Extreme deformations and/or huge clusters still allow one to separate clearly the different branches of the Mie plasmon mode as done, e.g., in the experiments on large Au nano-rods [SFW$^{+}$02], see also the examples of Sections 5.3.2 and 5.3.4.2 where deformation splitting in huge embedded or deposited clusters is used as handle for manipulations with intense laser beams.

The discussion of deformation splitting has dealt up to now with ground states of free clusters. The statement that Landau fragmentation overrules splitting only for $N > 40$ holds in these circumstances. The situation may differ for clusters on a substrate. As an example, we show in Figure 4.18 shapes (upper panels) and dipole spectra (lower panels) for the ground state (left) and first isomer (right) of $Na_8$ on a NaCl surface. The strong surface attraction produces a flat ground state like a pancake. The upper left panel shows a cut orthogonal to the surface. One sees the longest and the shortest cluster axis. The third axis has intermediate

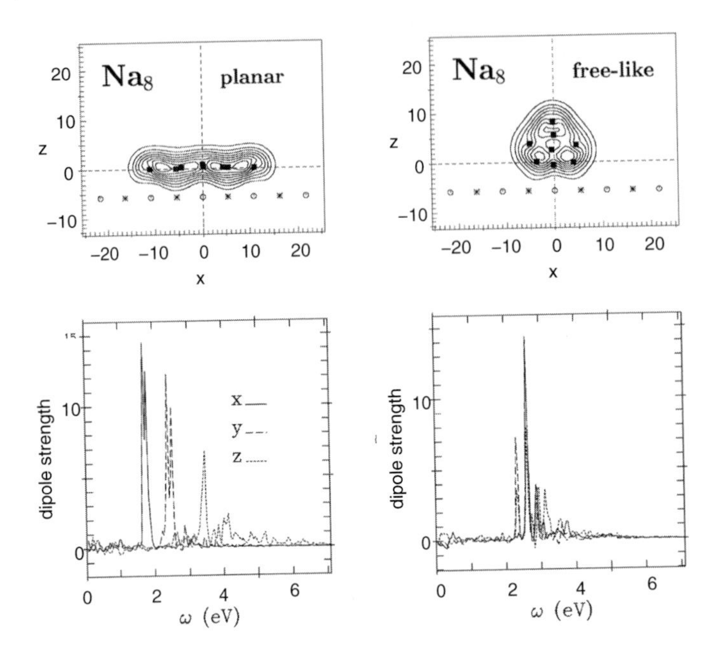


**Figure 4.18:** Upper part: Ground state (left) and isomer (right) configuration of $Na_8$ on a NaCl surface. The Na ions are indicated by stars, the electron density by equi-density lines. Lower part: Corresponding dipole strengths for modes in $x$, $y$, and $z$ direction as indicated. After [KCRS98].

length and is orthogonal to the plot. The ground state is thus strongly deformed and it is triaxial because it has axes with three different lengths. The corresponding dipole spectrum (lower left panel) thus shows a clean collective splitting into three different peaks. The highest energy peak is slightly fragmented. But this does not inhibit the clear distinction of the peaks. There is an interesting isomer for the system $Na_8$@NaCl. It corresponds to a free $Na_8$ cluster which is very gently dropped at the NaCl surface. This isomer is stable up to 1000 K. The stability is enhanced here by electronic shell effects. Indeed, $N = 8$ is a magic electron number for a three-dimensional shape while it is not magic for the flat configuration (the magic numbers for flat Na clusters are $N = 6, 12, 24, \ldots$ [KMR96]). The shape is shown on the upper right panel. The cluster keeps basically its original spherical shape. There is no quadrupole deformation. But the surface attraction creates a strong octupole (pear like) deformation. The dipole spectra show the same average peak position for all three modes. There is no deformation splitting. But all modes are strongly Landau fragmented. The strong octupole deformation breaks reflection symmetry. This, in turn, gives access to a band of energetically close $1ph$ states which were decoupled in the case of good parity. It requires a strong symmetry breaking to produce that large effect. The ground state (left panels) had also some violation of parity due to the surface. But this did not cause much fragmentation. We learn from this example that Landau fragmentation can become large also in small clusters if parity symmetry is massively broken. This arises for clusters on substrates and also in the case of thermal shape fluctuations of free clusters (see the discussion of thermal line broadening in Section 4.3.5 below).

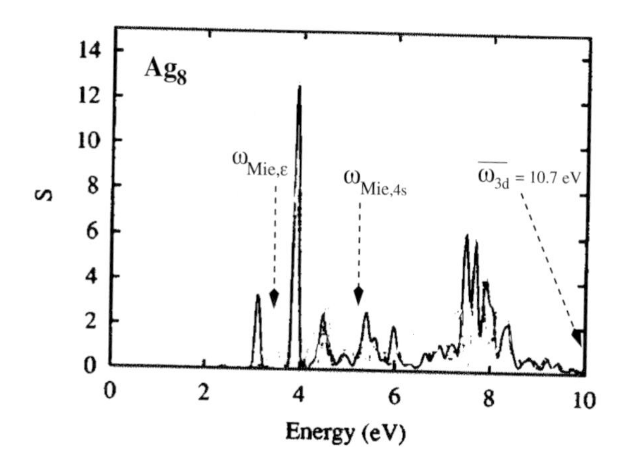


**Figure 4.19:** Photo-absorption spectrum of $Ag_8$ computed with TDLDA (full line). The positions of key frequencies are indicated by arrows: $\omega_{\mathrm{Mie,4s}}$ = the Mie frequency Eq. (1.18) using the density of $4s$ valence electrons; $\overline{\omega_{\mathrm{3d}}}$ = average excitation energy of the $3d$ dynamical polarization; $\omega_{\mathrm{Mie},\varepsilon}$ = the Mie frequency deduced from the strength Eq. (4.19) employing the dynamical dielectric response of bulk Ag. After [YB98].

## 4.3.4 Other materials

### 4.3.4.1 Noble metals

We have pointed out several times that metal clusters have dipole spectra with one pronounced surface plasmon peak which exhausts almost all the dipole sum rule. To be more precise, this holds for simple metals (alkalines, alkaline earths). There are more involved cases. Consider for example noble metals (Cu, Ag, Au). They have also one valence electron in the HOMO $s$ shell. But the next core state below that is a $d$ shell which is energetically not too well separated, see the example of Cu as opposed to Na in Figure 3.2. The $d$ shell is only loosely bound and thus easily polarizable. The ensemble of $d$ shells of all atoms in a cluster together build also a collective dipole mode where all core dipoles oscillate coherently. We thus have two strong dipole polarizabilities, one from the $d$ states and the other one from the valence $s$ electrons, setting a collective surface plasmon as in simple metals. These two modes interact with each other and recouple to two new hybrid modes. The optical absorption spectrum of a cluster from noble metals thus exhibits several regions of strong response. Figure 4.19 shows the case of $Ag_8$ as an example. One sees substantial strength below 4 eV, associated with a fragmented resonance whose average peak is located at around 3.5 eV. But this strength is not as isolated as it used to be for simple metals. Only 20% of the total strength is exhausted here and there resides appreciable strength above the Mie resonance (part of it for even larger energies, not shown on the plot). A simple analysis with the Mie formula in version Eq. (1.18) yields a surface plasmon resonance at 5.18 eV whereby this assumes that the $4s$ electrons are the only active electrons (taking $r_s = 3.09\,\mathrm{a}_0$). The observed low-energy strength is far below. One has to take into account that the $3d$ core electrons have a rather large dipole polarizability. They couple to the $4s$ surface plasmon and yield new frequencies. This was estimated in a simple two states model in [YB98]. Input are the frequencies of the $4s$ surface plasmon ($\omega_{\mathrm{Mie,4s}}$ in the figure), the average response frequency of the dynamical polarizability of the $3d$ electrons ($\overline{\omega_{\mathrm{3d}}} = 10.7$ eV), and an estimate for the dipole–dipole coupling ($\propto R^{-3}$). Solution of that simple $2\times2$ matrix problem (special case of the coupled dipole oscillators in Section 5.2.1.2) then yields one much lower frequency around 3.6 eV (and another one upshifted from the 10.7 eV). This recoupled frequency coincides very well with the mean of the

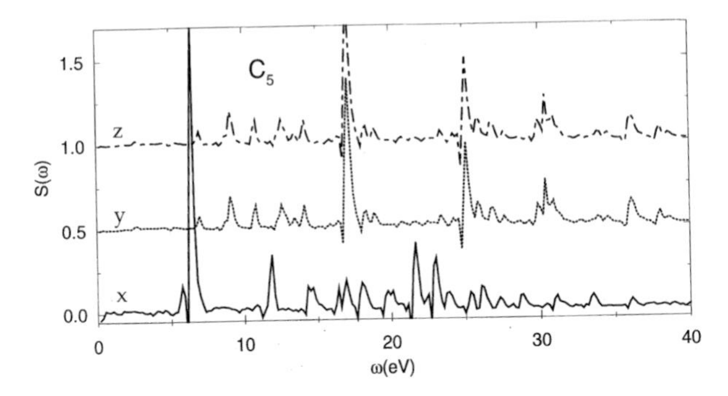


**Figure 4.20:** Optical absorption spectra for $C_5$ which aligns as a chain computed with TDLDA and subsequent spectral analysis. The modes along the three principal axes are shown as indicated. The $x$ direction points along the chain. After [WCF$^+$98].

two low-energy peaks from TDLDA. It is noteworthy that the Mie theory yields correctly that low frequency peak if one uses the form Eq. (4.19) which employs the macroscopic dielectric constant of the material. One indeed obtains directly the frequency of 3.6 eV (as indicated in the figure) when inserting the full dielectric function of bulk Ag. This is not surprising because the bulk $\varepsilon(\omega)$ embodies already the $4s$–$3d$ hybridization. The more involved composition of dipole resonances in noble metals requires reconsideration of the effects discussed above as, e.g., deformation splitting or Landau damping. Very little of that task has been worked on until now.

The coupling with the core polarization does also have consequences on the slope of the evolution of the actual resonance frequency with particle number as shown in Figure 4.13. The polarization shifts the frequency to red. But the surface layer of the cluster is free from that polarization effect [Lie87, TKMBL93]. There is thus less down-shift with increasing surface zone. This was the explanation of the positive slope seen in Figure 4.13.

#### 4.3.4.2 C chains

In spite of the covalent nature of binding in carbon clusters, electrons in such systems are known to possess a certain amount of metallic character. The optical absorption spectra thus show pronounced resonance peaks together with a swamp of non-collective peaks [WBT93]. This is not so surprising. An interesting aspect of C is that its particular binding properties allow the formation of stable chains over a broad range of sizes (a fact that was found first in vibrational spectra as recorded with radioastronomy of stellar dust, see e.g. [BHK89, CGC00]). The linear geometry delivers much different trends for a plasmon resonance. As in the Mie picture, we can model the plasmon mode as a collective oscillation of the electron cloud against the ionic background. But now we approximate cloud and background by the shape of a needle and consider the longitudinal vibration. The potential energy of the displacement $d$ along the chain is just the energy in an electric dipole field $\propto d^2/L$ where $L$ is the length of the needle. The spring constant is thus $\propto 1/L$ and the plasmon frequency becomes $\propto L^{-1/2} \propto N^{-1/2}$. The case is exemplified for the chain $C_5$ in Figure 4.20. The spectrum is significantly fragmented. The modes orthogonal to the chain axis show a series of peaks all having about the same strength. But within the modes along the chain ($x$ direction in the figure) one sees here that a strong resonance forms at the lower end of the frequencies. The resonance becomes more and more pronounced with increasing cluster size $N$ [BRS02]. The

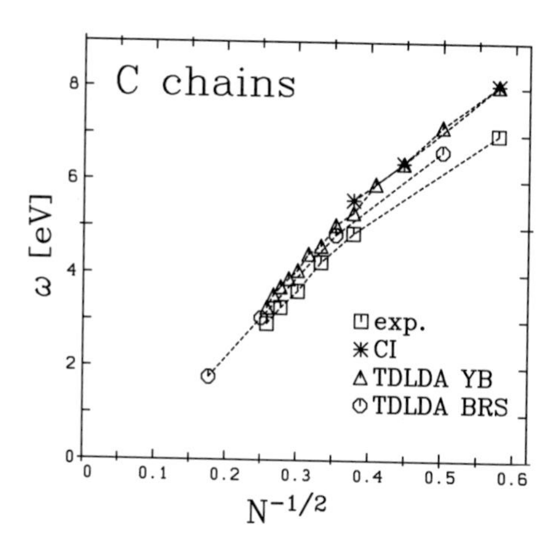


**Figure 4.21:** Systematics of longitudinal plasmon resonance frequencies for C chains with $N$ = 2–32. Results from various sources are compiled, experiments from [CG82, For$^+$96], CI calculations from [PK88, Kol95], TDLDA YB are TDLDA calculations from [YB97], and TDLDA BRS from [BRS02].

resonance is actually composed from the collective motion of the (metallic) $p$ electrons of C. The $s$ electrons are more localized but take part by their large dipole polarizability (in a way similar to the $d$ electrons in noble metals).

A summary of the resonance peaks is shown in Figure 4.21. Theoretical and experimental results are in agreement and the results line up nicely with the predicted trend $\propto N^{-1/2}$ for the longitudinal mode (a straight line on this figure, note the abscissa). The two TDLDA results use similar grid techniques but different (non-local) pseudo-potentials. The agreement, or slight discrepancies, give an impression about the impact of the pseudo-potential on resonance properties.

## 4.3.5 Widths

### 4.3.5.1 Landau fragmentation

We have already seen in the discussion of Figure 4.15 that Landau fragmentation distributes the dipole strength of the Mie surface plasmon resonance over energetically close $1ph$ states. In a first step, we thus discuss the impact of Landau fragmentation on the observed widths.

For this purpose we consider the trends of dipole spectra in the case of positively charged Na clusters: they are summarized in Figure 4.22. The position of the Mie surface plasmon resonance is drawn as a heavy line. Note that this is a result of the collective flow model such that the line correctly represents the position of the final plasmon peak, see the discussion of Figure 4.15. One sees a roughly linear decrease from the bulk value (at $N^{-1/3} \to 0$). This is due to the increasing importance of the surface and it is the electron spill-out at the surface which reduces the resonance frequency (see e.g. the estimate (4.17)). The linear trend turns to a flat one for small clusters ($N^{-1/3} \approx 0.3 - 0.5$, $N \approx 8 - 40$) due to shell effects. The deformation splitting of the resonance frequency is not shown in this figure for the sake of clarity. The figure does also show the distribution of the $1ph$ states up to $N \approx 60$. There are large spectral gaps for small clusters. They shrink with increasing cluster size and merge around $N \approx 60$ into one large pool of $1ph$ states from the lower band edge on. This is related to the formation of more or less continuous bands as demonstrated and discussed in Figure 1.11. The general trend is nevertheless overlayed by shell oscillations. Gaps are larger

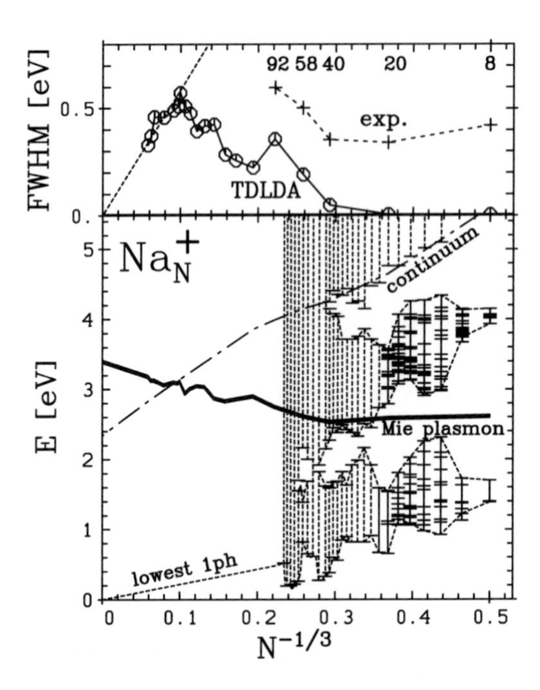


**Figure 4.22:** Trends of spectral properties in positively charged Na clusters versus system size. The lower panel shows excitation energies. The $1ph$ energies are indicated by horizontal bars for small clusters and low excitations. For large sizes they form dense bands of excitation which are indicated by vertical dotted lines. The bounds of the bands are connected by dotted lines. The bands merge for heavy clusters and only the lowest $1ph$ energy is shown. The average energy of the Mie plasmon is drawn as a full thick line. The electron continuum threshold is indicated by a dash-dotted line. The positions of the first electron shell closures are indicated by the magic numbers 8, 20, 40, and 58. The upper panel shows the full width at half maximum (FWHM) of the Mie plasmon resonance from TDLDA and is compared with experimental values from [RESH97]. The fine dashed line indicates a fit to the asymptotic $\propto N^{-1/3}$ for large $N$.

for magic clusters (numbers are indicated in the upper panel) and smaller for the deformed systems mid-shell. The relation between resonance frequency and $1ph$ states is decided on the strength of Landau fragmentation. One sees that the resonance frequency resides well inside a spectral gap of $1ph$ states for all small clusters up to about $N = 40$. This explains why Landau fragmentation plays only a minor role in these small systems and why one can deduce deformation cleanly from the splitting of the plasmon resonance. The resonance line dives quickly into a swamp of $1ph$ states for $N > 40$. Landau fragmentation then becomes inevitably active.

The upper panel of Figure 4.22 shows the trend of the width (i.e. FWHM) of the surface plasmon resonance as it can be read off from linearized TDLDA calculations [BR97]. Only results from spherical clusters are drawn such that we see basically the effect of Landau fragmentation. The corresponding width is practically zero for small clusters, reaches a maximum around $N \approx 1000$ and then turns into a decrease linear with $N^{-1/3}$. The decrease

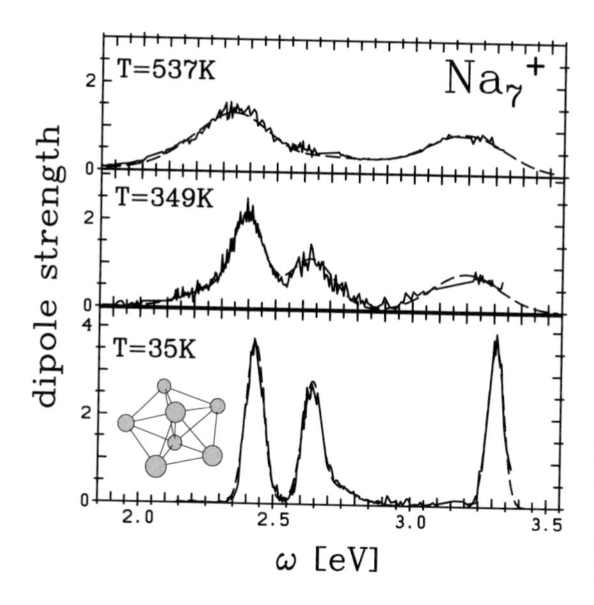


**Figure 4.23:** Photo-absorption strength for $Na_7^+$ at three different temperatures as indicated. The configuration is indicated in the lowest panel. Data taken from [ESS+95].

develops in spite of increasing density of $1ph$ states. What happens is that the coupling matrix elements shrink towards bulk material. The trend linear in $N^{-1/3}$ can actually be analytically derived in a model of plasmon waves scattered by the cluster surface (wall friction, see [YB92]). Figure 4.22 also shows experimental results for the FWHM [RESH97]. The trends are qualitatively reproduced by calculations. But there is a large offset. There is obviously an important mechanism for line broadening yet missing in the theoretical description.

#### 4.3.5.2 Temperature effects

A hint on the missing mechanism is given by Figure 4.23, where the optical response is shown in $Na_7^+$ clusters at various temperatures. The spectrum at low temperature shows three narrow and well separated peaks. We see the collective splitting of the strongly oblate $Na_7^+$ and a low energy peak, fragmented by one close $1ph$ state. The picture changes systematically for larger temperatures. The peaks grow broader and the lower double peak merges into one. It is the temperature which causes the large width (it ought to be remembered that the experimental widths in Figure 4.22 were deduced from spectra taken at 400 K). Thermal effects at the side of the electrons cannot induce such a width because 400 K is a small energy at an electronic scale. It is the thermal ionic motion which is responsible for that line broadening. The ionic excitation energies reach down to below 100 K for $Na_7^+$ [RS02]. A thermal excitation of 400 K is a lot on that scale. Many ionic eigenmodes are excited to rather large amplitudes. The cluster then undergoes substantial thermal shape fluctuations. We have seen in Section 4.3.3 that deformation has a strong influence on the dipole spectrum. Thus each member of the thermal ensemble with its different deformation contributes a different spectrum to the total optical response. These spectra all add up incoherently to rather broad peaks.

The superposition of spectra from different shapes is sketched in Figure 4.24. The test case is again $Na_7^+$ with ionic structure treated by local pseudo-potentials and electrons propagated

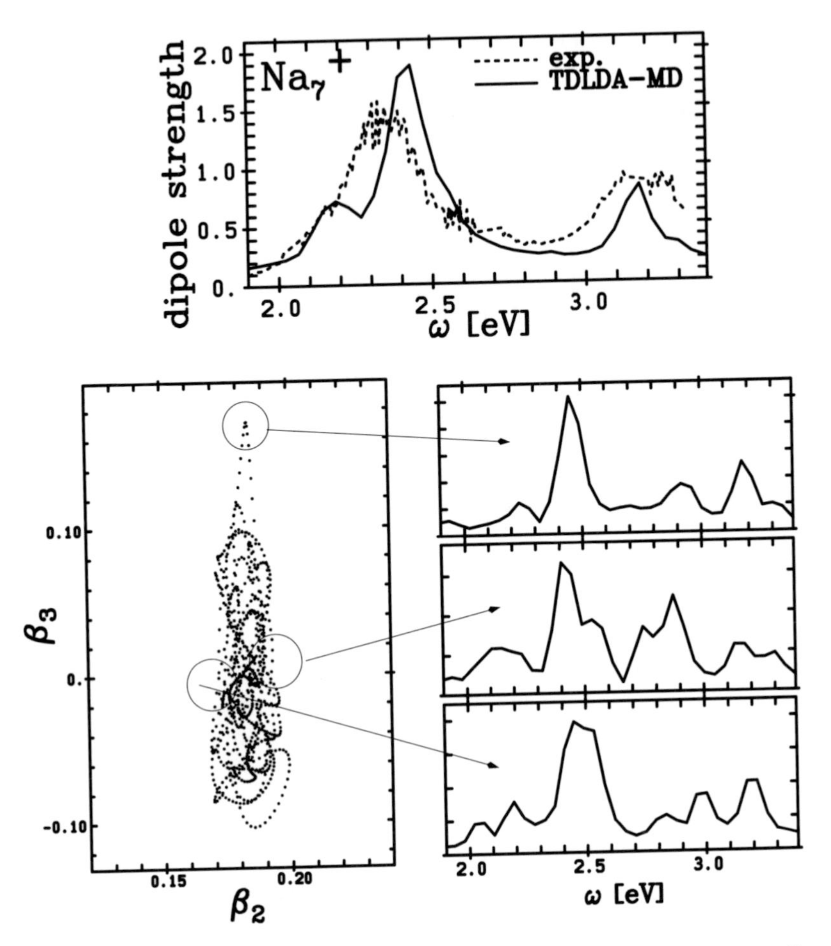


**Figure 4.24:** Upper panel: dipole strength for an ensemble of $\mathrm{Na_7}^+$ clusters at temperature $T = 540\,\mathrm{K}$ compared with experimental data from [ESS$^+$95]. Lower left panel: distribution of ionic quadrupole ($\alpha_{20}$) and octupole ($\alpha_{30}$) deformations for the ensemble. Lower right panels: dipole strengths for three selected members of the ensemble. The selection is indicated by arrows from the left panel. All spectra have been computed by TDLDA with ADSIC. The uppermost panel uses actually TDLDA-MD at given ionic kinetic energy to simulate the thermal ensemble. The spectra are down-shifted by 0.2 eV to allow better comparison with the experimental width.

by TDLDA including ADSIC (Appendix E.3). The thermal ensemble is simulated dynamically by starting TDLDA-MD from the ground state configuration and applying to the ions the appropriate thermal kinetic energy $E_{\mathrm{kin}} = 3(N-6)/2\,kT$ (note the six subtracted degrees of freedom for c.m. and angular motion). The velocities are chosen at random. The TDLDA-MD is then run for a very long time to explore the phase space of ionic configurations. Actually we went up to 7 ps. This procedure simulates, in fact, a micro-canonical ensemble which is usually a good approximation to the canonical ensemble. The approximation is a bit at the edge for this small cluster. But it should suffice for the schematic considerations here. The same holds for the somewhat short simulation time. The upper panel of Figure 4.24 shows the dipole spectrum accumulated over the whole ensemble and compares the theoretical dipole strength with the experimental data at the same temperature. The thermal width of the low energy peak is well reproduced. The upper energy peak looks a bit different. The experiment shows a broad but well concentrated peak while TDLDA-MD has one narrower peak with very broad tails. The lower panels of the figure try to illustrate the mechanisms of thermal line broadening. The left panel shows the distribution of cluster deformations (octupole versus quadrupole) in the thermal ensemble. The amplitude in octupole direction is much larger than along quadrupoles; this is a typical feature for all small Na clusters [MR95b, MLY01]. We know from the previous considerations (Section 4.3.3) that quadrupole as well as octupole deformations have a strong impact on the dipole strength distribution. The optical spectrum will thus look different at each one of these points in ionic phase space. The incoherent summation of all different spectra then yields the broad distribution as shown in the upper panel. Three examples for spectra at selected and fixed ionic configurations are shown in the lower right panels. They corroborate the picture of fluctuating strengths. The importance of shape fluctuations had been pointed out earlier in [BT89, YPB90], although restricted to quadrupole deformations. The necessity of octupole softness was worked out within the framework of the jellium model in [MR95b]. A recent exploration which has employed a detailed ionic background has confirmed the thermal broadening mechanism and the importance of the octupole deformation [MLY01].

#### 4.3.5.3 Electron–phonon coupling

It is to be pointed out that the thermal broadening mechanism deals with incoherent superposition from spectra taken as snapshots at quasi-stationary samples of the ensemble. There exists in addition a coherent line broadening from dynamical coupling of the Mie plasmon to ionic oscillations (the "phonons" of the cluster). A coherent superposition of the plasmon with one or more phonon modes yields sidebands to the plasmon peak with frequency shifts typical of ionic modes. This is demonstrated for the example of ${Li_4}^+$ in Figure 4.25. The upper panels show two peak regions at an enlarged scale. This reveals substructures, in the 10 meV range, which can be associated with molecular vibrations coupled to the plasmon oscillations, as corroborated by the complementing theoretical analysis (vibrations along BO surfaces). The existence of the coherent coupling to phonons has thus been proven, although Figure 4.25 does also show that the sidebands are rather weak. The phonon coupling is thus not very strong. The coherent coupling of ionic and electronic modes is automatically arranged when propagating them simultaneously (i.e. beyond the Born-Oppenheimer approach) for example with TDLDA-MD, see Section 3.4.1. An analysis of the electron–phonon coupling for $Na_n$ clusters produces the whole spectrum including phonon vibrations at once and leads to similar pattern [RS02] which confirms that electron–phonon coupling is visible, but only weakly.

#### 4.3.5.4 Even more widths ...

We ought to mention that we have already discussed a further possible broadening mechanism in connection with Figure 4.16. It is the effect of electron–electron correlations (coupling to $2ph$ and higher states). The outcome is similar to the case of electron–phonon coupling. The

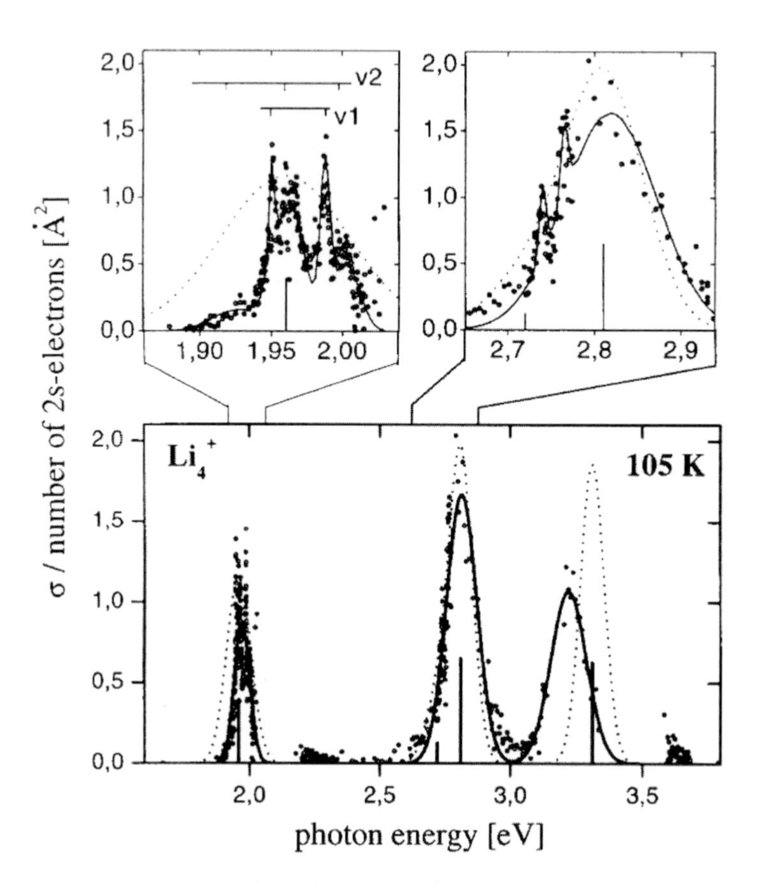


**Figure 4.25:** Illustration of the coherent coupling between the Mie surface plasmon and ionic vibration modes for $Li_4{}^+$ as test case. Lower panel: The optical absorption spectrum on a broad scale. Upper panels: Enlarged views of two small energy bands as indicated. The dots are the measured points. The full lines are a fit through the data, the vertical bars indicate theoretical results (of CI type), and the fine dotted lines show the same theoretical results folded with an appropriate Gaussian. The horizontal bars in the upper panels indicate vibrational frequencies as obtained from BO surfaces. From [ESH+02].

CI calculations show the $2ph$ states. But their strength is small and does not contribute much to the width of the plasmon. However, the dynamical electron–electron correlations are known to become important for strong excitations beyond the linear regime. The Ühling-Uhlenbeck collision term has been designed to account for that effect, see Section 3.3.5.3.

But all in all, Landau damping and incoherent thermal line broadening have been found to be the most important mechanisms creating the width of the Mie plasmon peak in the optical absorption spectra, i.e. in the linear regime.

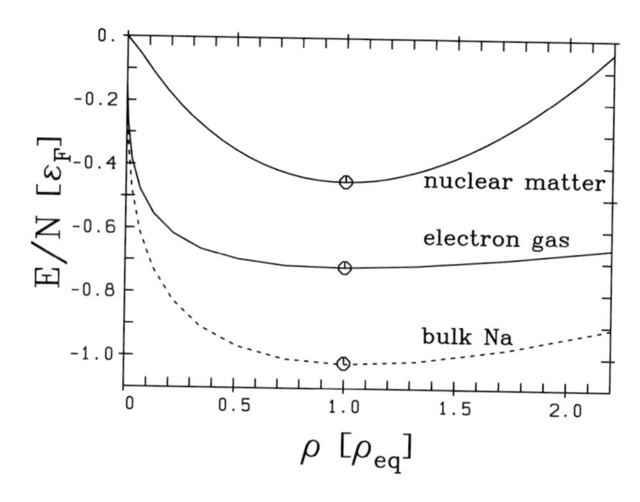


**Figure 4.26:** Binding energy per particle for symmetric nuclear matter, for the electron gas, and for bulk Na. The densities are expressed in units of the bulk equilibrium density of each system and the energies in units of the Fermi energy. They are $\rho_{\mathrm{eq}} = 0.17\,\mathrm{fm}^{-3}$ and $\varepsilon_{\mathrm{F}} = 38\,\mathrm{MeV}$ for nuclei, $\rho_{\mathrm{eq}} = 0.0037\,\mathrm{a}_0^{-3}$ and $\varepsilon_{\mathrm{F}} = 3.2\,\mathrm{eV}$ for Na and approximately the same for the electron gas.

## 4.4 Metal clusters and nuclei

The properties of metal clusters are to a large extent determined by the cloud of valence electrons. This holds even more so in the jellium model where the ionic background is structureless. The cluster becomes a drop of electron gas which is, in fact, a Fermion liquid [PN66]. Other examples of Fermion liquids are nuclear matter and liquid $^3$He. Finite nuclei and $^3$He are drops of Fermion liquid similar to metal clusters. The properties of all Fermion liquids have much in common. One thus expects many similarities between nuclei, $^3$He clusters, and metal clusters. There are, indeed, many such similarities. For example, the Nilsson model [Nil55] which had been developed as a shell model for nuclear structure calculations has been taken up to explain the shell structure of Na clusters long ago [Cle85]. There are also many analogies concerning collective resonance excitations [RWGB92, RWGB94, RS99]. The case of $^3$He droplets has been treated in [WR92, WR93] with much the same methods (density functional theory, shell structure, resonances) as nuclei and metals. We will concentrate the following discussion on a brief comparison of alkaline clusters (especially Na clusters) and nuclei. An early review comparing these two species can be found in [BC95] and a more recent discussion extending up to non-linear dynamics is found in [RS99].

### 4.4.1 Bulk properties

We start with the bulk properties. Figure 4.26 shows the binding energy per particle for infinite nuclear matter, the free electron gas and bulk Na (bulk parameters for Na and other alkalines are found in Table 1.5). The similarity lies in the fact that all three curves fit nicely into one plot when using the natural units of each system, i.e. expressing energies in terms of Fermi energy and densities in terms of equilibrium density. There are small differences in detail. The curvature of the $E/N$ curve for the free electron gas (= the incompressibility) is much smaller than the nuclear one, for example. Ionic structure effects which are incorporated in bulk Na enhance that again half way towards the nuclear matter curvature. But these are details. The main point is that all three systems are saturating in that they have one equilibrium point. As a consequence, finite drops have a radius $R \propto N^{1/3}$ and their binding energy is dominated by the volume energy $E_{\mathrm{bulk}}/N$ [Mye77]. This is indeed the case for Na clusters and for nuclei.

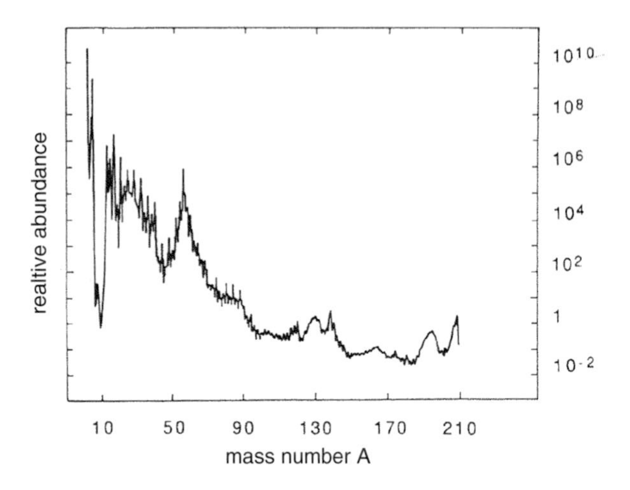


**Figure 4.27:** Nuclear abundances in the solar system, as a function of nucleon number $A$, drawn after [Vio90]. The curve has been "normalized" to the abundance of silicon ($A \sim 28$) taken equal to $10^6$.

### 4.4.2 Shell effects

The shell effects in metal clusters were first identified experimentally by measuring mass abundances, see Section 4.2.1. Abundances are also a key tool to mark the magic proton and neutron numbers in nuclei. The difference between metal clusters and nuclei lies in the experimental technique to identify these shell effects. Thermal ensembles of nuclei can hardly be generated in a laboratory on earth, at variance with the case of clusters. Nuclei are "naturally" produced in stars in processes involving possibly high temperatures, though. The thus formed nuclear species are then released to the interstellar medium by various mechanisms, in particular in explosive situations such as for example supernovae. The distributions of elements can finally be measured from the stellar abundances of elements. They are shown in Figure 4.27. One clearly sees the peaks associated with magic numbers. Corroborating information through laboratory experiments can be drawn from two-proton separation energies $S_{2p}(N_p, N_n) = E(N_p, N_n) - E(N_p - 2, N_n)$ and two-neutron separation energies $S_{2n}(N_p, N_n) = E(N_p, N_n) - E(N_p, N_n - 2)$ which are the analogue of the ionization potentials in clusters. Shell closures are then signaled by a jump in the separation energies. One finds the so called magic numbers $N = 2, 8, 20, 28, 50, 82$ for protons as well as neutrons and additionally $N = 126$ for neutrons. The sequence agrees with magic numbers of electrons in Na clusters up to $N = 20$. From then on the (harmonic oscillator) sequence changes due to the large spin-orbit force in nuclei [RS80].

A comparison of mean-field and associated shell sequence is shown in Figure 4.28 for the cluster $Na_{20}$ and the nucleus $^{40}Ca$ which has the same number of neutrons or protons, $N_p = N_n = 20$. The shell gaps are indicated by horizontal dashed lines and augmented by the corresponding magic numbers. The local potential wells look much the same and they alone would produce the same sequence of magic numbers. But the nuclear potential is strongly modified by the spin-orbit force (not presented in the plot). The consequence is visible, look for example at the nuclear spin-orbit splitting for the $1p$ shell (between $N = 2$ and 8) where the Na cluster shows only one $1p$ state. The splitting grows with angular momentum and reaches just above the $N = 20$ shell a point where the down-shifted spin-orbit partner, the

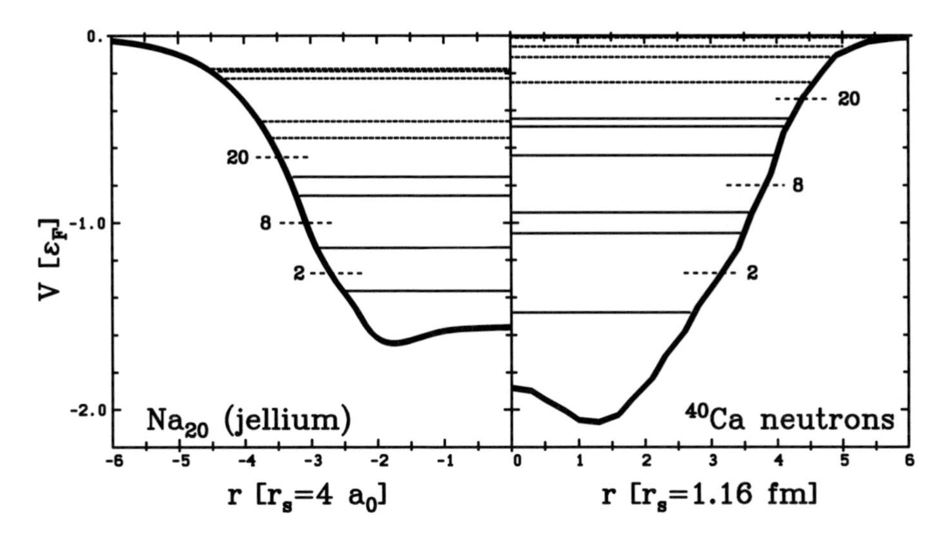


**Figure 4.28:** The mean-field potentials (heavy lines) and the bound single-particle energies (full lines = occupied, dotted lines = unoccupied) for a Na cluster (left part) and for a nucleus (right part). Length and energies are scaled in the natural units of each system. The total numbers of occupied states up to a given energy are indicated by horizontal dashed lines and associated numbers. The Na cluster was computed with the soft jellium approximation for the ionic background and the energy functional of [PW92]. The nuclear mean field was computed with the Skyrme force SkI4 [RF95] and the results are shown for the neutron part.

$1f_{7/2}$ state, is separated by a new gap. This starts the $N = 28, 50, 82$ sequence. After all, nuclei and metal clusters share the pronounced shell effects. They differ in two respects: first, nuclei have a strong spin-orbit force which leads to a different shell sequence for $N > 20$, and second, nuclei consist of a mix of two Fermion liquids while metal clusters mix one Fermion liquid with a completely different partner, the ionic background. The latter difference means that we can have doubly magic nuclei where $N_p$ and $N_n$ both are magic while we deal with the more subtle interplay of electronic and atomic shells in metal clusters, see Figure 4.8 and its discussion.

Shell effects also determine the shapes of Fermion systems, as discussed in Section 4.1.3. Thus the systematics of nuclear deformations should have similarities with metal clusters. This holds at least for small systems as demonstrated in Figure 4.29. The same sort of Jahn-Teller effect is at work in both sorts of "materials". There are, however, also marked differences between nuclei and metal clusters. The nuclear mean field has a very strong spin-orbit force which changes the shell sequence as compared to the cluster case. This becomes more important for larger systems. And a comparison like the one in Figure 4.29 becomes dangerous for $n > 20$. Moreover, there is a rather strong pairing interaction in nuclei which tries to reduce deformations.

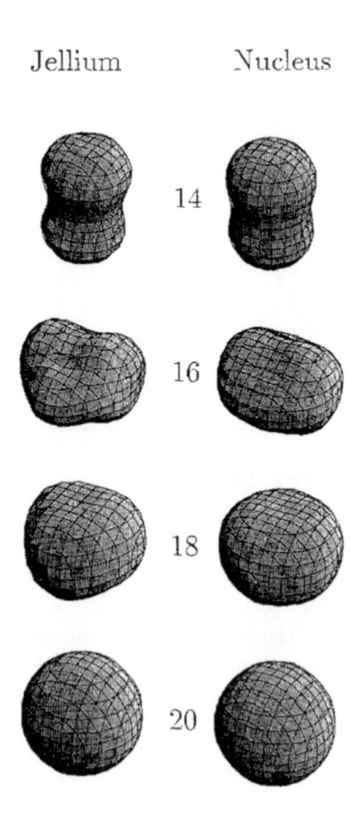


**Figure 4.29:** Comparison of shapes of Na clusters with nuclei in the range of $n = 14 \ldots 20$. The nuclei have the same numbers of protons and neutrons $N = Z = n$. From [KLM95].

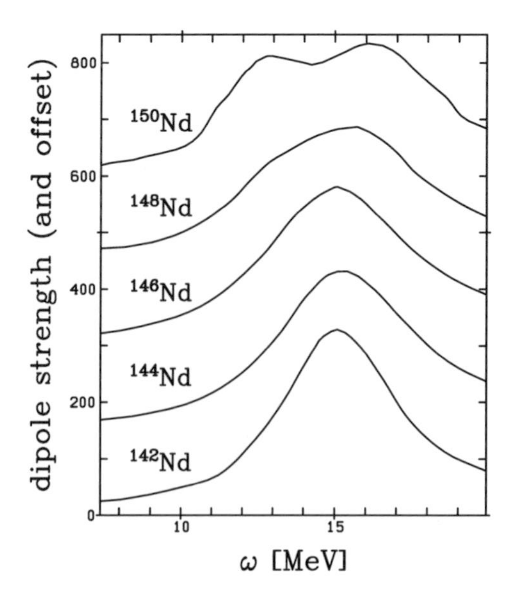


**Figure 4.30:** Nuclear isovector dipole strength distribution for the series of more and more Nd isotopes (from below): $^{142}$Nd, $^{144}$Nd, $^{146}$Nd, $^{148}$Nd, $^{150}$Nd. After [DB88].

### 4.4.3 Collective response

There is also a similarity in the excitation properties of nuclei and metal clusters. Both display well concentrated and dominant resonance peaks in the dipole spectra. In particular, the analogue of the Mie surface plasmon is the isovector-dipole giant resonance. The picture is the same as in the Mie model (see Section 4.3.1). The "isovector" in the name means that the centers-of-mass of the protons and neutrons oscillate against each other. It also implies that this mode strongly couples to the photon, much as the Mie resonance. The nuclear giant dipole resonance shows all the features discussed above for clusters: Landau fragmentation, $2ph$ correlations, dynamic coupling to surface modes (the equivalent of nuclear phonons), and thermal broadening. However, the relative strengths of these effects differ. There is no thermal broadening for nuclei excited from the ground state. Landau fragmentation is important for all nuclear sizes and $2ph$ correlations also play a non-negligible role [RTY86]. The coherent line broadening through resonance–phonon coupling can be significant [BB81]. In general, the nuclear dipole resonance is relatively much broader than a comparable Mie resonance in a cold cluster. Moreover, the deformation of small nuclei is not well defined due to huge (quantal) shape fluctuations in the ground state. The deformation splitting of the resonance can thus only be seen in heavy nuclei. Figure 4.30 shows experimental results of the photoabsorption strength for a series of Nd isotopes. This sequence is interesting because it starts at $^{142}$Nd with a magic neutron number ($N = 82$) and moves away from it when adding more and more neutrons. Consequently, the series starts with a spherical $^{142}$Nd and swaps soon to well deformed nuclei. This transition can clearly be read off from the structure of the dipole resonances in Figure 4.30, just as was seen for clusters in Section 4.3.3. There is a single peak in the lowest line and a nice deformation splitting in the uppermost. Many other similarities for dynamical response in nuclei and metal clusters are discussed in [RWGB92, RWGB94]. A broader overview of comparable features is given in the review article [BC95] and an overview specialized on dynamical features including the non-linear regime is given in [RS99].

### 4.4.4 Fission

Fission is also a feature common to both nuclei and metal clusters. But again the details differ. The fragmentation channels depend on an interplay between the Coulomb force, the surface tension, and shell effects. The case of cluster fission has been reviewed thoroughly in [NBF$^+$97]. Estimates using a droplet model (fissility parameter) hint that clusters preferably decay into very asymmetric fragments while nuclei stay always close to symmetric fission. Nevertheless, shell effects can overrule that to a certain extent such that symmetric fission could also be observed in clusters [NBF$^+$97]. Figure 4.31 compares typical situations for clusters and nuclei. The left panel shows barriers for cluster fission (see also Figure 3.17). As a rule, minima there correspond to peaks in mass distribution. It is obvious that this $K_{26}{}^{++}$ cluster strongly prefers asymmetric fission into a trimer $K_3{}^+$ and the remaining $K_{23}{}^+$. Note that trimer emission is then analogous to $\alpha$ decay in nuclei. The magic, thus especially stable, $K_9{}^+$ also plays a role, but has not been identified so clearly in experiments. The nuclear result (right panel) shows that the fragment distribution is clearly concentrated around symmetric fission. Small fragments are so rare that it is not even worth of extending the mass scale more than $\pm 30\%$ around the equal partition. The extreme asymmetric wing of cluster

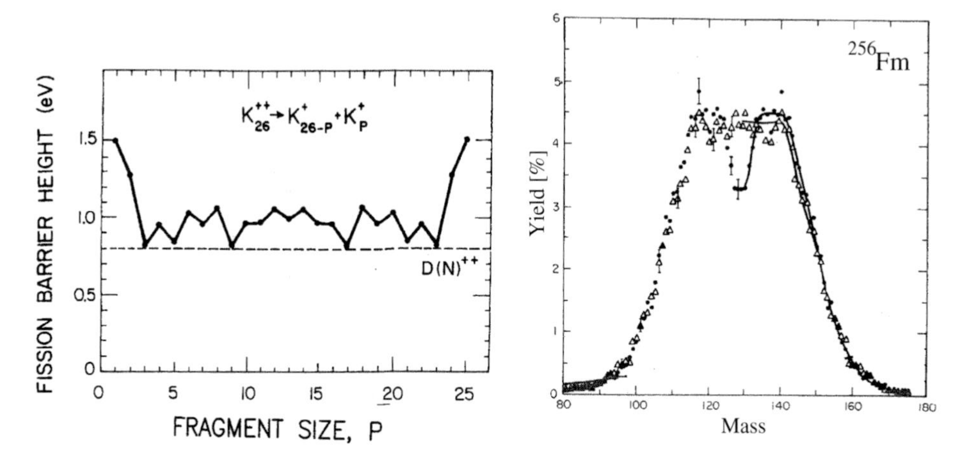


**Figure 4.31:** Distribution of fission fragments for fission. Left panel: estimated barriers for fission of the cluster $K_{26}{}^{++}$ versus fragment size, from [NBF+97]. Right panel: fission fragment yields for spontaneous fission of $^{257}$Fm, after [VH73].

fission is emission of a charged trimer which dominates in most experiments. The analogue in nuclear fission is $\alpha$ decay which is usually discussed as a separate issue. It is a key decay mode of heavy nuclei, often more important than (nearly symmetric) fission. This feature has something in common with clusters.

### 4.4.5 Cluster versus nuclear time scales

The comparison of nuclear and cluster time scales is also quite enlightening. In order to make it more telling we use reduced units in terms of Fermi gas characteristics of both systems, following the strategy used above to compare time scales in alkalines (Section 1.3.2). We thus again consider the basic time $r_{s,0}/v_F$ and energy $\epsilon_F$ scales, as built from the Wigner-Seitz radius $r_s$ for clusters and from the parameter $r_0$ entering radius systematics in nuclei ($R \sim r_0 A^{1/3}$, with $r_0 \sim 1.12$ fm). In clusters, and for the sake of simplicity, we restrict the analysis to the case of Na, thus taking for the Wigner-Seitz radius $r_s = 4a_0$, which leads to $r_s/v_F = 0.2$ fs and $\epsilon_F = 3.2$ eV. In nuclei the corresponding values read $r_0/v_F = 3.3$ fm/c and $\epsilon_F = 40$ MeV. As in Section 1.3.2 times are plotted as a function of temperature, which again is mostly a matter of convenience and does not necessarily imply a full thermalization of the system under any circumstances.

With this system of reduced units we can now compare nuclear and sodium time scales. This is done in Figure 4.32 for various relevant times: plasmon (giant dipole resonance) period, ionic (nuclear fission/fragmentation) time scale, electron (neutron) evaporation time, electron–electron (nucleon–nucleon) time scale. The comparison calls for several comments. The first impression is a relative similarity between electronic and nuclear time scales, with comparable dependence (or independence) of times on temperature. But when looking more into the detail one can see that the hierarchy of time scales is pretty different between Na and nuclei. Grossly speaking nuclear time scales are more similar to each other than cluster ones.

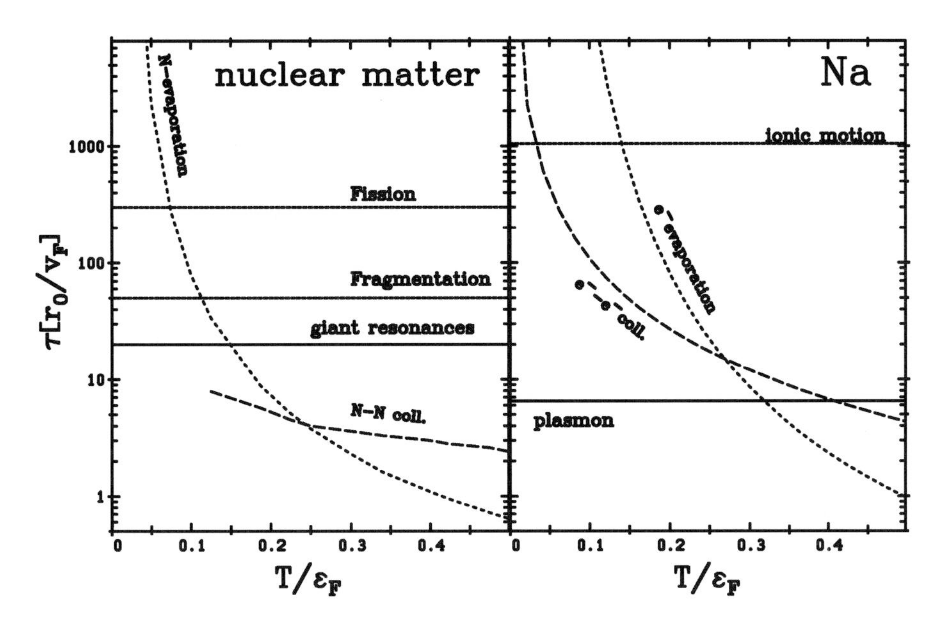


**Figure 4.32:** Comparison of relevant times scales in nuclei and clusters. Reduced units are used in both cases to allow a relevant comparison (see text for details). The various times plotted are: plasmon (giant dipole resonance) period, ionic (nuclear fission/fragmentation) time scale, electron (neutron) evaporation time, electron–electron (nucleon-nucleon) collision time scale.

In other words there is a natural hierarchy of time scales in clusters, while nuclear times are more mixed up. This has important implications in particular at the level of theoretical descriptions. In nuclei one can hardly identify slow degrees of freedom, except maybe in very special cases like fission. In cluster physics electron time scales are an order of magnitude smaller and thus more responsive than ionic ones. They are thus to be accounted for in priority in cluster dynamics. One should nevertheless note that the separation of electronic and ionic time scales tends to shrink in strongly non-linear situations (see discussions in Chapter 5). Differences between nuclear and cluster hierarchies of time scales do not only concern the hierarchies by themselves but also the times with respect to each other. It is in particular noticeable how electron–electron interactions come into play as compared to nucleon–nucleon ones. The former become dominant for much higher temperatures than the latter, which means that mean field methods can probably be used at much higher excitation energies in clusters than in nuclei, a welcome feature in view of the theoretical difficulties such effects raise (Section 3.3.5.3). Similarly, thermal emission comes into play much earlier in nuclei than in clusters, a feature again reflecting the more "conflicting" nature of nuclear time scales as compared to cluster ones.

## 4.5 Conclusions

The study of systematic trends and gross properties of clusters has provided us a handful of important information. As already mentioned at many places in this book, clusters are scalable objects. Systematics thus constitute a fundamental access to their properties and provide guidelines to understand the way clusters are built and respond to external perturbations. This is especially true in the case of metal clusters which exhibit particularly generic features; they have thus constituted a large part of the body of results presented in this chapter.

In the first part of the chapter we have described the typical observables one may theoretically access, in order to establish a proper link with experimental results. We have then focused the presentation on structure questions, discussing in particular the famous shell effects, both at electronic and ionic levels. Stepping into the realm of dynamical responses, we have also discussed in detail the optical response, especially in the case of metal clusters, in which it plays a key, dominating role and in which it exhibits systematic behaviors. We have discussed the dependence of the optical response on size, shape and temperatures, insisting in particular on the impact of these parameters on the width of the optical peaks. The last part of the chapter was dedicated to a comparison between metal clusters and nuclei, two finite fermion systems which are known to exhibit several similar behaviors. We have indeed seen that similarities do exist, both at the level of structure and dynamical properties. Differences in the detail nevertheless remain, and they have also been pointed out when necessary.

# 5 New frontiers in cluster dynamics

The previous chapters have presented the basic features of clusters and associated experimental as well as theoretical methods. They have summarized the state of the art as it has been accumulated over the past two decades. All that could give the impression of a well understood and closed subject. The opposite is true. There are many open questions and promising directions for future research. This holds particularly in the realm of cluster dynamics. But also structure questions remain of actual interest because they are to be explored ahead of dynamical problems. Furthermore, the impressive technological developments of lasers and laser facilities have opened up many new directions of research in truly dynamical situations. And, as we have seen at several places in this book, lasers offer a particularly flexible tool of investigations of cluster properties. This chapter will thus discuss a large variety of very recent and near future investigations, mostly in the domain of truly dynamical problems. The style of the presentation will naturally change as compared to previous chapters, as is appropriate for open ends. We will put together a (for sure non exhaustive) mosaic of examples more or less briefly explained and refer the reader to appropriate, more dedicated references. Of course these new developments often, but not always, represent extensions of previous experience, as described in previous chapters. We shall thus rely on the basics as detailed in the previous chapters and complement new techniques where necessary.

There is a multitude of new frontiers in cluster physics. We divide them into three domains: close to ground state, semi-linear dynamics, and massively non-linear processes. Before going into the details, we will introduce the three groups briefly by one example for each domain.

The seemingly most conventional part concerns investigations close to the ground state. This covers structure and low-energy dynamics in the regime of linear response. Most progress has been made in the past at that frontier, and yet there is still a rich ground for new investigations, mainly triggered by new or developing experimental techniques. We will discuss a broad selection of such developments in Sections 5.1 for structure and 5.2 for linear response. Figure 5.1 shows here as a starter one typical example, the assessment of single-electron levels in clusters by means of photoelectron spectroscopy (PES) [WavI02]. In this experiment, one lifts a bound electron at once into the continuum using one photon with frequency safely above emission threshold. The excess energy is converted into kinetic energy of the released electron. One thus can read off the original single-electron binding energy from the final kinetic energy which provides direct mapping of PES into bound single-particle energies. This method has long been used for negatively charged clusters where the emission threshold is in the range of visible light, see [MEL$^+$89] and Sections 2.2.3 and 2.3.8. Better light sources are required for neutral or positively charged clusters, and are now becoming more and more available. The example of Figure 5.1 was performed with UV light from frequency tripled

*Introduction to cluster dynamics.* Paul-Gerhard Reinhard, Eric Suraud

ISBN: 3-527-40345-0

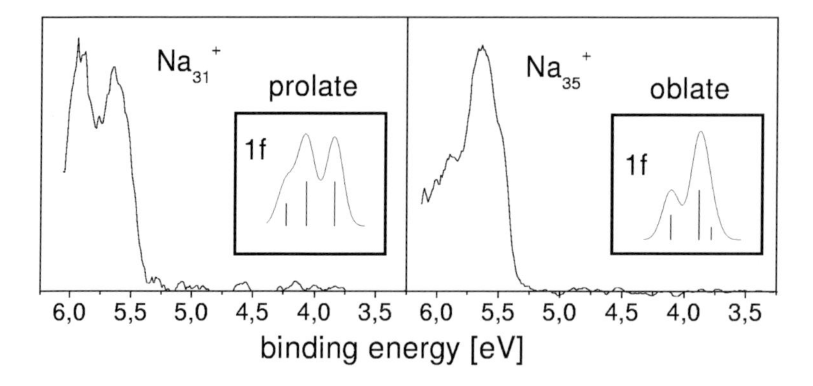


**Figure 5.1:** Experimental photoelectron spectra (PES) of positively charged Na clusters with UV light of a frequency tripled laser, from [WavI02]. The insets show three last bound levels as estimated by a deformed shell model [Cle85] with their multiplicity and the smoothed spectral strength derived thereof.

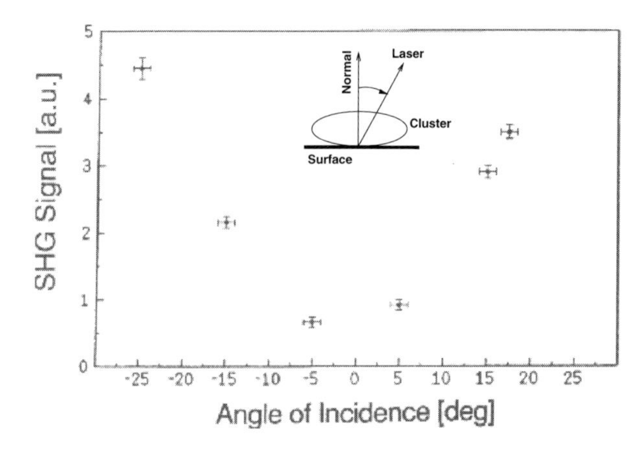


**Figure 5.2:** Dependence of the SHG output signal on the angle between laser beam and surface-normal. Test cases are large Na clusters ($R > 100$ nm) on LiF surface. The wavelength of the incident laser pulse was 1064 nm. After [GBD+95].

intense laser light. This technique will most probably be even further improved. Even more flexible sources are to be expected from free-electron lasers and synchrotron radiation. This will pave the path towards more and more precise information on electron structure in clusters.

Investigations beyond the linear regime are related to the development of readily available femtosecond lasers. Just beyond linear phenomena comes what we call the semi-linear regime. Non-linear effects come into play, as e.g. multi-photon processes, but the cluster remains basically intact, or at least as a whole for a sufficiently long time to allow further analysis. With a view on Figure 3.3, we associate with the semi-linear regime laser intensities up to about $10^{13}\,\mathrm{MeV/cm^2}$ (with a dependence of the threshold on the material, of course). There exist already quite a few studies in that direction. Most of them could be qualified as promising in the sense that they provide extremely interesting results which increase the appetite for more. In other words, the field is still much in a developing stage. We will discuss selected test cases in Sections 5.3 and 5.4. As one typical example, we show in Figure 5.2 results from second harmonic generation (SHG) on clusters [GBD+95]. The SHG is an interesting feature

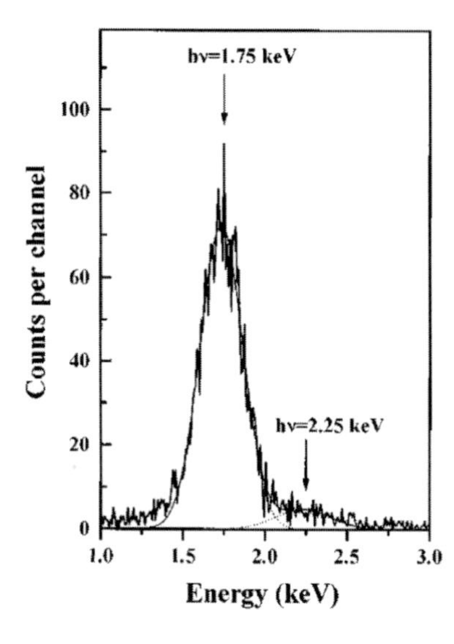


**Figure 5.3:** X-ray spectrum as generated by irradiation of Kr clusters with intense laser beams. After [DLS$^+$97].

as such and it is often used as an analyzing tool to obtain a background-free signal from cluster response. It is used, e.g., to monitor growth of clusters on surfaces [BJR00, BR01]. We will see in Section 5.3.4 that, for example, SHG helps to suppress background signal in pump-and-probe experiments. The example here demonstrates a principal feature. SHG is easily possible only for clusters deposited on a surface. One needs to break reflection symmetry to obtain second harmonics and the surface provides a perfect ("natural") source of such a parity breaking. Note that this acts only in the direction perpendicular to the surface while the clusters remain more or less symmetric in their extension parallel to the surface. It is obvious that the intensity of SHG thus depends on the angle of the incident laser light. And that is demonstrated in Figure 5.2. There is a minimum of the SHG signal if the light shines orthogonal to the surface (angle $= 0$). The polarization axis of the light then lies in plane with the surface and there is too little symmetry breaking in that direction. With increasing angle the part of the polarization along the symmetry breaking direction increases and thus the SHG signal increases.

The non-linear regime covers violent processes where destruction of the cluster is a key part of the mechanism. In the strongly non-linear domain the cluster may even not be the object of the investigation itself, but rather provide a laboratory for "not cluster oriented" investigations. This is typically the case for irradiation by very intense laser beams. That case will be discussed in Section 5.5. The hefty reactions strongly couple all degrees of freedom in the system. Consequently, all sorts of reaction products emerge, amongst them very energetic ions and sometimes even neutrons from nuclear fusion. Figure 5.3 exemplifies another reaction product, namely X-ray emission from the Coulomb explosion of Kr clusters after an intense laser pulse. The X-rays stem from atomic transitions amongst the core states in highly ionized Kr atoms. The width is caused by incoherent superposition of X-rays from different ionization stages.

The above discussed examples are just spotlights on the multitude of ongoing works and we shall go into more detail below. Still, the three above experimental examples should not give the impression that there are only experimental challenges ahead of us. As explained at many places in the book, the many-body problem in clusters is very involved, in particular when dealing with time dependent problems. Although we are already able to understand a lot of physically interesting situations, the theoretical problems leave plenty of space for developments as we are still far away from a fully dynamical description of an experiment. We have seen that DFT offers a robust and flexible tool of investigations in various dynamical domains. It thus probably provides a good starting basis for further studies of cluster dynamics when keeping in mind the still existing insufficiencies of the approach. Without entering into details, it is hence interesting to quickly list a few important directions along which the theory should make progress. The first aspect concerns the nature of the material constituting the cluster under study. While experimentalists may consider a bunch of various materials, theorists are often bound to the simplest ones, e.g. alkaline metals with bias on sodium or rare gases. It is obvious that an important step forward would consist in a proper dynamical account of core electrons, for example in noble metals, or even in covalent clusters. This represents a first challenge. A question which is somehow similar concerns improved modeling of cluster-substrate interfaces. Most theoretical studies up to now have been performed for free clusters. The interaction with a substrate or a matrix adds a layer of complication but gives way to a wide variety of applications to ongoing experiments. Last, but not least, one should insist again on the effect of dynamical electron electron correlations which increase with increasing strength of the perturbation. As discussed in Section 3.3.5.3, a way to account for dynamical correlations is to complement the effective mean field by a collision integral in the spirit of kinetic theory. Such developments are still in their infancy in cluster physics. First attempts are available in semi-classical approaches. The question needs to be considered at a quantal level, as is already done in other fields of physics. This is one more theoretical challenge for cluster dynamics, in particular in relation to ongoing experiments with intense laser beams.

## 5.1 Structure

Although cluster structure is not the particular topic of this book, it is worth pointing attention to the fact that many new developments are in store even in that most traditional branch of cluster physics. By far not all conceivable materials are thoroughly understood. For example, a seemingly everyday metal as Mg analyzed with standard observables poses ongoing puzzles concerning its electronic shells [DDTMB01, SRS02] and concerning its metallicity [AJ02, TTXB02]. Many more interesting perspectives open up when considering further observables or novel production schemes. This section presents a few selected examples for such ongoing investigations. Section 5.1.1 deals with a pure structure question but gathers its new features from the modified cluster–cluster interaction on a surface. Section 5.1.2 deals with a very particular sort of cluster, namely He droplets. These constitute a peculiar material in itself and they serve as extremely useful low-temperature laboratory for producing and analyzing other clusters or molecules. The following sections discuss other observables, such as polarizabilities in Section 5.1.4, heat capacities in Section 5.1.3, and magnetization in Section 5.1.5. Another observable is the conductivity of clusters. This is very fashionable due to its promising perspectives for applications. We refer here to the discussions in previous chapters, namely in Sections 2.3.6 and 4.1.6.

### 5.1.1 Fractal growth

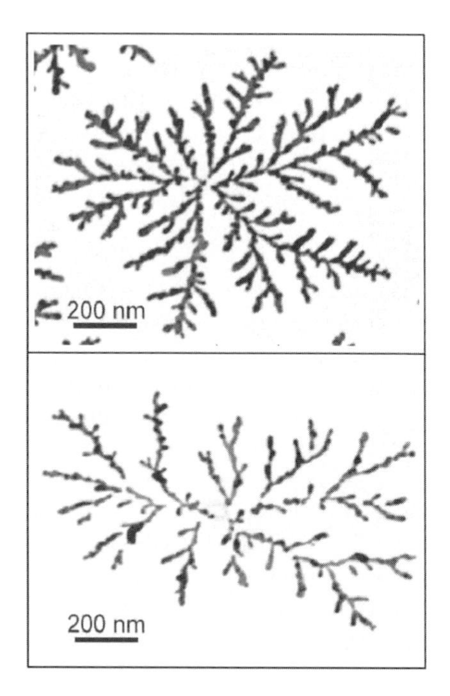


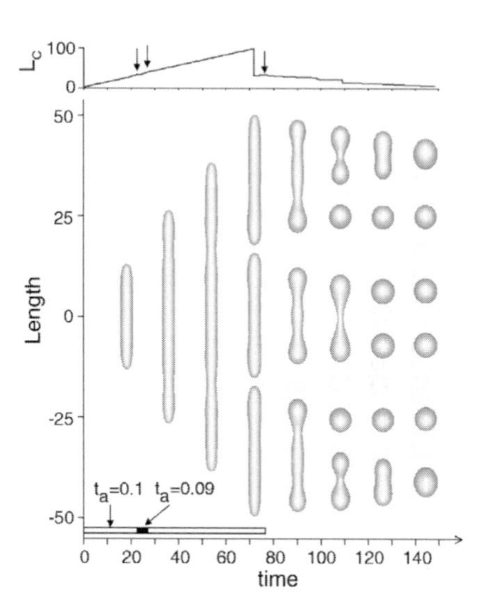


**Figure 5.4:** The left panels show scanning electron microscope (SEM) pictures of large aggregates of Ag clusters on graphite. Upper: sample with average coverage of 6 mono-layers. Lower: average coverage of 2 mono-layers. The right panel shows MD simulations for growth of the Ag islands and their disassembling through instability in the course of time. Cluster deposit continues up to 79 time units. Disassembling takes over after switching off the supply of fresh Ag clusters. Composed from [BCC+02].

The left part of Figure 5.4 shows the distribution of Ag clusters on a graphite surface after low energy (0.05 eV/atom) deposit of Ag clusters with typical size of $N = 150$, i.e. radius around 2 nm. At first glance, one is impressed by the obviously fractal structures which appear. The mechanism beyond that has been eludicated by MD simulations [BCC+02] whose results are sketched in the right part of Figure 5.4. During deposit, the Ag clusters have a tendency to coalesce and to form long cylinders. But these cylinders develop an intrinsic hydrodynamical instability if they grow beyond a critical length. They decay into shorter pieces until a stable size is reached. The final state thus reached depends on the history of growth and termination. Long exposure to the cluster beam and subsequent larger coverage lead to broader and longer branches while lower coverage yields more fragmented structures.

Although the patterns look like structures, we see here a result of fragmentation dynamics of clusters. The attachment to a surface and fixation by final oxidation provides these snapshots of fractal structures. The analogue process for free clusters is fission. For a detailed discussion of fission for metal clusters see the review article [NBF+97]. Actual experiments deal mainly with fission of small clusters [BCC+94] while multi-fragmentation and hydrodynamical instabilities develop for larger metal clusters. Collecting the clusters on a surface allows one to assemble the needed huge cluster sizes for these fragmentation studies. The dynamics thus observed is, of course, much different from those of free metal clusters, because it becomes a 2D (instead of a 3D) process. An extreme example for fragmentation of free clusters is the laser driven Coulomb explosion as discussed in Section 5.5. The regimes in between small and huge clusters as well as between moderate and extremely high excitations have yet to be explored.

### 5.1.2 He droplets

Bulk He is a very special material for which a broad literature exists, see e.g. the books [BK78, KN02]. Its binding is extremely weak with a binding energy per atom of –2.5 K ($\simeq$ –0.22 meV). As a consequence, bulk He never reaches a solid state. Instead it becomes a quantum liquid near zero temperature, a Fermi liquid for $^3$He and a Bose liquid for $^4$He. On the other hand, the short range atom-atom repulsion is huge such that one deals with a highly correlated system. Therefore, liquid He has been a critical testing ground for all sorts of highly developed many-body techniques, as e.g. variational Jastrow correlations [PPW86] or the hyper-netted chain method [Kro00]. A mean field description is possible when employing effective energy-density functionals, see e.g. [SGB$^+$91] for $^3$He and [DRHPT90] for $^4$He. These functionals are motivated by similar ideas as energy functionals for electronic systems. However, the higher degree of correlations requires much more elaborate energy functionals which invoke folded densities and a strong momentum dependence as well [WR91] in order to reproduce the empirical polarization potentials for normal modes in liquid He [AP78]. The same methods have been used for the description of finite He droplets (clusters). Present day computing facilities allow fully correlated calculations even for not so small He clusters [KZ01]. But large scale surveys on finite systems are better off with the more efficient density functional methods [SGB$^+$91, WR92, WR93].

The He clusters are more than interesting clusters of a Fermi liquid. They serve as an extremely useful low-temperature laboratory for molecular and cluster physics experiments [TVW01]. For example, Na atoms stay at the surface of a He droplet due to the strong electron repulsion inside. But they tend to gather and to form surface compounds. And these are preferably found in fully spin aligned states. The spin saturated combinations have much larger binding energy and thus the energy gained in fusion to a molecule (cluster) overrules the very weak surface binding and removes the freshly composed system [SV01]. This is exemplified in Figure 5.5 showing the strength of the fluorescence signal from small Na clusters on a He surface. The weak binding to the He droplet leaves the electronic spectra of the Na clusters almost unchanged such that they can be easily identified from the known spectra of the various feee clusters. This allows the assignment as shown in the figure. The signals from spin-saturated clusters are obviously much suppressed and thus these clusters are basically not present while one sees clear response from the spin-aligned species. The existence of fully spin-aligned larger Na clusters can be expected, but they have yet to be identified and explored. A theoretical survey of their properties can be found, e.g., in [ARS02a]. There are many other materials for which He clusters serve as a laboratory, for a long list of examples see the review [SV01]. One example from cluster structure experiments is the production of Mg clusters by collecting a huge amount of Mg atoms in the droplet and then triggering clus-

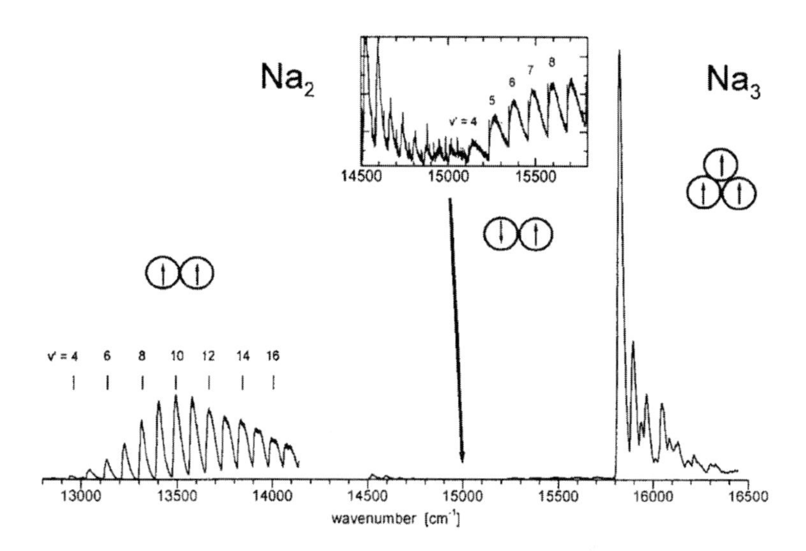


**Figure 5.5:** Laser-induced fluorescence spectra of Na doped He droplets. The spectral range (12900-16500 cm$^{-1}$ $\equiv$ 1.60-2.06 eV) is chosen to cover crucial electronic transitions in $Na_2$ and $Na_3$. Blocks of spectra can be distinguished and associated with specific transitions as indicated. Large signals from fully spin-polarized $Na_2$ and $Na_3$ can be identified while spectra from corresponding spin-saturated systems are much suppressed. After [SV01].

ter growth by heating away the He droplet [DDTMB01]. Most other examples reach into the study of dynamical processes, as e.g. the systematics of medium effects on the optical spectrum of embedded Ag clusters as discussed in Figure 5.10 or the pump-and-probe experiments to explore the coupled dynamics of $Na_2$ on the surface of a large He droplet as discussed in Figure 5.36.

### 5.1.3 Heat capacity

Clusters are distinguished from molecules in that they are "scalable", i.e. for a given material there exists virtually any cluster size up to the bulk. This feature has been exploited to understand the bulk limit with respect to many observables, e.g. the emergence of electronic energy bands with increasing cluster size, see Section 1.2.1.2. A key observable is, of course, the total energy. Its dependence on temperature yields the specific heat from which basic thermodynamic properties, as e.g. transition temperatures and latent heat, can be deduced, see Section 2.3.9 for the basics of such measurements. There are several intriguing problems involved. At the theoretical side, one has to check how thermodynamical concepts can be extended to finite systems, for a thorough discussion see [Gro01]. It turns out that the microcanonical ensemble is to be preferred for finite and isolated clusters. At the experimental side, it is anything but trivial to map binding energy versus temperature with sufficient resolution to pin down critical points. And last but not least, the emerging trends pose an interesting puzzle yet awaiting a fully microscopic answer.

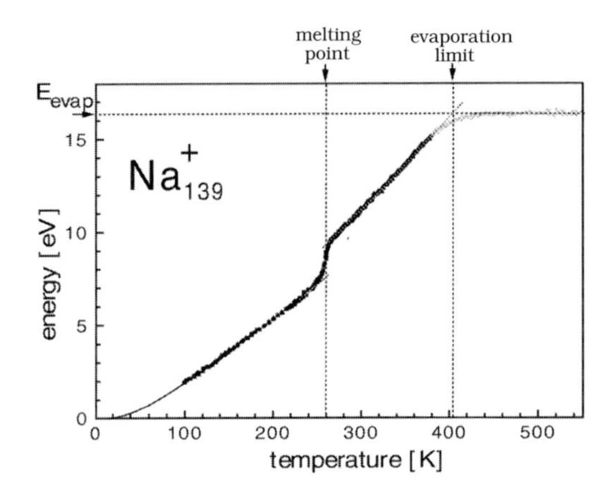


**Figure 5.6:** Internal energy of a $Na_{139}{}^{+}$ cluster as a function of temperature. The melting temperature and the transition into the evaporation limit are indicated. After [SHD$^+$01].

Figure 5.6 shows results from measurements of the caloric curve of Na clusters. We repeat briefly the experimental method to obtain the caloric curve $E_{\text{int}}(T)$ (see also Section 2.3.9): a special cluster source delivers a size- and temperature-selected beam of $Na_n^+$ clusters. The clusters are irradiated with an intense laser pulse whose frequency lies below direct electron emission threshold. The absorbed energy will thus be converted into internal heat which with appropriate delay gives rise to evaporation of monomers. The final state corresponds to an evaporative ensemble and becomes independent of the initial conditions. Only the total energy determines the emerging fragment mass spectra. The experiment starts at some low temperature $T_1$. One irradiates and keeps a record of the mass spectrum. Then one increases the initial temperature until $T_2$ where the same pattern reappears. Although the initial $T$ was higher, the same final state could be reached because one photon less was absorbed. The difference $E_{\text{int}}(T_2) - E_{\text{int}}(T_1)$ is then identified with the photon energy $\hbar\omega$. Scanning the scale of temperatures with a variety of different photon frequencies allows one to compose the whole $E_{\text{int}}(T)$. A typical result is shown in Figure 5.6. One clearly sees a pronounced step in the $E_{\text{int}}(T)$ curve. This is interpreted as the transition from solid to liquid state of the cluster and the height of the jump then corresponds to the latent heat at the corresponding melting temperature.

Figure 5.7 shows the systematics of melting temperatures with system size. At first glance, one sees that these are all significantly smaller than the bulk limit. This is not so surprising because surface melting starts at lower temperature [LBW89] than bulk melting and because the surface becomes increasingly important for smaller clusters. At second glance, one realizes that there are strong fluctuations of melting temperature with cluster size. Note, e.g., the pronounced peaks at $N = 57$ and 142. One wonders how this could be related to magic numbers (as discussed, e.g., in Section 4.2.1). However, the electronic shell closures arise with $N = 59$ and 139 somewhat away from these peaks. On the other hand, there are ionic shells which close at $N = 55$ and 147, again slightly off the wanted point. But note that the peaks reside in both cases just in between an electronic and an ionic shell closure. It is possible that we see here a cooperative effect of both closures which is maximal between them. A fully detailed microscopic evaluation of these trends has not yet been worked out because the task is extremely demanding: one has to simulate an ensemble of clusters at finite excitation

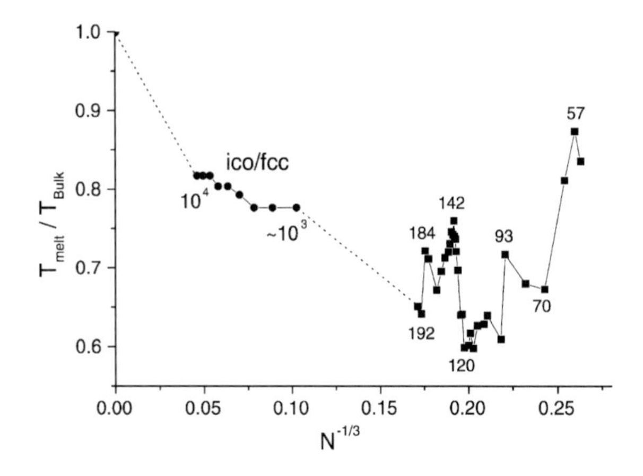


**Figure 5.7:** Trend of melting temperature $T_{\rm melt}$ (in units of bulk melting $T_{\rm Bulk}$) with system size drawn versus $N^{-1/3}$ such that the bulk limit is reached at the left side. The figure is drawn from [SHD⁺01]. The data marked by ico/fcc are taken from [Mar93] representing results for large Na clusters which already exhibit signs of crystalline structure (see also Figure 4.8).

energy, and worst of all, one needs to deduce the entropy from that. Qualitative features, on the other hand, are well understood theoretically [Gro01]. For example, certain clusters display the interesting phenomenon of a negative heat capacity [GFF⁺02, SKH⁺01]. Altogether, the case of heat capacities shows once more that there is a lot of interesting physics at finite temperature close to the ground state and many open ends for future research.

### 5.1.4 Static polarizability

The electric polarizability is a most prominent feature of molecular systems. Basic facts and measurements of it were already discussed in Sections 1.3.3.2 and 2.3.3. The dynamical polarizability belongs to the regime of linear response and optical absorption is related to its imaginary part. The static polarizability can be considered as the limit of zero frequency of the dynamical polarizability. But it is an observable of its own, measured differently from optical absorption and, at the theoretical side, allowing a purely stationary treatment. We thus discuss it here in the section on structural properties (although some response is invoked).

The (dipole-)polarizability $\alpha_D$ of a finite system is defined from the dipole moment $\mathbf{D} = e\mathbf{r}$ induced by an external homogeneous electrical field $\mathbf{E}$, i.e.

$$\mathbf{D} = \alpha_D \mathbf{E} \quad . \tag{5.1}$$

It is, in general, a tensor. It becomes diagonal in systems with reflection symmetry about three orthogonal axes. And it shrinks to one relevant number in the case of a fully spherical symmetric system. The measurement of the polarizability for free clusters with their random orientations naturally produces a spherical average. Calculated polarizabilities have thus to be averaged over all spatial orientations before comparison with data.

The theoretical calculations proceed in two basically different ways. The conceptually simpler access is to apply directly the definition Eq. (5.1). This implies several steps:

1. A weak external potential $\eta\mathbf{D}\cdot\mathbf{E}$ is added to the Hamiltonian where $\eta$ is a sufficiently small parameter.

2. The static solution is sought for some small $\eta$ and $-\eta$.

3. The resulting dipole momenta $\mathbf{D} = \langle e\mathbf{r} \rangle$ are computed for each $\eta$.
4. Finally, we find the tensor polarizability as $\alpha_{D,ij} = (D_i(\eta) - D_i(-\eta)) / (2\eta E_j)$ (where $i, j = x, y, z$).

This simple technique can be applied to any method for calculating ground states, see Chapter 3. A diligent choice of the numerical parameter $\eta$ is nevertheless required. It should be as small as possible to explore for sure the linear regime, but it should be large enough to leave sufficient precision in the finite difference evaluation. An alternative method is possible if one has access to the spectrum of excited states, as e.g. in some CI calculations (see Section 3.3.2.2) or in DFT when RPA (=linearized TDLDA) is performed (see Appendix G). The perturbation is again $\hat{\mathbf{D}} \cdot \mathbf{E}$. It yields a perturbed state $|\delta\Psi_0\rangle$ which at lowest order of perturbation theory can be written as

$$|\delta\Psi_0\rangle = \sum_{N \neq 0} |\Psi_N\rangle \frac{\langle \Psi_N | \hat{\mathbf{D}} | \Psi_0 \rangle \cdot \mathbf{E}}{E_N - E_0}$$

where $\{|\Psi_N\rangle, E_N\}$ is the spectrum of excited states of the ground state system. From that expression one obtains the induced dipole moment in first order as $\mathbf{D} = 2\Re\{\langle \Psi_0 | \hat{\mathbf{D}} | \delta\Psi_0 \rangle\}$ and the polarizability which reads in detail

$$\alpha_{D,ij} = 2 \sum_{N \neq 0} \frac{\langle \Psi_0 | \hat{D}_i | \Psi_N \rangle \langle \Psi_N | \hat{D}_j | \Psi_0 \rangle}{E_N - E_0} \quad . \tag{5.2}$$

This shows that the polarizability is dominated by the lowest and strongest dipole transitions. In metal clusters, this is usually the Mie plasmon which dominates, because the $1ph$ states below resonance have only very small strength.

Polarizabilities were amongst the first observables to be looked at in the early studies on metal clusters [KCdHS85]. But they are not so easy to measure and thus data about polarizability of clusters are still rare. Nonetheless, they constitute a key feature of clusters which require continued studies, the more so as there are several aspects which are not yet fully settled. For example, there was a long standing problem with the trends in the polarizability of Na clusters; all models have underestimated its values for the very soft clusters around $N \approx 14$ [GCKS95, KBRB00]. The case has found only recently a convincing explanation [KAM00]. One needs to take care of the actual temperature during measurement and to simulate an appropriate thermal ensemble. This provides perfect agreement between DFT and experimental results. And it indicates that one should control temperature sufficiently well in experiment and theory. Another example is polarizability in extreme situations, as e.g. in long chains, which has proven to be a particularly critical test case. It resists a treatment by standard LDA and calls for advanced methods as the extension of LDA to current-density functionals [vFBvL+02]. All that shows that polarizability is a matter of actual and future research. Old data often still call for explanation and new materials always require a first inspection of their polarization properties, as e.g. for carbon chains [BBG+02]. Particularly interesting here are combinations of two materials because different responses of both can give rise to surprising effects. Figure 5.8 shows results for an example of a $Na_n$ cluster attached to a $C_{60}$ cluster. Two different scenarios are conceivable here: first, the Na atoms are tightly covering the $C_{60}$ and

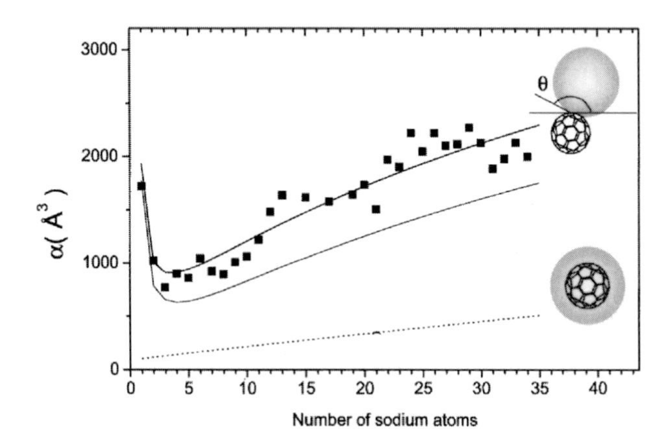


**Figure 5.8:** Polarizability of $Na_n$–$C_{60}$ as a function of the size of the $Na_n$ cluster. The full squares are results of measurements. The dotted line is the prediction from a model where the Na atoms cover the $C_{60}$ as a metallic surface layer (see lower inset). The two other lines come from a model of a $Na_n$–$C_{60}$ super-molecule (see upper inset) taken at two different relative angles. From [DAR+01].

respond to the external field like a metallic surface layer, or second, the $Na_n$ cluster maintains a memory of its own identity and oscillates as a whole with respect to the carbon cluster. The figure shows the predicted polarizabilities from both scenarios, the lower curve (dotted) corresponds to the model of a metal layer while the upper curve (full lines) is related to the model of a molecule of two clusters. Both scenarios are sketched in the insets. The experimental results (black squares) clearly rule out the model of a metallic surface layer. Altogether, the example demonstrates how polarizability delivers crucial information on the structure of a system. It should finally be noted that a $Na_n$ cluster attached to a $C_{60}$ cluster represents, to some extent, a prototype of a cluster deposited on a surface, in spite of the fact that the curvature of the surface is quite large, as compared to the cluster size. Still, it is interesting to keep that aspect in mind in relation to the physics of deposited clusters.

### 5.1.5 Magnetic properties

Cluster magnetism is another aspect of structure which is still much investigated. Its treatment requires, though, elaborate experimental and theoretical methods. (A first example from a theoretical perspective was given in Figure 3.12.) This makes development slower. On the other hand, the topic has enormous practical implications, e.g. for nano-structured storage techniques. This brings much motivation into the field which counterweights easily the complications. Figure 5.9 shows the magnetic response of Fe clusters as an example. The Fe clusters are produced in a special source yielding a high fluxes of clusters. To that end, a dense Fe vapor is produced by a cathode arc discharge. This is handled further with a He carrier gas, as usual. The Fe clusters are mass selected with a variance of less than 10% and then implanted into a Ag matrix on a Si(111) substrate. A magnetic field is applied to the sample and the emerging magnetization is measured by the electro-optical Kerr effect, i.e. from the rotation of the polarization through the magnetic moments. For large clusters, one expects a ferromagnetic behavior as in the bulk. Smaller clusters will develop super-paramagnetism, i.e. a continuous magnetization with all atoms oriented in the same direction. This happens if the energy contained in a change of magnetization direction (which is proportional to the number of atoms in the cluster) comes down to the order of the thermal energy $kT$. At room tempera-

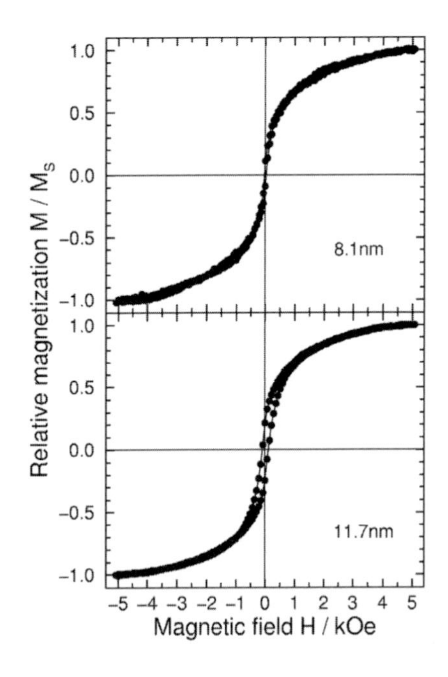


**Figure 5.9:** Hysteresis of Fe clusters at two different sizes as indicated by the the mean diameter given in the panels. The measurements were done at room temperature with samples of Fe clusters embedded in a silver matrix. From [MSK$^+$01].

ture, this point is expected for Fe clusters at a diameter of about 11 nm (= 210 $a_0$). Figure 5.9 shows results from two samples close to that diameter. The larger cluster still has some hysteresis while the smaller cluster has practically none. The transition to super-paramagnetism is thus nicely demonstrated [BSM$^+$00, MSK$^+$01].

There are, of course, numerous studies on this demanding and promising topic. A simplified theoretical description is possible with the extended Hubbard model [ADDP98, ODDP99]. Small magnetic clusters are accessible to fully fledged DFT calculations [KB99, OPC98] which are, however, one order of magnitude more expensive than those for spin-saturated clusters. For a review, mostly from a theoretical perspective, see [PB99]. More practical questions are addressed (amongst several other topics) in [Bin01].

An interesting combination of magnetism with questions of heat capacity (see Section 5.1.3) is discussed in [SHTM99]. It turns out that relaxation times are very slow for the spin degrees of freedom. Thus one can see a time dependence at the scale of seconds for the heat capacity of magnetized clusters [SHTM99]. This allows one to study the dependence of the heat capacity as a function of temperature and applied magnetic field.

## 5.2 Observables from linear response

### 5.2.1 Optical absorption

#### 5.2.1.1 The environment of embedded clusters

Optical response has been a key observable since the early days of cluster physics, see the extensive discussion in Section 4.3. It will maintain its key position also in future investigations

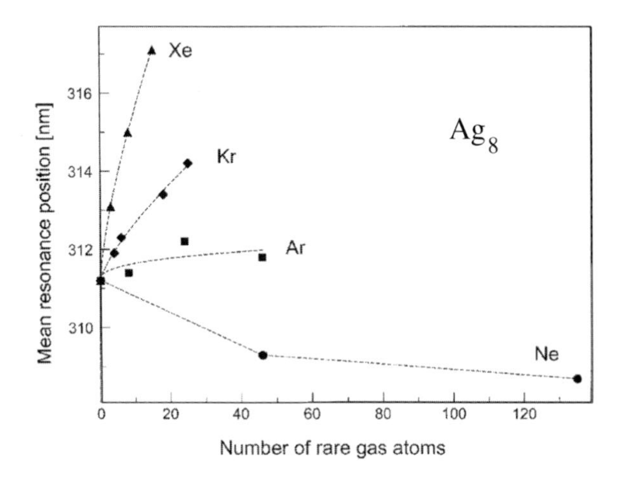


**Figure 5.10:** Mie resonance frequency for $Ag_8$ surrounded by layers of different rare gases and embedded finally in a liquid He drop. The frequency is drawn versus number of attached rare gas atoms. From [DTMB02].

because any step into new combinations of materials has first to consider the basic spectral properties. Moreover, there are still many details of known data which deserve deeper discussions. Examples are the spectral signatures of electron-vibration coupling visible in small metal clusters, see Figure 4.25 in Section 4.3.5.

Much space for new developments also exists for embedded or deposited clusters. The combination of different materials multiplies the range of possibilities. The interface of the cluster with the surrounding material and the polarization interaction considerably modify the dynamical response of the cluster. This was already discussed to some extent in Section 3.4.4.2. We continue this here by presenting an example from a recent study in which the rare gas surroundings have been systematically varied [DTMB02]. The result for the Mie plasmon wavelength of $Ag_8$ embedded in different layers of different rare gases is shown in Figure 5.10. There is a noticeable trend with the size of rare gas layers and the slopes thus strongly depend on the embedding material. Trends and slopes are produced by two counteracting effects: first, the valence electrons of the clusters are repelled by rare gas atoms, which compresses the cluster and thus leads to shorter wavelengths of the Mie plasmon, and second, the rare gas atoms build a strong polarization interaction with the cluster dipole which tends to produce larger wavelengths. As a rule of thumb, the core repulsion is about the same for all rare gases while the polarizability grows strongly with increasing system size, as can be read off from table 3.1 (the atomic radius is related to the range of repulsion). The core repulsion dominates in Ne thus leading to the trend towards shorter wavelengths in Figure 5.10. The polarizability takes over for the heavier atoms which reverses the trends of wavelengths, the more so the larger the rare gas atom.

Altogether, the results from Figure 5.10 and model for the interaction of a cluster with a rare gas environment as outlined in Section 3.4.4.2 exemplify the crucial role played by the polarization interaction. This was also visible in the extended Hubbard model discussed in Section 3.4.5.3. The dipole interaction with the environment can (or has to) be added as well to other descriptions as, e.g., the tight-binding approach [DS96]. But not only outside dipoles take part in the dynamics. There are several cluster materials where the dipole interaction with their own ionic cores is also important. Sodium is particularly inert in that respect. But the optical response of heavier alkalines (K, Cs, Rb) is somewhat influenced by the dipoles of the cores [YB98]. This effect is much stronger and unavoidable in noble metals, see Section 4.3.4.1. One can easily, and one should, extend the description by an electron-core dipole model in a fashion similar to the dynamical substrate coupling. This amounts then to the approach as exemplified for Ag clusters in [SR97].

#### 5.2.1.2 Arrays of clusters

As we have seen in Section 5.2.1.1, the polarization interaction with the environment is crucial for embedded or deposited clusters. It is even more important for another interesting compound: arrays of clusters. There are meanwhile several techniques to shape regular assemblies of clusters, e.g. by etching [WGC+98], by electrochemical processing [MKPE98], or with the help of organic templates [EZK98]. These are often discussed as promising future switching devices with bias on their conductivity [JBD+00, LLC+00]. But the optical properties of arrays of clusters are also very interesting. The strong polarization interaction between the Mie plasmons in each cluster allows one to shape their optical properties, e.g. to deliberate shifts, broadening or special transport properties, see e.g. [RP99, FAL+02]. We will discuss them here as a simple and instructive example for the polarization interaction. Each cluster is represented as a one dipole oscillator, similar to the atoms or core electrons in the previous section. A frequency and a dipole mass are assigned to the harmonic oscillator to reproduce the Mie plasmon frequency and associated dipole strength. The clusters are assumed to have no electronic contact and to communicate only by dipole-dipole interaction. The energy of the model system thus reads

$$
\begin{aligned}
E_{\rm array} &= \sum_i \frac{\mathcal{M}_i}{2}\dot{\mathbf{D}}_i^2 + \sum_i \frac{\mathcal{M}_i\omega_i^2}{2}\mathbf{D}_i^2 \\
&\quad + \frac{e^2}{2}\sum_{i\neq j} \frac{\hat{\mathbf{D}}_i\cdot\hat{\mathbf{D}}_j - 3\mathbf{e}_{ij}\cdot\hat{\mathbf{D}}_i\mathbf{e}_{ij}\cdot\hat{\mathbf{D}}_j}{|\mathbf{R}_i-\mathbf{R}_j|^3} \quad , \qquad (5.3a)\\
\mathbf{e}_{ij} &= \frac{\mathbf{R}_i-\mathbf{R}_j}{|\mathbf{R}_i-\mathbf{R}_j|} \quad , \qquad (5.3b)\\
\mathcal{M}_i &= \frac{1}{\omega_i D_{{\rm ex},i}^2} \quad , \qquad (5.3c)
\end{aligned}
$$

where $D_{{\rm ex},i}^2$ is the dipole strength of the Mie plasmon $\omega_i$ in cluster $i$ and $\mathbf{R}_i$ denotes its (fixed) spatial position. Spring constant $\mathcal{C}_i = \mathcal{M}_i\omega_i^2$ and mass $\mathcal{M}_i$ are adjusted to reproduce the given frequency $\omega_i$ and dipole strength, $D_{{\rm ex},i}^2 = \mathcal{M}_i\omega_i = \sqrt{\mathcal{C}_i\mathcal{M}_i}$ of oscillator $i$. The third term in Eq. (5.3a) accounts for the dipole interaction between the clusters. It will lead to coupled oscillations combining coherently dipole deformation from several clusters and shifting the frequencies $\omega_i$ of the isolated clusters.

The model (5.3) is still very general and can be applied to many different situations, besides arrays, also to irregular ensembles of metal clusters of different sizes in a He droplet or in glass. For regular arrays, we can specify it one step further. All clusters are the same and thus $\omega_i = \omega_0 =$ constant and $\mathcal{M}_i = \mathcal{M}_0 =$ constant. The coupling is then better reformulated in terms of a dimensionless strength $\eta$. To that end, we express all lengths in terms of

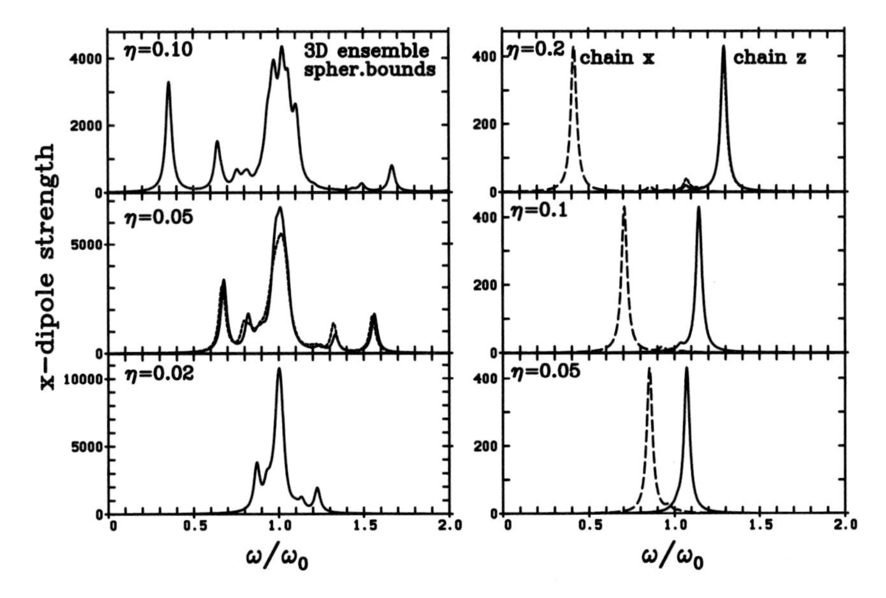


**Figure 5.11:** Dipole response for arrays of coupled dipoles for various strengths of dimensional coupling $\eta$. Left panels: Equidistant three-dimensional array within spherical bounds; the fine dotted line in the middle panel stems from a variant where all grid positions have been moved randomly within about 30% of the grid spacing. Right panels: One-dimensional chains along direction of the dipole ($x$-direction, dashed lines) and orthogonal to it ($z$-direction, full lines).

the Wigner-Seitz radius $r_s^{(\mathrm{array})}$ of the arrangement. Energies are expressed in terms of the frequency $\omega_0$ of one isolated cluster. Similarly, the dipole moments are rewritten in units of the dipole strength $\tilde{\mathbf{D}} = \mathbf{D}/D_{\mathrm{ex}}$ of one cluster. This yields the renormalized expression

$$\tilde{E}_{\mathrm{array}} = \frac{1}{2}\sum_i \dot{\tilde{\mathbf{D}}}_\mathrm{i}^2 + \left(\omega_0^2 \tilde{\mathbf{D}}_\mathrm{i}^2\right) + \eta\frac{\omega_0^2}{2}\sum_{\mathrm{i}\neq\mathrm{j}} \frac{\tilde{\mathbf{D}}_\mathrm{i}\cdot\tilde{\mathbf{D}}_\mathrm{j} - 3\mathbf{e}_\mathrm{ij}\cdot\tilde{\mathbf{D}}_\mathrm{i}\mathbf{e}_\mathrm{ij}\cdot\tilde{\mathbf{D}}_\mathrm{j}}{\left(|\mathbf{R}_\mathrm{i}-\mathbf{R}_\mathrm{j}|/\mathbf{r}_\mathrm{s}^{(\mathrm{array})}\right)^3} \quad , \tag{5.4a}$$

$$\eta = \frac{e^2 D_{\mathrm{ex}}^2}{\omega(r_s^{(\mathrm{array})})^3} = \frac{\alpha_D(\omega=0)}{(r_s^{(\mathrm{array})})^3} \quad . \tag{5.4b}$$

This is a classical system of coupled harmonic oscillators. One obtains the new frequencies $\omega_N$ and dipole strengths in a standard fashion: from diagonalizing the curvature matrix $\partial_{D_{i\nu}}\partial_{D_{j\mu}}\tilde{E}_{\mathrm{array}}$ (where $\nu,\mu \in \{x,y,z\}$). Figure 5.11 shows results for two limiting cases, a one-dimensional chain and a three-dimensional array. Finite probes have been considered, the chains with a length of 10 clusters and the 3D ensemble contains clusters within a sphere of 7.5 $r_s^{(\mathrm{array})}$. The three-dimensional array is rather symmetric and shows the same dipole spectrum in all three basic directions. Chains are different. It matters in which direction one explores the dynamical polarization. The interaction in Eqs. (5.3) or (5.4) tries to align the dipoles along the chain. Thus it is attractive for polarization in the direction of the chain and repulsive orthogonal to it. Accordingly, the frequency of the $x$-mode in Figure 5.11 is lowered as compared to $\omega_0$ while the orthogonal $z$-mode is blue-shifted. The shift grows monotonically with the coupling strength. One can observe in the right uppermost panel a faint dotted line almost identical with the full line for the $z$-mode. This represents a variant of the model where the clusters have been displaced stochastically by about 30% in the average from their equidistant positions. This serves to simulate slight disorder as could be induced, e.g., by thermal motion. The effect is obviously very robust against this disorder. It is noteworthy that the chains produce a clean shift of the resonance position with very little broadening or fragmentation. The situation is much different for a three-dimensional array, see left panels of Figure 5.11. Whatever way we excite the system, there is always a fraction of aligned dipoles and another fraction of orthogonal ones. This dichotomy leads to a substantial fragmentation of the spectra. The leading lowest-frequency and highest-frequency peaks show trends much similar to the down-shifted and up-shifted modes of the chain. They represent probably the extremes of aligned and anti-aligned coupling. There are several more modes in between, representing so to say this or that sort of compromises. The spectral spreading increases with coupling strength (note that the net coupling also depends on the dimensionality such that the $\eta$ are not directly comparable between chain and three-dimensional array). The fine dotted line in the middle left panel is again result from a stochastic displacement by 30% in the average. And once more, we see that the spectra are robust against slight disorder.

These results for limiting cases allow one to estimate the effect for other scenarios. For example, a two-dimensional array will show a clean and strongly blue-shifted mode in a direction orthogonal to the array-plane while the modes in-plane will be fragmented around the base frequency. Another typical example is the stochastic mix of metal atoms and small metal clusters of different sizes which assemble during the collection phase in a He droplet or a rare gas matrix. One expects from the above results a strong broadening with even less structure than seen in Figure 5.11. This has the interesting consequence that laser excitation of such an ensemble will always find a resonance frequency to couple to, even if the laser is far away from the typical Mie plasmon frequency of the given metal. A further example is a metal cluster embedded in a glass matrix. One observes a substantial broadening of the Mie plasmon after irradiation with a strong laser pulse [SKBG00]. What happens is that the strong heating of the cluster due to irradiation leads to evaporation of metal particles into the vicinity of the cluster. They gather there to many small metal clusters. Applying the above model with one strong dipole mode (large center cluster) and many small dipoles at stochastic distances leads indeed to a diffusion of the originally sharp Mie plasmon peak. There are many more conceivable scenarios which could be understood in terms of this simple model of coupled dipoles. This allows one to understand the basic features of optical response in term of shifts, fragmentation and broadening. This as such gives rise to interesting investigations and it serves as doorway for the non-linear processes from which switching devices may be built.

#### 5.2.1.3 Tailoring colors with semiconductor clusters

A certain size dependence of the optical response is well known from metal clusters, see Figure 1.17. It is due to surface effects on the Mie plasmon (electron spill out, core polarization). Pronounced size effects are also seen in the optical response of semiconductor clusters [All96]. We consider here the example of embedded $Ag_2S$ clusters as discussed in [BLGC02]. The environment is provided by Ca zeolite A micro-crystals. The clusters are grown and kept in small cavities of the zeolite crystal. This establishes, in fact, an array of embedded $(Ag_2S)_n$

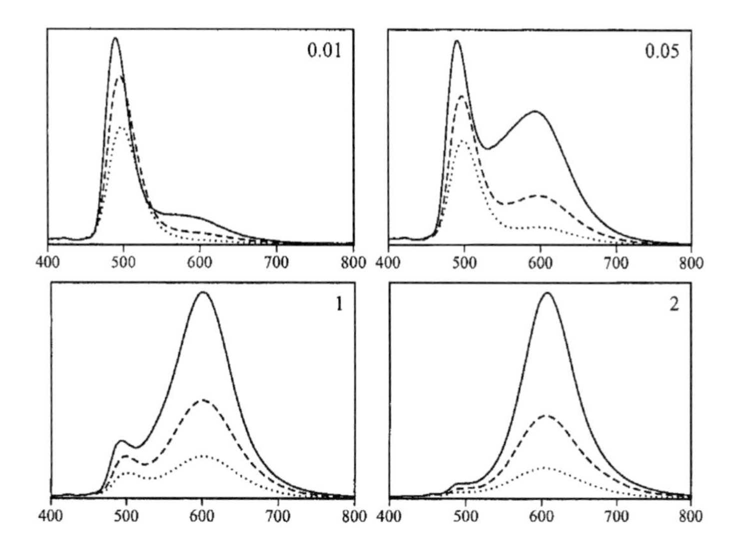


**Figure 5.12:** Photoluminescence spectra of $Ag_2S$ clusters in the cavities of Ca zeolite A for various densities and sizes of the clusters. The spectra are drawn versus wavelength in nm. The number in the right upper corner of the panels gives the average number of $Ag^+$ ions per cavity which determines the size of the $Ag_2S$ clusters and their density over the zeolite cages. Spectra are taken at three temperatures: full lines = 78 K, dashed lines = 173 K, and dotted lines = 223 K. After [BLGC02].

clusters. The average size of the clusters is determined by the supply of AgS during growth. The largest cluster which fits into one cage is $(Ag_2S)_4$. Figure 5.12 shows the luminescence spectra of the embedded $Ag_2S$ clusters for various cluster sizes. The left upper panel shows the smallest Ag content where probably $Ag_2S$ molecules (monomers at the level of clusters) prevail. The luminescence is dominated by a blue-green line corresponding to an electronic dipole transition in the molecule. With increasing cluster size, this molecular line is slowly superseded by a broad line in the range of orange-red light which finally takes over. The orange-red luminescence is most probably caused by $Ag_8S_4$ clusters, the largest sample which fits approximately into the cavities. A fully microscopic description of these trends has not yet been achieved because the $(Ag_2S)_n$ clusters possess an enormous number of isomers, not to mention the complications added by including the interface interaction [BLGC02]. One also needs to account for the mode coupling in the array of clusters, see Section 5.2.1.2, which is probably responsible for the broadening of the observed cluster emission line. The interesting aspect for practical applications is that the average cluster size, and with it the color, can be easily tuned by the chemical conditions during cluster growth.

#### 5.2.1.4 Luminescence in biological environments

The combination of clusters with biological material is a very promising line of development in view of the great interest on bio-molecules in general. We had already one example in Section 5.2.1.2 in connection with the fabrication of oriented arrays of metal clusters [MKPE98]. The proteins could serve here as convenient templates. In the other direction, clusters can

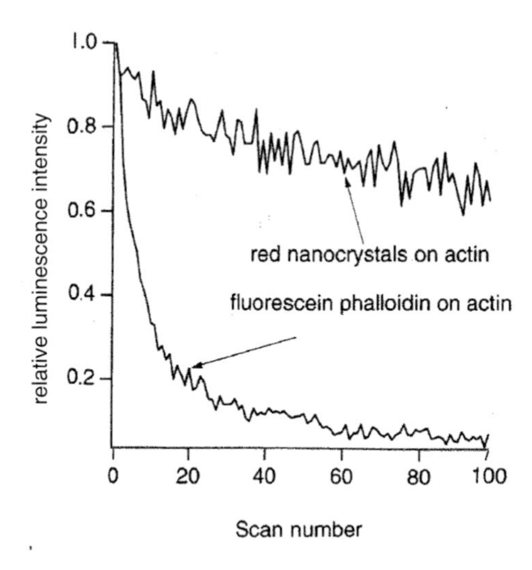


**Figure 5.13:** Sequential-scan stability of the photoluminescence signal of fluorescent markers in a biological fiber (actin). Compared are fluorescein phalloidin as a standard marker with a semiconductor cluster. Fluorescein was excited at 488 nm and the cluster at 353 nm. After [JMG$^+$98].

serve as powerful probes for biochemical status or reactions. For example, the construction of a biochemical sensor-chip from a surface composed of proteins and metal clusters was discussed in [MPBS01]. The chemical status of the environment can be recorded optically from the change of resonance spectra. Another example of the possible usefulness of clusters for bio-systems is discussed in [JMG$^+$98]. It was found that clusters in biological fibers are much more stable against photochemical destruction than ordinary fluorescent molecules. This is demonstrated in Figure 5.13. The stability of a standard fluorescence marker, fluorescein phalloidin, is compared with that of a semiconductor cluster. The stability of the fluorescence signal with repeated scans is obviously much larger for the cluster. This suggests a possible use of clusters as biological markers [JMG$^+$98]. A more recent and further reaching application of such markers has been reported in [DSN$^+$02].

## 5.2.2 Beyond dipole modes

### 5.2.2.1 Other multipolarities

The optical response of metal clusters is overly dominated by the dipole plasmon as we have seen at several places in this book. But the cluster as such carries a rich variety of other modes. For example, other multipoles such as monopole or quadrupole and various magnetic modes, or ionic vibrational modes. The theoretical description of the electronic modes proceeds in a way very similar to the dipole case, see Chapter 3 and the examples of applications in Chapter 4. The next step beyond the dipole is to look at the other modes with natural parity, for a discussion of them see e.g. [RGB96b]. There is, in principle, a surface plasmon mode for each multipolarity $L$ with frequency $\omega_L = \omega_{\rm pl}/\sqrt{2L{+}1} = \omega_{\rm Mie}\sqrt{3/(2L{+}1)}$ where $\omega_{\rm pl}$ is the volume plasmon frequency. However, the phase space of interfering $1ph$ states grows with $L$ and frequency such that these modes are strongly Landau fragmented. The monopole mode ($L$=0) resides in the frequency range of the volume plasmon. It is also much fragmented

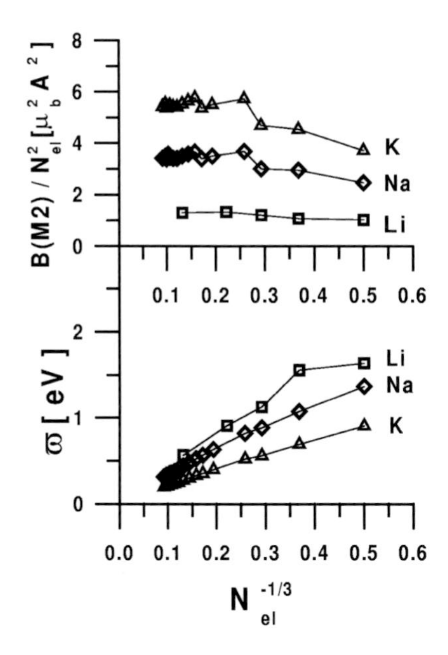


**Figure 5.14:** Properties of the twist mode in spherical and positively charged alkaline clusters drawn versus system size in the form $N_{\rm el}^{-1/3}$. Lower: Main peak position. Upper: Transition strength. From [MNdSC+00].

for finite clusters. The larger spectral width and the dominance of the dipole mode hinder an experimental exploration of these modes, see the discussion in Section 5.2.2.3.

#### 5.2.2.2 Magnetic modes

New and interesting features arise when proceeding to modes with unnatural parity. These are related to magnetic transitions for which two different sources exist: spin flow or orbital flow. The dominant orbital magnetic mode is related to M1 transitions. It is often called the scissors mode because its strongest part can be viewed as angular oscillations of the electron cloud against the (deformed) ionic background. The scissors mode is excited and analyzed in terms of the orbital angular momentum $\hat{\mathrm{l}}$ which plays here the role of the dipole operator for the Mie plasmon. Scissors modes are a general feature of deformed Fermion systems. They have also been observed, for example, in atomic nuclei [BRS+84, Iud97]. They have been discussed for metal clusters in early studies using collective excitation models [LS89, LS91] and later on in microscopic models [NKdSCI99, RNS+02]. They appear at low energies in the region of the lowest $1ph$ states. Their energy drops with system size as $N^{-1/3}$ while the strength increases.

Next come the M2 modes. A collective picture of M2 excitations is the twist mode related to helical flow pattern [MNdSC+00] and similar to the quadrupole torsional deformation mode of a globe [Lam82]. In contrast to the scissors mode, the twist is fully present also in spherical clusters. There it is the leading orbital magnetic mode. Similar to the scissors, the M2 strength is dominated by low-lying $1ph$ states which are only very little shifted by the residual interaction. The largest M2 strength stems from the $1ph$ excitation connecting single particle states with the highest multipolarity. These turn out to be amongst the lowest excitation energies as well. One can thus characterize the twist mode very well by a peak energy and an associated strength. The trend of both these quantities with system size is shown in Figure 5.14. The peak excitation energy drops as $N^{-1/3}$, as is typical for $1ph$ energies. The strength grows as $N^2$. Note that we use a scaled strength to cope with this dramatic trend. The conditions for observation are thus best for very large clusters, but not too large to discriminate the mode from vibrational excitations. A promising compromise lies around $N \approx 10^4$.

Up to here we have discussed orbital magnetism. There is another branch of magnetic modes caused by spin transitions. These are, of course, coupled to orbital magnetism in the case of strong spin-orbit force as is typical for magnetic materials. Spin modes are independent from orbital magnetic modes if the spin-orbit force is negligible as is the case, e.g., for simple metals. One can distinguish here a spin-dipole mode which is excited by a dipole shift of the electronic spin-up and spin-down partition in opposite directions. The spin-dipole modes in metal clusters have been investigated in [SBBN93, SL97, MCRS96, KEGC$^+$99, RNS$^+$02]. Particularly interesting samples are small clusters with triaxial ground state, as e.g. $Na_{12}$ or $Na_{16}$ (see Section 4.1.3). These have often low-lying isomers with net spin of two [KMR95, KFR97]. Such naturally spin-polarized systems exhibit cross talk between dipole and spin-dipole modes which, in turn, offers a chance for detection of low-lying spin-dipole modes. Spin-dipole and scissors modes are dominated by low-lying $1ph$ transitions with very little shift through residual interaction, leaving the energy $\propto N^{-1/3}$. Thus they reside in the same energy range which makes a unique identification even more involved. There are, however, different trends of the excitation strength with system size. The strength of orbital modes grows with system size while spin modes do not. They become relatively unimportant for large clusters which are then the ideal laboratory for studying the scissors mode. The spin-dipole should be looked at in small clusters and may be easiest seen in systems with a spin polarized ground state.

#### 5.2.2.3 Experimental access

Thus far the sorting of non-dipole modes is simple from a theoretical perspective. The problem resides at the experimental side. Mere photon absorption couples almost exclusively to the dipole transitions. More elaborate methods are required, as e.g. Raman spectroscopy or inelastic electron scattering. These more subtle methods suffer from weak signals. This disadvantage can be counterweighted by performing the experiments with deposited or embedded clusters to provide a higher density of scatterers and thus bearable counting rates. Still, there has not yet been a clear assessment of any one of the above discussed electronic modes beyond the dipole. Progress has been made already for measuring vibrational modes of clusters, though. Figure 5.15 shows an example where Raman spectroscopy gives access to the vibrational monopole (=breathing mode). The sidebands caused by the coupling of the electronic dipole to the monopole are clearly visible (note that 1 cm$^{-1}$ = 1/8 meV) as well as, on an enlarged scale, the quadrupole vibrations. It is hoped that the technique of Raman spectroscopy can be further refined to give access also to the non-dipole electronic modes of clusters. This requires sources with high photon frequencies (UV up to X-ray range) which will become more readily available in future. Careful analysis of angular distributions and polarization should allow one to disentangle modes with natural parity (monopole, quadrupole) from magnetic ones. Inelastic electron scattering has also a high potential for detailed analysis because

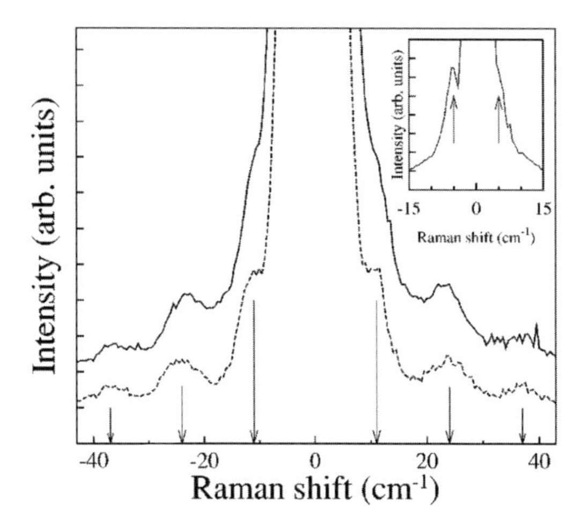


**Figure 5.15:** Low-frequency Raman spectra of Ag clusters in glass. Two photon frequencies were used: full line = 458 nm, dotted line = 515 nm. The arrows indicate the positions of the sidebands which can be associated with breathing (monopole) vibrations. The inset shows sidebands associated with quadrupole vibrations. From [PSD+01].

one has several parameters to control the experiment (energy, energy loss, angular distribution, optionally polarization). In fact, it was electron scattering which enabled the detection of scissors modes in nuclei [BRS+84]. However, the practical problems in the realm of cluster physics are enormous because low electron energies are required which are much harder to handle experimentally, see Section 5.4.4. An option may also be electron energy-loss spectroscopy (EELS) [IM82]. It is widely used in solid state physics and allows, e.g, to assess the volume plasmon. It could allow one to see the monopole modes in finite clusters.

### 5.2.3 Photoelectron spectroscopy

As mentioned in the introduction to this chapter, photoelectron spectroscopy (PES) is one of the basic tools to eludicate the electronic structure of clusters. The underlying mechanism is very simple: one photon of a well prepared frequency $\omega$ removes one electron from a bound state with single electron energy $\varepsilon_{\rm bound}$ and lifts it to the continuum. The final kinetic energy $\varepsilon_{\rm kin}$ allows one to conclude on the energy of the initially bound state mediated by the simple relation Eq. (2.1) specialized to a one photon process, i.e.

$$\varepsilon_{\rm kin} = \hbar\omega_{\rm las} + \varepsilon_{\rm bound} \quad .$$

Although conceptually very simple, PES requires well controllable sources with high frequencies. And these are rather recent (if not future) achievements.

UV light is well suited to explore the valence electrons in clusters, see Figure 5.1 and the discussion around it [WavI02]. The various core states can be resolved with X-rays. These are a standard analyzing tool in other systems such as solids, molecules or atoms. And there exist, of course, also several investigations for clusters, particularly in connection with the brilliant synchrotron radiation, see e.g. [YSR+02, HKR+02]. Figure 5.16 shows an example taken with classical X-ray sources [DKSM01]. A sample of Au clusters deposited on a $SiO_2$ film has been produced by depositing on glass, coating with $SiO_2$ and subsequent thermal annealing. The sample has been analyzed by spectroscopy and microscopy. It consists of nearly spherical Au clusters. The cluster size and density depends on the amount of Au supplied initially. The

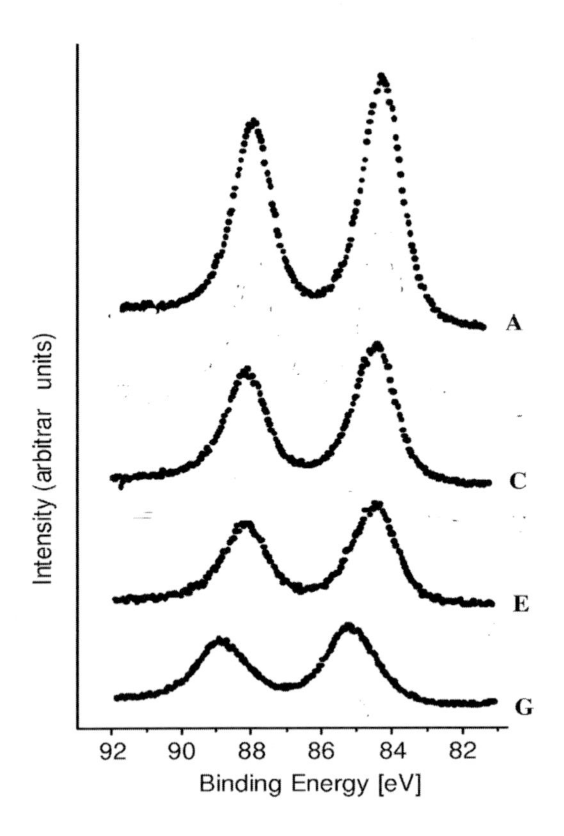


**Figure 5.16:** Photoelectron spectra of Au clusters on $SiO_2$ taken with X-rays at a frequency of 1.25 keV. The spectra are shown in the window of the $4f$ core levels of Au. Results for different concentrations of Au (and with it different clusters sizes) are shown in decreasing order (A labels the highest concentration). After [DKSM01].

figure shows the two levels $4f_{7/2}$ and $4f_{5/2}$ almost unperturbed from which one concludes that nearly clean Au is present with negligible oxidation. There are clear trends with system size. The core binding and the width of the levels increase with decreasing Au content (A to G in Figure 5.16). This results from various effects. Charging during X-ray impact enhances the core binding energy with a trend $\propto R^{-1} \propto N^{-1/3}$ thus yielding more binding for smaller clusters (less Au content). The trend in Figure 5.16 is stronger than that and further mechanisms acting on the surface of the cluster may play a role. The trend of the width hints also at surface effects. Last but not least, there is a polarization interaction of the many Au cores in a cluster which contributes to the shift (see the discussion of arrays of dipoles in Section 5.2.1.2). Note that this interpretation was only possible with the help of other analyzing tools (spectroscopy, microscopy). This is a typical procedure when disentangling effects in such complex structures.

PES is a still developing tool. The very versatile synchrotron radiation and free-electron lasers are being steadily improved. Besides the already good frequency definition, time-resolved measurements are coming up with the development of short X-ray pulses (for technical progress in free-electron lasers see e.g. [And$^+$00] and for a first cluster application [Wab$^+$02]). This will allow one, e.g., to record the effect of ionic motion on core levels or a mapping of time-dependent single-electron energies in the valence shell. PES is then combined with the typical scenario of pump-and-probe analysis, see Section 5.3.4.

## 5.3 Laser excitations in the semi-linear regime

The notion "semi-linear" concerns the regime of moderate laser intensities. The cluster, although much shaken in some cases, remains basically intact, or at least metastable, say as one single piece of matter for some time. The laser frequency thus still plays a decisive role in this regime. At the lower energy end, one may still sort the processes in terms of numbers of photons. At the upper end, one merges into the "field dominated" processes where intensity gains importance while frequency looses. This semi-linear regime covers a large variety of interesting dynamical scenarios, as e.g. second-harmonic generation or pump-and-probe analysis. We will discuss in this section several selected examples. Most of them deal with rather recent results and leave an open end which demonstrates that the field is much in a developing stage.

With a brief look back at Figure 3.3 in Section 3.2.1, we associate the semi-linear regime with laser intensities typically in the range between $10^8\ \mathrm{W/cm^2}$ and $10^{14}\ \mathrm{W/cm^2}$, depending on the material. However, intensity alone is not the only criterion. The frequency also plays a key role in determining the reaction path. The field strength is fully pulling on the cluster electrons in the low-frequency limit while the effective field strength looks smaller for high frequencies because the field oscillations tend to cancel. This aspect is quantified by the Keldysh adiabaticity parameter [Kel64]

$$\gamma = \sqrt{\frac{2E_{\mathrm{IP}}\omega_{\mathrm{las}}^2}{I}} \qquad (5.5)$$

where $E_{\mathrm{IP}}$ is the ionization potential of the system, $\omega_{\mathrm{las}}$ the frequency and $I$ the intensity of the laser, all expressed in natural units. The case of $\gamma \ll 1$ is considered as the case of strong fields where direct ionization through tunneling prevails. The case $\gamma \gg 1$ signals the regime of weak perturbations where multi-photon ionization dominates. The trends are understandable. High frequency means strongly oscillating fields which average out for slow processes, as e.g. tunneling. On the other hand, high frequency means high photon energy which favors photon dominated processes. The estimate of perturbation strength by the Keldysh parameter is still rather global because it does not account for the actual response of the system. Laser frequencies near strong eigenmodes of the system induce resonant response with subsequent strong field amplification [RS98]. Here we have to expect plasmon enhanced multi-photon emission. The same consideration applies, of course, when taking the laser intensity as measure of perturbation strength. Field amplification can lower the limits near resonant frequencies.

### 5.3.1 Electron emission

In Section 5.2.3, we have discussed at length photoelectron spectroscopy (PES) from ionization through one-photon processes. This requires photon frequencies above the ionization threshold. Nonetheless, electron emission is also found for frequencies below threshold due to multi-photon processes. This is called above threshold ionization (ATI). It has been much studied in the context of atomic physics [PKK97, Lam$^+$98, JDK00]. It is, of course, also a very interesting topic for cluster physics. The situation is somewhat different because clusters are much more complex. This often yields a more diffuse picture. But it also gives access to new phenomena.

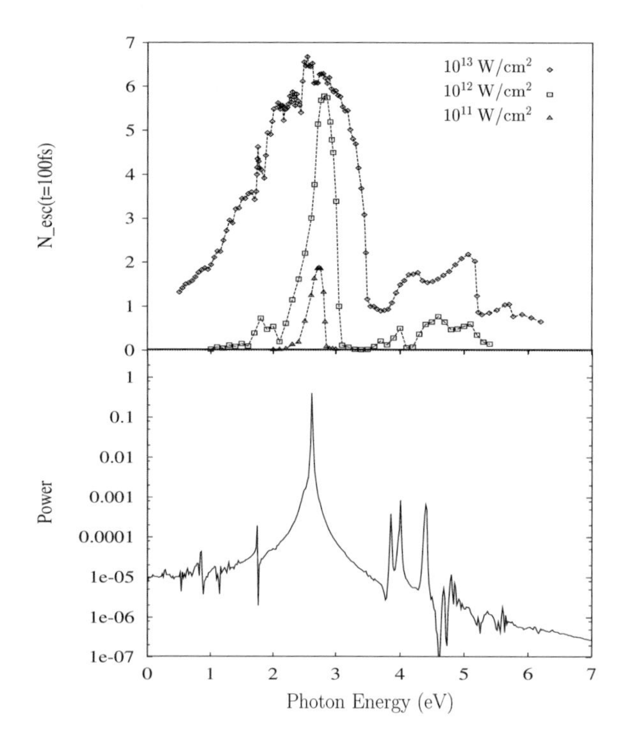


**Figure 5.17:** Results from TDLDA for electron emission and dipole response in $Na_9^+$ described with a spherical soft jellium background (see Section 3.2.3.2). Upper panel: Net ionization $N_{esc}$ versus frequency of the laser pulse, for Gaussian pulses with FWHM=25 fs and different intensities as indicated. Lower panel: Spectral distribution of dipole power obtained from spectral analysis (see Section 4.1.7). From [URS97].

#### 5.3.1.1 Trend with frequency

The response of a cluster depends very much on the excitation frequency, see e.g. the extensive discussions in Section 4.3. Strong response is associated with large energy absorption from the laser pulse (see e.g. the discussion in Section 1.3.3.2). Energy is needed to emit electrons. And thus one expects that direct electron emission depends on excitation frequency in a manner much similar to optical absorption. This is exemplified in Figure 5.17 for $Na_9^+$ with a soft jellium ionic background. The upper panel shows the photoelectron yield versus laser frequency for a variety of intensities. This is to be compared with the dipole spectrum in the lower panel. And indeed, electron emission takes place wherever optical absorption shows peaks, first of all in the vicinity of the strong Mie plasmon peak, but also near the secondary peaks at higher frequencies. Low intensities produce narrow emission peaks and a photoelectron distribution much similar to the optical absorption. The distribution broadens for higher intensities. This is an understandable consequence of the enormous perturbation of the system which can be read off from the huge number of electrons emitted from the cluster. More than 80% of the electrons are removed at the maximum. It is to be noted that the mere TDLDA description becomes incomplete at the highest excitation energies. The strong perturbation does not only lead to strong electron emission but also to a substantial internal heating of the remaining electron cloud. Thermalization through electron-electron collisions becomes an important competitor to direct electron emission [GRS03]. Notwithstanding this precaution, the above TDLDA results still serve well to show the basic trends.

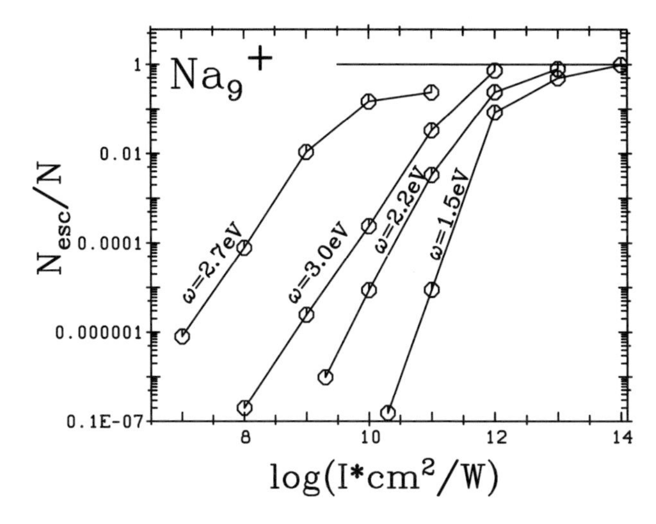


**Figure 5.18:** Laser intensity dependence of the net ionization $N_{\mathrm{esc}}$ relative to total electron number $N$ after laser irradiation with a laser pulse having $\cos^2$ profile with FWHM = 50 fs and varying frequency as indicated. Test case is $\mathrm{Na_9}^+$ described within TDLDA. The upper limit $N_{\mathrm{esc}}/N = 1$ is indicated by a horizontal line. Results are drawn on doubly logarithmic scales.

The above example shows that dissipation broadens the response peak. In that case (one-body) dissipation from electron emission was the leading mechanism. Electron-electron collisions becomes an important dissipation channel for highly excited clusters (see Figure 1.15). The semi-classical VUU scheme, as outlined in Section 3.3.5.3, allows a detailed computation of that effect. An example was shown in Figure 3.14. As a consequence of electron electron collisions, the emission peak (at plasmon frequency) in the ionization yields versus frequency acquires some additional broadening where peak ionization is reduced by about 25 % and much enhanced (by a factor of 3 and more) in the tails far off resonance.

#### 5.3.1.2 Trend with intensity

Multi-photon ionization by definition requires non-linear processes. This has immediate consequences for the trend of ionization probability with laser intensity $I$. The average number of emitted electrons can be estimated as

$$N_{\mathrm{esc}} \propto I^{\nu} \quad , \quad \nu = \mathrm{int}(E_{\mathrm{IP}}/\hbar\omega_{\mathrm{las}}) \tag{5.6}$$

where "int" means the integer part. The trend can be deduced from simple considerations. The minimum number of photons required for ionization is $\nu$. The amplitude for emission is then to be computed in $\nu$-th order perturbation theory and thus depends on the field amplitude as $E_{0,\mathrm{las}}^{\nu}$ (for an extensive discussion of multi-photon perturbation theory see [Fai87]). The total ionization probability is then $\propto E_{0,\mathrm{las}}^{(2\nu)} \propto I^{\nu}$. These trends are exemplified in Figure 5.18 showing ionization versus intensity for various laser frequencies in the $\mathrm{Na}_9^+$ metal cluster. The ground state of $\mathrm{Na}_9^+$ was determined by LDA for the electrons and simulated annealing of the ionic configuration (see Chapter 3). The laser excitation was described by TDLDA with frozen ionic positions. The LDA yields an ionization threshold of 5.9 eV which is a bit too low (see Appendix E.3). But it suffices for the present demonstration of the $I^{\nu}$ law (5.6). The relation between intensity and field strength is given in Appendix A.1. In practice, an intensity of $I = 10^{14}\,\mathrm{W/cm^2}$ corresponds to a field strength of $E_{0,\mathrm{las}} = 0.107\,\mathrm{Ry/a_0}$. The figure is drawn on doubly logarithmic scales. A trend like $I^\nu$ is distinguished by a straight line in that frame and the slope is directly proportional to the minimum photon number $\nu$. We see indeed straight lines in the perturbative regime of small emission. The slopes for the various frequencies behave exactly as predicted in this regime. The highest frequency of 3 eV needs two photons to come over the threshold of 5.9 eV and the straight section in the yield has indeed a slope of 2. The lowest frequency of 1.5 eV needs 4 photons and we find a slope of 4. The curves deviate from the straight trends for larger field strengths. We merge here into a non-perturbative and field dominated regime where photon counting is obsolete. The "critical" intensity (bending from a straight line towards the asymptotics) strongly depends on the frequency due to resonant field amplification. But the turning point lies always at about the same electron yield. The critical point is located approximately at an ionization of $N_{\rm esc} \sim 1$ which is an obvious limit where serious destruction of the system sets in. Experimental investigations of these trends for clusters are rare. One example is [SKvIH98].

#### 5.3.1.3 Electron spectra from multi-photon ionization

The kinetic-energy spectra of electrons emitted in one-photon processes map the level scheme of bound one-electron states, as has been discussed in Section 5.2.3. Multi-photon ionization (ATI) will also map the level scheme, but now with the full relation (2.1) i.e.

$$\varepsilon_{\rm kin} = n_{\rm phot}\hbar\omega_{\rm las} + \varepsilon_\alpha$$

where $\alpha$ is the bound single-particle state from which the electron is removed and $n_{\rm phot}$ is the number of photons actually involved. The minimum number $\nu$ of photons as given by Eq. (5.6) will dominate the spectrum. But higher order photon processes are also conceivable with decreasing weight. The expected picture is thus a mapping of the level scheme repeated in bins of $\hbar\omega_{\rm las}$ with strong decrease for increasing $\varepsilon_{\rm kin}$. This can indeed nicely be seen if the single-particle energies are well separated and their span remains below the photonic energy step $\hbar\omega_{\rm las}$, a situation which requires small clusters and high photon frequency [PRS00]. The high level density for larger clusters leaves mainly one broad bump reminding one of a valence band (for an example with huge clusters see [MPT$^+$00]). But the repeated sequences corresponding to increasing photon order will persist.

Figure 5.19 shows ATI spectra in $C_{60}$ for various laser conditions [CHH$^+$00]. The uppermost panel for the shortest pulse nicely confirms the expectation. The spectral resolution of a 25 fs pulse is, of course, low, such that one cannot resolve the single-electron states. But the repetition of the "valence band" with higher order photon processes is well visible. Quite as expected one also observes a dramatic decrease with photon order (or equivalently, increasing electron kinetic energy). The IP of $C_{60}$ is about 7.6 eV [LNR$^+$91] such that at least five photons are required for an electron emission with the laser used here ($\hbar\omega_{\rm las}$ = 1.6 eV). The pattern becomes more diffuse for longer pulses. This is explained as thermal emission. The idea is that the energy absorbed from the laser can easily be distracted to internal heating of the cluster. Thermal excitation energies of the order of 20–400 eV are estimated under these experimental conditions [CHH$^+$00]. The absorbed energy is released in part by thermal electron emission which, of course, produces smooth and exponentially decreasing kinetic-energy spectra. The estimated thermal contribution is indicated by the dashed lines in Figure 5.19.

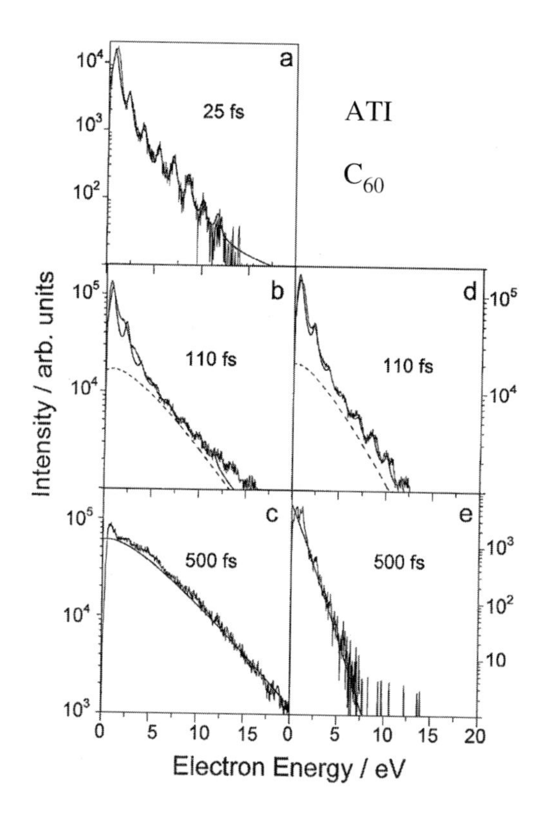


**Figure 5.19:** Photoelectron spectra from $C_{60}$ irradiated by a laser with wavelength of 790 nm ($\equiv 1.6$ eV) and various pulse lengths as indicated. The laser intensity was $I = 8 \times 10^{13}\,\mathrm{W/cm^2}$ for all left panels and reduced to $I = 3{\times}10^{13}\,\mathrm{W/cm^2}$ for panel d and $I = 8{\times}10^{12}\,\mathrm{W/cm^2}$ for panel e. Curves (full lines) are fitted to data with sum of Lorentzian peaks over thermal background. Dashed lines represent estimated thermal contributions. After [CHH⁺00].

A mix of ATI spectra and thermal emission is seen at intermediate time and almost complete thermal emission for the longest pulses. The left panels were produced for constant intensity and varied pulse length. This increases the fluence ($\propto$ intensity $\times$ pulse length) with increasing pulse length. The transition to thermalization could then be induced either by pulse length or just by brute force, i.e. more energy absorption. As countercheck, the right panels represent a series with equally varied pulse length but constant fluence. The transition to thermal pattern looks precisely the same as in the left panels. That allows one to conclude that predominantly the pulse length is decisive for the branching between direct ATI and thermal emission. This is plausible because one expects a competition between relaxation times, namely the time scale for direct emission (typically 2–10 fs) and the electron-electron collision time, see Figure 1.15 in Section 1.3.2. A similar interpretation in terms of ATI versus thermal emission was also given to the experiments of [SKvIH01].

The smoothing by thermal emission is the correct interpretation of the example in Figure 5.19 where large intensities, large photon numbers and large amounts of stored energy are involved. We ought to remember that there is a competing mechanism producing smooth spectra while remaining in the regime of ATI. This has been discussed in [PRS00]. Assume pulses which stay sufficiently short that thermalization is suppressed. Now consider increasing intensity. The ionization will increase. This, in turn, enhances the ionization threshold as $N_{\mathrm{esc}}e^2/R$ where $R$ is the cluster radius. The upshift follows almost immediately the time evo-

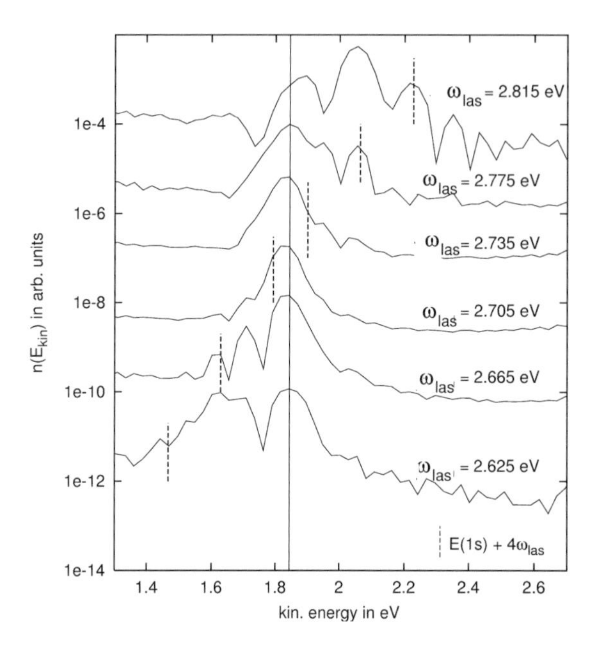


**Figure 5.20:** PES for $Na_9{}^+$ and different laser frequencies, as indicated. The energy window concentrates on four photons excitation out of the $1s$ state. The short dotted vertical lines indicate the energies of the direct peaks for a four-photon process, the long vertical line indicates the energy $E_{GS} + 4\hbar\omega_{\rm plas}$ of a four-plasmon process which is independent of laser frequency. The plasmon frequency is $\hbar\omega_{\rm plas} = 2.7\,{\rm eV}$ in $Na_9^+$. After [PRS01].

lution $N_{\rm esc}(t)$ of ionization because it is an electronic process. The sliding threshold smears the ATI spectra and produces eventually a completely smooth pattern. The deceiving feature is that such spectra also fall off exponentially. This can be understood from the $I^\nu$ law Eq. (5.6). Following that trend, the peaks of the ATI spectrum line up $\propto \exp\left(-\nu \log\left(I/I_0\right)\right)$. The envelope thus follows an exponential trend $\propto \exp\left(-c\varepsilon_{\rm kin} \log\left(I/I_0\right)\right)$. The smeared spectra approach the envelope and thus also exhibit an exponential decrease, although only direct emission is involved.

The regime of fully resolved ATI spectra also carries a lot of interesting mechanisms. There is, e.g., the competition between direct and sequential processes which has been discussed in atoms [KL99]. To consider the case of clusters in that respect, let us briefly sketch the amplitude for a two-photon excitation from initial state 0 to final state $f$,

$$T(\omega_{\rm las}) \propto \sum_n \frac{\langle \Psi_f|\hat{D}|\Psi_n\rangle}{E_f - E_0 - 2\hbar\omega_{\rm las}} \frac{\langle \Psi_n|\hat{D}|\Psi_0\rangle}{E_n - E_0 - \hbar\omega_{\rm las}} \tag{5.7}$$

where $\hat{D}$ is the dipole operator. Only a direct two-photon ionization takes place if none of the intermediate states $n$ is in resonance with the laser, i.e. if $E_n - E_0$ stays far from $\hbar\omega_{\rm las}$ for any state. But a sequential process is possible if one intermediate state can be reached by the laser frequency. For then, the excitation energy can be stored in the resonant state and the system can wait for the second photon to gather its final excitation stage. Clusters provide a far denser spectrum of intermediate states than atoms, which makes the situation less transparent. Metal clusters add a new spectral feature, the prominent and strong Mie plasmon resonance. An interesting competition thus emerges if the laser frequency comes close to the plasmon frequency: multi-photon versus multi-plasmon excitation.

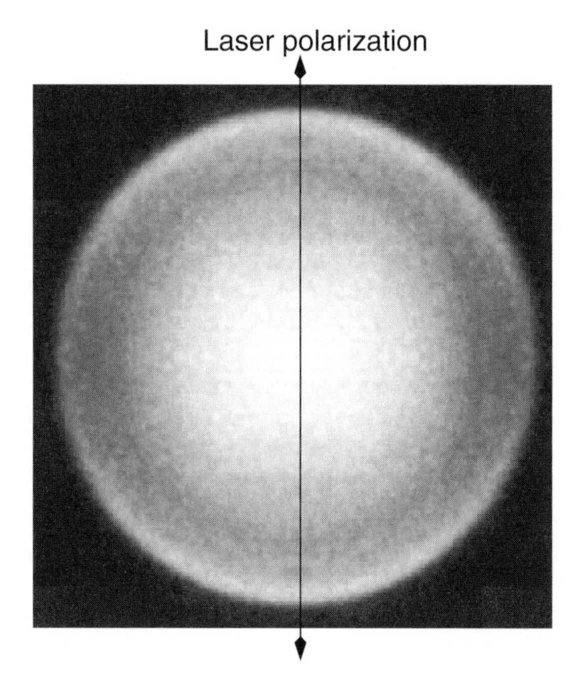


**Figure 5.21:** 2D equi-density plot of kinetic energy spectra and angular distributions from a $W_4^-$ cluster anion after irradiation with a laser of frequency 4 eV. The emission angle is mapped in polar coordinates while the kinetic energies grow with radial distance from the center of the plot. The gray scale indicates flux; high emission shines white. The laser intensity was low so that only one photon processes are to be expected. The pulse width was in the ns range. The laser polarization points along the vertical axis. After [BPBB01].

The relation (2.1) for the kinetic energy spectrum is then generalized once more to

$$\varepsilon_{\text{kin}} = n_{\text{phot}} \hbar\omega_{\text{phot}} + n_{\text{plas}} \hbar\omega_{\text{plas}} + \varepsilon_{\text{bound}} \tag{5.8}$$

where $n_{\text{plas}}$ counts the number of Mie plasmons involved and $\omega_{\text{plas}}$ their frequency. The mix is demonstrated in Figure 5.20 showing results from a theoretical analysis in $Na_9^+$ by means of TDLDA [PRS01]. Take, for example, the spectrum for the largest frequency (uppermost line). There is a clear triple peak in a narrow energy window. Analyzing the wavefunctions allows one to assess that all three peaks are related to emission out of the $1s$ state in $\text{Na}_9{}^+$. The energetic separation is about 0.23 eV, suspiciously close to twice the energy difference $\hbar\omega_{\text{las}} - \hbar\omega_{\text{Mie}}$. The vertical solid line indicates the position of a four-plasmon process from the $1s$ state ($n_{\text{plas}}$ = 4). The left peak is very close to that. The small dashed line indicates the position of a four-photon process ($n_{\text{phot}}$ = 4) and the right peak coincides very well. The middle peak can then be interpreted as a mixed two-plasmon-two-photon process ($n_{\text{plas}}$ = 2 and $n_{\text{phot}}$ = 2). It is obvious that the peak structure should depend sensitively on the relation between laser and plasmon frequency. The variation of laser frequency in Figure 5.20 confirms that. One clearly sees how the four-photon peak moves relative to the four-plasmon peak. One may miss the odd combinations $n_{\text{las,plas}} \in \{1, 3\}$. They are suppressed for the short pulses considered here and appear for longer pulses with their better frequency resolution. After all, this theoretical example shows that lots of interesting details could be worked out by high resolution measurements of ATI on clusters.

#### 5.3.1.4 Angular distributions

Besides the kinetic energy, one can also measure the angular distribution of the emitted electrons. This also carries a lot of interesting information. Figure 5.21 shows combined PES and angular distributions obtained from irradiation of a $W_4{}^-$ cluster anion with a low-intensity ns laser at a frequency of 4 eV. The anion has a low ionization threshold around 1.6 eV as compared to the monomer evaporation threshold which is larger than 7 eV. This favors thermal ionization over monomer evaporation. The competition remains, though, between direct (one-photon) and thermal electron emission. The extremely long laser pulse gives thermalization through electron-electron collisions good chances such that one expects a large contribution from thermal emission. This is indeed seen in Figure 5.21 where light grey indicates large emission and dark grey low emission. The broad central spot can be associated with thermal emission. Emission is visibly isotropic and the kinetic energy spectra (not directly visible in the figure) confirm the correct trend $\propto \sqrt{\varepsilon_{\rm kin}}\,\exp\left(-\varepsilon_{\rm kin}/T\right)$ [BPBB01]. But at larger kinetic energies (larger radial distances in the plot) one spots a non-isotropic emission pattern directed along the laser polarization axis. This is clearly a signal from direct emission. Thus we see here again a mix of thermal and direct emission, similar as was analyzed in Figure 5.20. But the interpretation of Figure 5.21 depends much less on theoretical modeling because we have additional information from the angular distributions.

A full description of such processes should account for electron-electron collisions to cover both regimes, direct emission as well as thermal evaporation. This can be done by virtue of the VUU scheme as outlined in Section 3.3.5.3. The semi-classical approach requires sufficiently high excitation. But that is the typical situation if thermalization is to play a role at all. Figure 5.22 compares angular distributions from pure mean-field calculations with VUU ones. (Unlike the previous figure, the distributions are here integrated over final kinetic energy.) First, we see again (as in Figure 3.14), that TDLDA and Vlasov-LDA agree nicely in that excitation regime. Both show a clearly peaked emission with direction along the laser polarization, similar as in the direct emission part of Figure 5.21. The collision term in VUU distracts a large amount of electrons from the direct emission part. The delayed emission of thermalized electrons has lost any memory of the original polarization axis and subsequently one obtains a much smoother angular distribution, as seen on the VUU curve in Figure 5.22. The distribution is not perfectly isotropic here. There remains still a sizeable fraction of directly emitted electrons for the chosen conditions. But one can easily imagine that the branching between direct emission and thermalization sensitively depends on the actual excitation. Systematic studies of these influences could deliver valuable information on the underlying dynamics. They have yet to be worked out, theoretically as well as experimentally. This holds the more so for the combined analysis of kinetic energies and angular distributions as exemplified in Figure 5.21.

#### 5.3.1.5 Resolving the electronic potential by scanning ionization

Kinetic energies and angular distributions provide, of course, valuable information on structure and dynamics of a cluster as outlined in the previous subsection. But even a global quantity as net ionization (total number of emitted electrons) can become extremely useful when some systematics is added to display trends. The most obvious is to record the trend

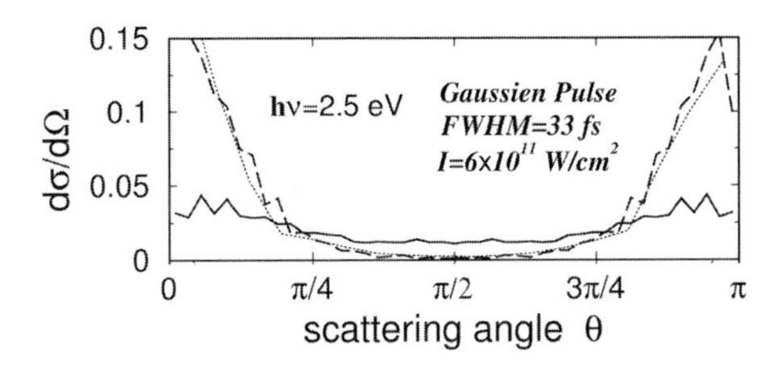


**Figure 5.22:** Angular distribution of emitted electrons computed at three different levels of approach: quantum mechanical TDLDA (dotted), its semi-classical analogue Vlasov-LDA (dashed), and VUU (Vlasov-LDA with collision term, full line). Test case is $Na_{41}^{+}$ with ionic structure. Laser parameters are indicated in the figures. The angle is defined relative to the laser polarization axis. After [GRS03].

of photoemission yield versus photon frequency. It turns out that such a conceptually simple measurement allows one to gather information on the spatial extension of a cluster [FR97]. This will be discussed in this subsection.

Consider the transition amplitude (in the limit of large wavelength) $T_{\mathbf{k}i} \propto \langle\varphi_{\mathbf{k}}|\mathbf{E}_0\cdot\hat{\mathbf{p}}|\varphi_i\rangle$ for photoexcitation from a specific initial single particle state $\varphi_i$ to an outgoing wave $\varphi_{\mathbf{k}}$ with asymptotic momentum $\mathbf{k}$. The (laser) photon polarization is given by $\mathbf{E}_0$. This expression is given in the velocity gauge. A schematic estimate is easily obtained when switching to the acceleration picture (sometimes called the acceleration gauge), changing in $T_{\mathbf{k}i}$ basically $\hat{\mathbf{p}} \longrightarrow \nabla U$ where $U$ is the local mean-field potential felt by the electrons. Let us, furthermore, replace $\varphi_{\mathbf{k}}$ by a plane wave, namely assume $\varphi_{\mathbf{k}} \longrightarrow \exp(\imath\mathbf{k}\cdot\mathbf{r})$ as a simplistic ansatz for the outgoing wave. For then we obtain

$$T_{\mathbf{k}i} \propto \int d^3r\, e^{\imath\mathbf{kr}}\nabla U\varphi_i \overset{\sim}{\propto} \varphi_i(r_{\mathrm{surf}})\int d^3r\, e^{\imath\mathbf{kr}}\nabla U \propto -\imath\mathbf{k}\varphi_i(r_{\mathrm{surf}})\int d^3r\, e^{\imath\mathbf{kr}}U \quad (5.9)$$

assuming that $\nabla U$ is strongly peaked at the cluster surface and that $\varphi_i$ is rather soft as compared to it. The estimate (5.9) means that the photoionization yield is strongly related to the Fourier transform of the mean field. On the other hand, the Fourier transform of a localized function is known to show an oscillatory structure. These are, to begin with, oscillations when changing the outgoing momentum $k$. But this momentum is uniquely related to the photon frequency $\hbar\omega_{\mathrm{las}}$ as $k = \sqrt{\hbar\omega_{\mathrm{las}} - |\varepsilon_i|}$. And this finally maps to oscillations of the photoionization yield as a function of frequency from which one can deduce, in turn, information about the global extension of the mean field [FR97].

The experimental realization requires a broad range of arbitrarily tunable photon frequencies. Synchrotron radiation provides an appropriate source. Furthermore, one has to tag specific single electron states because an average over many states would smear out the searched pattern. Figure 5.23 shows results for photoionization of free $C_{60}$ clusters out of the HOMO level, or alternatively the HOMO-1 state (which is the next highest occupied state having different nodal structure). One nicely sees the expected oscillations with photon frequency. The lines through the data are computed from a simple shell-model potential for $C_{60}$ which is radially symmetric, zero inside and outside the cluster, and has its deepest well at the mean radius

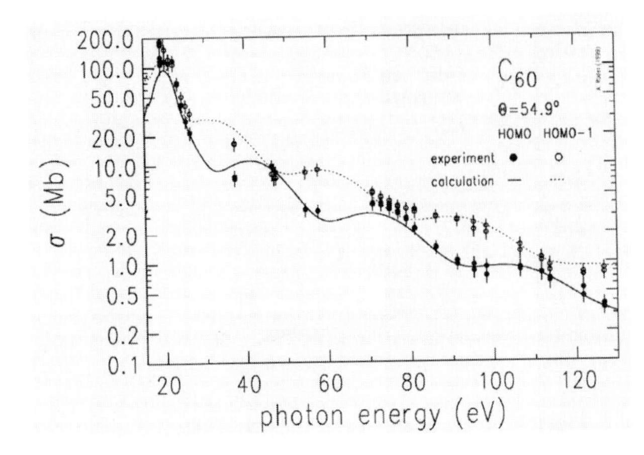


**Figure 5.23:** Cross section for photonionization out of the HOMO or HOMO-1 levels of $C_{60}$, as indicated. After [BGR00].

of the ionic cage [BGR00]. The two different single electron states produce about the same step-size of oscillations but place minima and maxima at different photon frequencies due to their different nodal structure.

This analysis of frequency dependence of photoionization yield is not confined to clusters. It applies to any well localized electron system. For example, it is argued in [HCN+03] that the pattern observed in semiconductor quantum dots [FIM+00] can be explained by the same mechanism as discussed above.

## 5.3.2 Shaping clusters

Section 5.3.1 dealt with moderate excitations, just enough to see multi-photon processes but safely away from destruction of the cluster. Here we go to the opposite limit employing intense laser pulses just enough to destroy certain clusters or at least to modify them dramatically. The intensity is at the upper end of the semi-linear regime (typically $10^{14} - 10^{16}\ \mathrm{W/cm^2}$), but still low enough to prevent violent processes (Coulomb explosion, nano-plasma).

Figure 5.24 demonstrates the dedicated shaping of cluster deformation by intense laser pulses (from [WBGT99], see also more recently [VBST02]). Silver clusters are grown on quartz by assembling from vapor. With increasing growth time, increasing coverages are reached which, in turn, yield increasing average cluster size. Thus the number of mono-layers (ML) indicated in the plot serves as a rough indicator of cluster size. The typical cluster radius varies between 2 nm at 4 ML to 12 nm at 23 ML where the clusters are strongly oblate, deformed by the attractive surface interaction. The optical absorption spectra of the deposited clusters are analyzed by shining on the substrate light from a Xe arc lamp and recording reflection under an angle of 45°. The resulting spectra are shown on the left panel. There are two well separated Mie plasmon peaks. This indicates that the deposited clusters are deformed (on the relation between spectra and deformation see Section 4.2.2.2). The attractive interface interaction favors oblate clusters that are less extended perpendicular to the surface. The left panel of Figure 5.24 shows that the deformation grows with cluster size. The spectra of the deformed clusters are more weighted towards the low-frequency (= long-

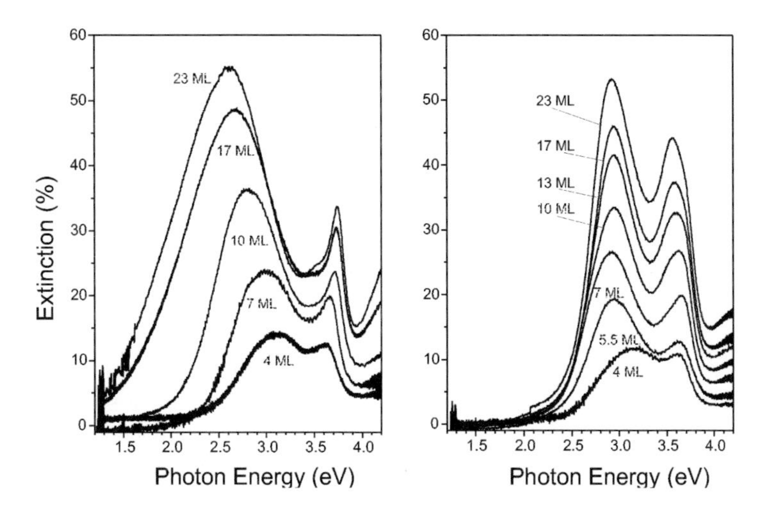


**Figure 5.24:** Optical extinction spectra of large Ag clusters on quartz substrate for various coverages as indicated (ML=mono layer). Left panel: Spectra from unperturbed growth. Right panel: Spectra from growth combined with irradiation by a laser of wavelength 530 nm ($\equiv$ 2.3 eV) and fluence 400 mJ/cm$^2$. From [WBGT99].

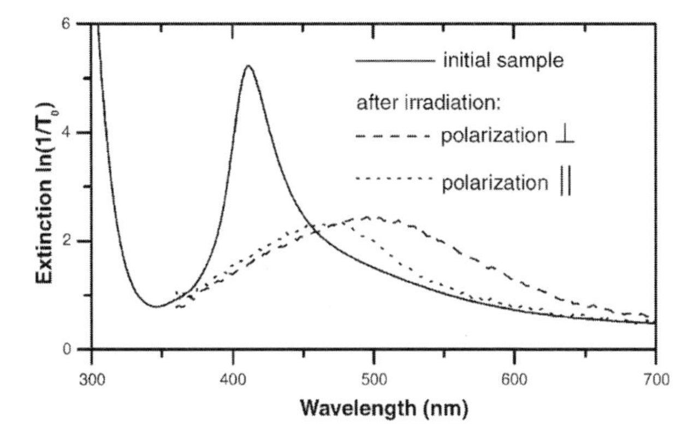


**Figure 5.25:** Optical absorption spectra of Ag cluster in a glass matrix. Full line: spectrum after preparation from ion exchange ($Na^+$ into $Ag^+$) and annealing. Dashed and dotted lines: spectrum long after irradiation with an intense laser pulse having wavelength of 400 nm ($\equiv$ 3.1 eV) and FWHM of 150 fs. From [SKBG00].

wavelength) end, as expected for oblate shapes (Section 4.2.2.2). In a second setup, a strong laser pulse in that lower-end region (wavelength = 530 nm $\equiv$ frequency = 2.3 eV, fluence = 400 mK/cm$^2$) was applied repeatedly during growth. The resulting spectra are shown in the right panel. The collective splitting of the spectra is obviously limited, and with it the deformations. The laser has "burned" away all strongly deformed clusters because they were in resonance with the laser, acquired a lot of energy and evaporated until they run out of resonance. As a result the same (small) deformation can be maintained for all cluster sizes. The mechanism allows tailoring of samples of deposited clusters. Here we have seen how deformation can be limited. Proper tuning of the "burning" laser (optionally with a sequence of different pulses) also allows a size selection or other wanted properties.

Figure 5.25 shows an example of how a strong laser pulse can modify large embedded Ag clusters. The change is again read off from a change in the optical spectrum. Large Ag clusters in glass are produced by inserting the glass into molten Ag salt. The Ag ions diffuse into the glass and substitute the Na ions therein. With slowly decreasing temperature, the Ag ions coagulate into embedded Ag clusters (for a theoretical MD simulation of that process see [TSG97]). These clusters are preferably spherical and their diameter ranges from 5 nm near the glass surface to 50 nm in the interior. Accordingly, these clusters show one clean Mie plasmon peak as seen from the solid line in Figure 5.25 (the peak at lower wavelength in the left corner of the plot is due to the optical response of the glass matrix). The sample is then irradiated with a strong laser (wavelength = 400 nm ≡ frequency = 3.1 eV, fluence 10 mJ/cm$^2$). After a sufficiently long pause for full relaxation of the shaken system (about 1 s), the optical spectra are recorded again. Two laser shots with different polarizations have been performed on two different samples. The results in Figure 5.25 show a substantial redshift and broadening, independent of the laser polarization. The strong broadening suggest that a somewhat diffuse environment has been produced by the violent laser excitation. The most probable scenario is that the laser heats the cluster up and ejects a lot of electrons initially. The electrons are stuck in the vicinity of the cluster due to the poor conduction of glass. The heat stored in the cluster leads to evaporation of monomers, Ag ions, and larger fragments. These diffuse slowly through the glass. The ions are neutralized by the available free electrons. The rigidity of the glass matrix confines these diffusion processes to a neighborhood of the original cluster from which a major fraction survives at its place. At the end, we have large Ag clusters surrounded by a halo of small Ag clusters. The Mie plasmon in each cluster constitutes a typical situation of coupled dipoles as has been discussed in Section 5.2.1.2, now with one strong dipole and a statistical mix of small dipoles coupled at various distances to it. It can be shown that this leads to a red shift (through collective coupling and effectively increased extension) combined with broadening of the spectra (due to the diffuse mix of dipoles). A deformation of the central cluster cannot be unambiguously deduced from the spectra, though, because of the small clusters surrounding the central one. But the width after irradiation is so large that some deformation (with subsequent spectral splitting) cannot be excluded. Up to here, the example has demonstrated once more how a cluster can be modified by dedicated strong laser pulses. The dynamical mechanisms on the way to the final result have also been tracked in [SKBG00]. It will be discussed in connection with pump-and-probe experiments in Section 5.3.4.

### 5.3.3 Ionic effects in laser pulses of varied length

Modern lasers allow short pulse lengths in the ps range down to a few tens of fs and a flexible choice of them. These times are in the range of important time scales in clusters, particularly the scale for ionic motion (see Section 1.3.2). Variation of the pulse profile thus will explore ionic effects. A most flexible variation is undertaken in pump-and-probe scenarios which will be discussed at length in Section 5.3.4. But a simple variation of pulse length already reveals a lot about the underlying ionic motion as we will see in this section.

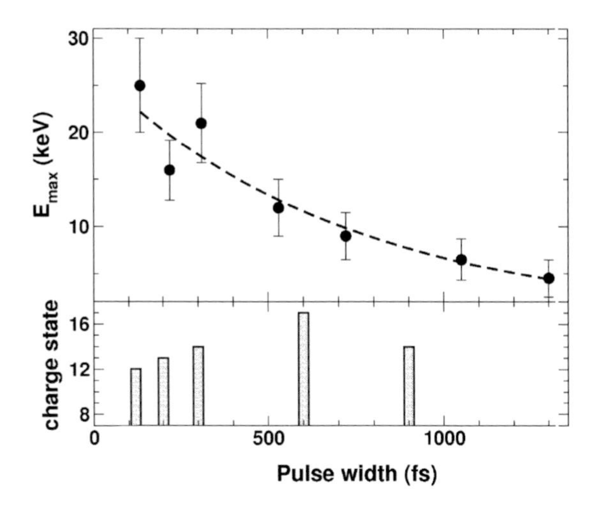


**Figure 5.26:** Coulomb explosion of small Pt clusters induced by a laser with frequency 1.54 eV and fluence of about $I\,T_{\rm pulse} = 1.2 \times 10^{18}\,{\rm Wfs/cm^2}$. Experiments have been done with systematically varied pulse length $T_{\rm pulse}$. Results are drawn versus pulse length. Lower panel: Net charge state reached. Upper panel: Maximum kinetic energy of emitted ions. From [TDF$^+$01].

#### 5.3.3.1 Experimental results for Pt clusters

Exposure of a cluster to a sufficiently intense laser beam will always lead to Coulomb explosion with total destruction. Such violent processes with huge laser intensities in the range of $10^{18}\,{\rm W/cm^2}$ and above follow their own rules almost independent of the initial cluster material. This case will be discussed from a general perspective in Section 5.5. But some Coulomb explosion can also be observed at more moderate intensities around $10^{15}\,{\rm W/cm^2}$ and these processes are still very sensitive to the cluster material and the details of the laser pulse, see e.g. the experiments of [TDF$^+$01, LVC$^+$02] on clusters of heavy metals, Pt and Pb. Figure 5.26 summarizes the findings of experiments on small Pt clusters (sizes somewhat below $N$=50) in pulses with varied pulse length [TDF$^+$01]. The fluence of the laser, $I\,T_{\rm pulse}$, was kept constant. Intensities thus range around $10^{15}$ to $10^{16}\,{\rm W/cm^2}$, depending on pulse length. As can be seen from the lower panel of Figure 5.26, high charge states are reached in any case. The interesting feature is that the final charge state depends on the pulse, having a maximum at $T_{\rm pulse} = 600$ fs. A first clue on the underlying mechanism is given in the upper panel. The kinetic energy of the final fragments drops monotonically with pulse length. It is produced from the Coulomb pressure at the stage of strong ionization (= charging). Low kinetic energy means a large starting radius while high kinetic energy a small one. This suggests the following mechanism: Let us assume that the laser is strong enough to remove initially a few electrons from the cluster. The thus released Coulomb pressure will drive the cluster into a slow expansion at ionic time scales (hundreds of fs). The cluster then changes its resonance conditions as a function of time. Dramatic ionization may arise at a later stage if the laser is still irradiating and if the cluster moves into resonant coupling with the laser (i.e. if the laser frequency matches the plasmon frequency, see e.g. Sections 1.3.3 and 5.3.1.1). Nothing special will happen if the laser is switched off before that point. Figure 5.26 shows that the ionic expansion, proceeding always at its typical slow time scale, drives the cluster into optimal conditions around a time of 600 fs. Shorter pulses are less efficient because they do not reach the optimum time, even though the initial ionization is larger due to larger intensity (note that fluence $= IT$ is kept constant). Longer pulses become inefficient again because

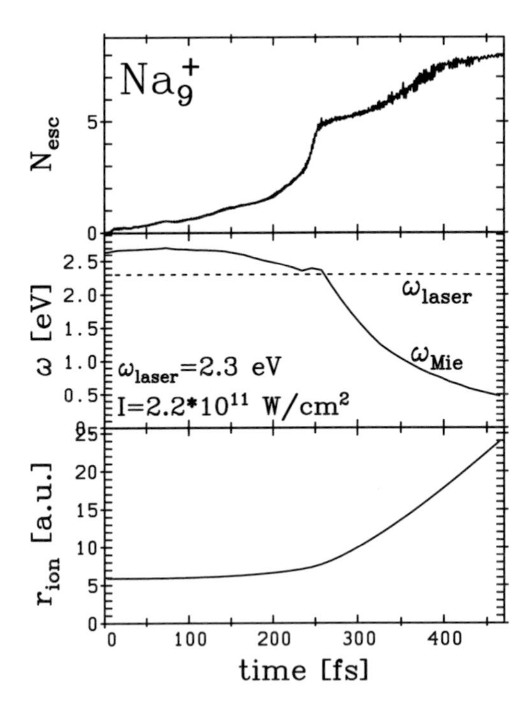


**Figure 5.27:** Time evolution of key observables for $Na_9{}^+$ irradiated by laser pulses with varying intensity as indicated. All pulses have frequency $\omega = 2.3\,\text{eV}$ and a pulse length of 1000 fs. Lowest panel: Ionic r.m.s radius. Middle panel: Frequency of the estimated average Mie plasmon peak; the horizontal straight line indicates the laser frequency. Uppermost panel: Number of emitted electrons $N_{\text{esc}}$. After [RS01a].

the initial ionization is weaker and the Coulomb pressure does not suffice any more to deliver the required expansion. The question remains to what process these "resonant" conditions relate. Two very different mechanisms have been discussed: resonant enhancement due to a red-shifted Mie plasmon and force-assisted enhanced tunneling rates. These will be discussed in the following two subsections.

#### 5.3.3.2 Plasmon enhanced ionization model

It was discussed in Section 5.3.1.1 that ionization yield follows the trends of optical absorption. The latter has a pronounced resonant structure in metal clusters dominated by the Mie plasmon. And the Mie plasmon frequency Eq. (1.18) depends on the radius as $\omega_{\text{Mie}} \propto \rho^{1/2} \propto R^{-3/2}$. The Mie frequency thus moves into resonance with the laser (having initially a frequency below the resonance) because expansion of the cluster yields a red-shift [RGB96b]. This mechanism was studied in detail for Na clusters in [RCK$^+$99, SR00, RS01a] using TDLDA-MD (as presented in Chapter 3). Figure 5.27 illustrates the mechanism. The uppermost panel shows the ionization as a function of time. The initial ionization increases with intensity but maintains a moderate slope in any case. The initial ionization triggers a more or less slow Coulomb expansion depending on the intensity, see the lowest panel. The middle panel shows the mean Mie plasmon frequency estimated according to the actual ionic radius (lower panel) and charge state (upper panel) as $\omega(q,R) = (2.1\,\text{eV} + q\,0.2\,\text{eV})\,(R/R_0)^{-3/2}$ where $R_0$ is the ionic radius of the cluster in the ground state. The fast initial ionization induces a slight blue-shift. The subsequent ionic expansion drives a red-shift $\propto R^{-3/2}$. Sooner or later, the actual Mie plasmon frequency crosses the laser frequency (horizontal line in the

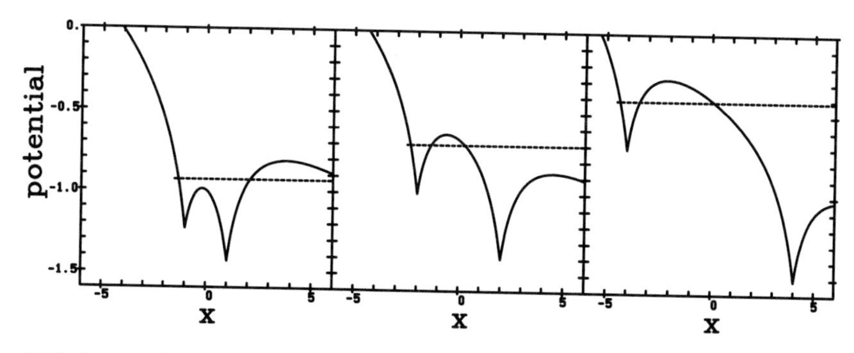


**Figure 5.28:** Schematic view of force-assisted tunneling. Shown is a soft Coulomb potential for two atoms at a given distance plus the potential slope exerted from a homogeneous electrical field. The fine dotted horizontal line indicates the energetic position of a core level. The three panels show the same setup for different molecular distances, growing from left to right.

middle panel). The thus established resonant conditions dramatically increase ionization as seen from the associated steep slopes in the uppermost panel. Thus far the figure exemplifies the resonant enhancement mechanism. The pulse length was kept constant to keep things simple. Now imagine constant fluence, i.e. a pulse length inversely proportional to the intensity. The lowest intensity has a long sustained pulse. But the forces are very small such that initial ionization remains low and resonant conditions are never reached. The highest intensity would then be associated with a pulse width of only $T = 100$ fs. This produces large initial ionization. But the time is too short to "see" the resonant conditions during the pulse. Thus the final ionization is low in both limits and has a maximum in between, in qualitative agreement with the lower panel of Figure 5.26. For a more detailed analysis see [RS01a].

#### 5.3.3.3 Field assisted tunneling model

The experimentally studied case of Pt, however, is more involved than the simple metal Na used as an example in Section 5.3.3.2. The Pt atoms contain weakly bound core electrons (as in all noble metals, see the example of Cu in Figure 3.2). These take part in the dynamics, first by polarization and later they may be released from the atoms to join the valence cloud. One thus needs to check the interplay of forces releasing and recapturing core electrons. This was done in a rate equation model in [SR02b]. The basic underlying idea stems from molecular physics [SIC95, ZB95]. We first discuss this effect schematically for the case of a dimer. Consider an atom in a laser field. In an adiabatic picture, the instantaneous field adds a linear potential slope. This renders the core electron asymptotically unstable. But it is still confined by a strong potential barrier with correspondingly very long tunneling time. Two atoms at far distance experience the same situation, each one separately. For a sketch of this situation see the right panel of Figure 5.28. Now move the atoms towards each other. The attractive Coulomb field from the atom at the lower potential side enhances the pulling force on the atom at the higher end and effectively lowers the tunneling barrier, see the middle panel of Figure 5.28. Tunneling rates thus increase with decreasing distance. There is a counteracting effect for even narrower packing. The potential difference between the attracting atom and the

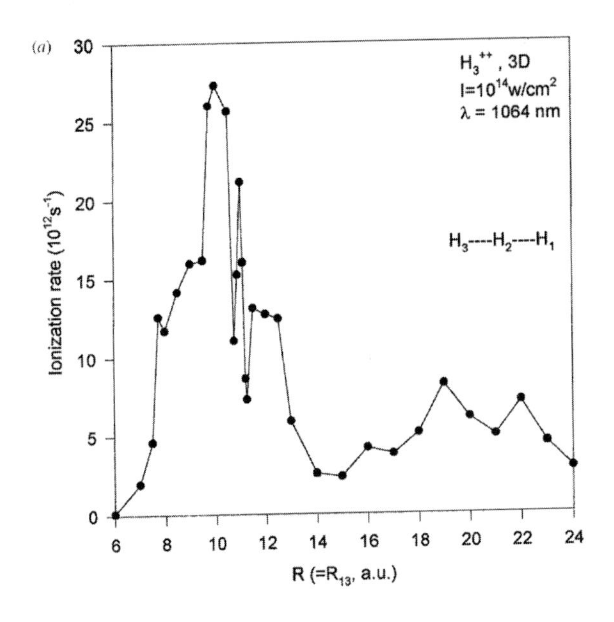


**Figure 5.29:** Ionization rate versus molecular distance for a linear $H_3^{2+}$ molecule. Shown are results of the fully three-dimensional solution of the time-dependent Schrödinger equation for the one active electron. The laser parameters are as indicated. From [YZB98].

donating atom decreases with decreasing distance. This enhances the outer potential barrier and hinders again the release of the electron, see left panel in Figure 5.28. There is obviously an optimum molecular separation at which force-enhanced tunneling rates have a maximum.

This effect has been studied and confirmed for molecules [SIC95, ZB95]. Figure 5.29 shows a result for the linear $H_3^{2+}$ molecule. The molecular distance between the two H atoms was kept fixed during each calculation of laser excitations. The resulting ionization is drawn versus this distance. There is obviously a strongly "resonant" peak at a distance of 10 $a_0$. It is due to an optimal cooperation of the external field from the laser and the effective Coulomb field from the neighboring molecular centers. It is likely that this force assisted tunneling also applies to the more complex situation of clusters. This was checked in a model combining molecular dynamics for the valence electrons with quantum mechanically computed tunneling rates for the core electrons [SR02a]. The external laser field is added as a classical time-dependent external field as usual. The ionization is quantified in terms of the energy absorbed from the laser field. The electronic dynamics is simulated for fixed ionic configuration. Starting from the equilibrium configuration, a series of different stages of expansion have been produced by rescaling the ionic positions. The results are summarized in Figure 5.30. The left panel shows the result for a rare gas cluster $Ne_{16}$. A softened Coulomb potential was used for the atom adjusted to provide appropriate atomic levels. One clearly sees a maximum of energy absorption at an interatomic distance which is about twice the equilibrium value. To countercheck the mechanism, the process has been simulated at two different laser frequencies. The above schematic discussion makes it clear that the force-enhanced ionization mechanism depends only on force fields and by no means on the laser frequency. The left panel of the figure shows indeed that the predictions are independent of frequency. As a further countercheck, the situation of a simple metal cluster has been simulated by softening the atomic potential so much such that no core electrons remain and all are valence electrons,

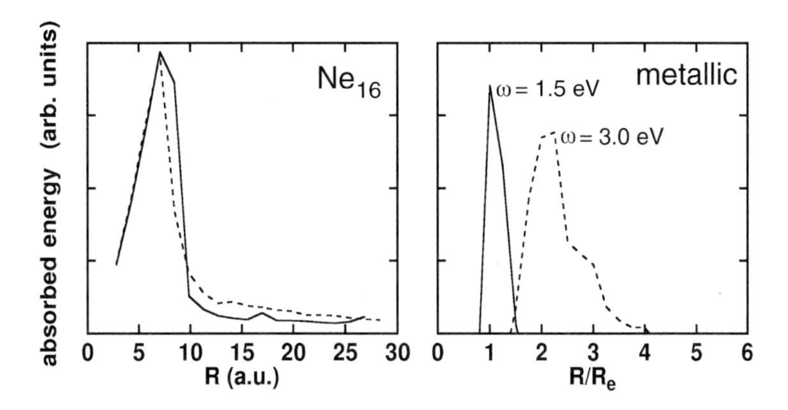


**Figure 5.30:** Energy absorbed from the laser field for clusters with N=16 at different stages of expansion, resulting from a classical MD simulation augmented by microscopically computed tunneling rates. The left panel shows results for a rare gas cluster $Ne_{16}$; the extension is characterized in terms of the average interatomic distance. The right panel is obtained by smoothing the atomic Coulomb well such that effectively a metal cluster is simulated; the extension is given here as ratio to the initial radius. After [SR02b].

mobile throughout the whole cluster. For such a situation, we expect a revival of the plasmon resonance enhanced ionization which, of course, strongly depends on the laser frequency. The right panel shows the results from that variant of the model. And indeed one sees the strong dependence of the resonance radius on the laser frequency.

The question remains as to which which one of the two mechanisms applies to the experiments on Pt of [TDF$^+$01, LVC$^+$02]. The laser intensities of $10^{15}$ to $10^{16}\,\mathrm{W/cm}^2$ are rather high. We have seen in Figure 3.3 that this is already in a force dominated regime where plasmon frequencies are not sensitively explored. The experiments yield also high ionizations which can only be explained if also core electrons are released. That would tend to favor the second mechanism of force-enhanced ionization. But more measurements, e.g., with varied laser frequency are required to settle this question finally.

### 5.3.4 Pump and probe analysis

Femtosecond (fs) lasers give access to time-resolved studies of molecular reactions. The standard tool are here pump-and-probe experiments. An initial short laser pulse (pump) excites the system into a definite state. This triggers a time evolution with accompanied changing electronic response. This change is explored with a probe pulse sent after a certain time delay. Scanning delay times allows one to map the time evolution of the system. The prototype example is a dimer molecule. The pump pulse lifts the electronic state from the ground state Born-Oppenheimer surface into an excited surface. The molecule is not in equilibrium any more for the excited surface. Thus ionic motion along the excited Born-Oppenheimer surface takes place. The energetic distance to the ground state surface changes with time and so changes the resonance conditions for de-excitation by the probe pulse. This is, of course,

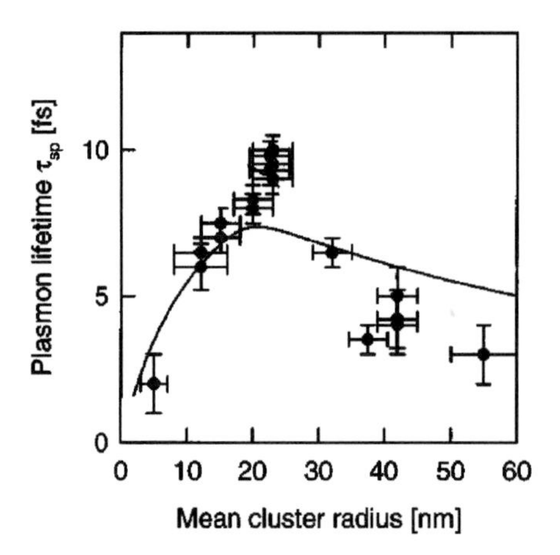


**Figure 5.31:** Life-time of the Mie plasmon for Na clusters on an insulator surface as a function of cluster size. The life-times are determined from pump-and-probe analysis recording the autocorrelation function of the SH signal from the probe pulse and comparing that with the free autocorrelation function. Laser pulse parameters are: wavelength = 570 nm, intensity = $2\times10^9$ W/cm$^2$, FWHM = 50 fs, and polarization perpendicular to the surface. The line through the data indicates a theoretical estimate, see text. From [KWSR97].

only one amongst a huge variety of pump-and-probe scenarios. One alternative setup is, e.g., to measure the lifetime of the excited state by tuning the probe pulse to ionization of the molecule. The multitude of variants grows with the complexity of the molecule. Altogether, femtosecond spectroscopy is an extremely exciting and rich field of physics and/or chemistry, for an overview see [Zew94]. And the field is still much under development. A recent achievement is, e.g., the dedicated shaping of laser pulses to switch chemical reactions in predefined directions [DDK$^+$02].

It is obvious that pump-and-probe analysis will also be an invaluable tool in cluster physics. Very small clusters allow scenarios much like in simple molecules which can still be characterized in terms of molecular dynamics along Born-Oppenheimer surfaces, see e.g. the experiments on trimers [LVWW99, HG00] and associated theory [HPBK$^+$98]. Clusters in general are acceded with various different strategies. In the following, we will discuss a few examples under a wide variety of conditions (deposited, embedded, or free clusters) and with respect to many different observables (electronic relaxation, ionic motion, response of the substrate).

#### 5.3.4.1 Deposited clusters

Subtle experiments on clusters, as e.g. pump-and-probe analysis, are easier for deposited or embedded clusters because this allows a higher density of scatterers. Most pump-and-probe studies on clusters are performed under such conditions. With laser pulses of width around 50 fs, they are best suited for resolving ionic time scales similar as in most molecular experiments. However, with a careful analysis of the probe signal, one can also conclude on electronic processes. One class of such experiments determines the electronic relaxation time of the Mie plasmon in metal clusters. An early measurement with Ag clusters on surfaces can be found in [SNHG$^+$92]. Similar measurements for other material combinations and cluster sizes are found in [KWSR97, KWSR99, MPT$^+$00]. We present here an example from this sort of experiment to demonstrate the richness of pump-and-probe experiments. Figure 5.31 shows results for the life-time $\tau_{\mathrm{Mie}}$ of the Mie plasmon in Na clusters attached to an insulator surface [KWSR97]. Cluster sizes have been monitored with the intensity of the second

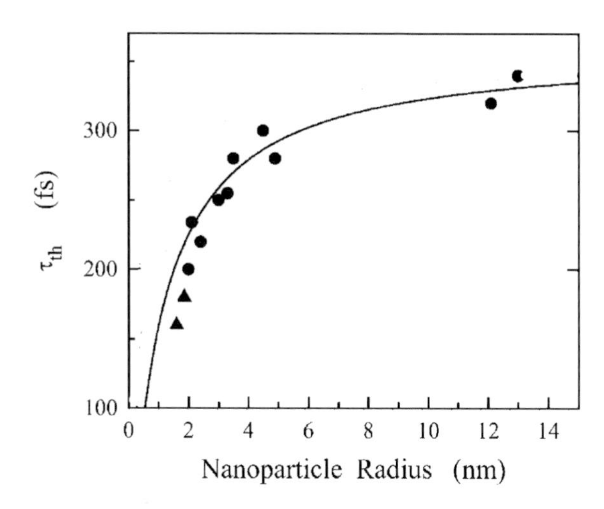


**Figure 5.32:** Relaxation times of electronic thermalization for Ag clusters in $BaO\text{-}P_2O_5$ (dots) or $Al_2O_3$ (triangles) matrices drawn versus average cluster size. The solid line results from a simple theoretical estimate accounting for electronic spillout and $d$-localization effects. From [VCF+00].

harmonic (SH) signal. The relaxation times are determined by pump-and-probe for samples with thus determined average cluster sizes where the effect of the probe pulse is monitored from the SH signal emitted by the cluster. The SH autocorrelation function for the Na sample is compared with that of a frequency doubling beta-barium borate (BBO) crystal for sake of synchronization. This allows one to deduce the life-times with a precision of about $\pm 1$ fs. The results shown in Figure 5.31 lie in the range as shown for free Na clusters in Figure 4.22. The main relaxation mechanism is here Landau fragmentation as discussed in Section 4.3.2. A simple estimate for large clusters is $\tau_{\text{Mie}} \approx (v_{\text{F}}/\lambda_{\text{el}} + v_{\text{F}}/R)^{-1}$ where $v_{\text{F}}$ is the electronic Fermi velocity, $\lambda_{\text{el}}$ the electronic mean-free path in bulk Na, and $R$ the cluster radius. The solid line in Figure 5.31 corresponds to that estimate and it fits the data very well.

Although we have concentrated here on presenting the analysis of electronic properties, we ought to mention that the same experimental setup of [KWSR97] did also allow the measurement of the electron-phonon relaxation time. It was determined to be around 1.1 ps.

#### 5.3.4.2 Embedded clusters

As with deposited clusters, there are many investigations using pump-and-probe techniques for embedded clusters. We discuss here two very different examples, one dealing with the time scale for electronic thermalization and another one measuring relaxation processes in the environment at ionic scales.

Figure 5.32 shows the electronic thermalization times for embedded Ag clusters as a function of cluster size [VCF+00]. The pump pulse was chosen with wavelength of 950 nm ($\approx$ 1.3 eV) in the infrared. The pulse width was 25 fs. It produces a $1ph$ excitation around the Fermi energy $\varepsilon_{\text{F}}$. The excited state will thermalize and at the end one finds a broadened Fermi distribution for the occupation of the single-electron states. The change in occupation numbers is out of equilibrium and widely spread shortly after the excitation. It contracts to a sharply confined region around $\varepsilon_{\text{F}}$ in the thermalized new equilibrium. The laser was tuned to have low intensity such that at most one single photon is absorbed. This deposits little energy and thus concentrates the changes to a very narrow region around $\varepsilon_F$. The re-occupations around $\varepsilon_F$ are explored with a probe pulse with wavelength of about 315 nm ($\approx$ 4 eV), i.e. in the UV. The pulse is obtained from frequency tripling of the original infrared pulse (sacrificing pulse width which is now about 60 fs). This allows appropriate synchronization and well defined delay times. The high frequency is chosen to coincide with the Mie plasmon frequency of Ag clusters. The changes in the occupation induce changes in the dynamical polarizability $\alpha(\omega)$ of the Ag clusters. The $\alpha(\omega)$ is recorded by scanning frequencies around the plasmon pole and that for systematically varied delay times. The occupation induced shift in $\alpha(\omega)$ peaks at $\omega = 4$ eV and the height of the peak as a function of delay time allows one to read off the thermal relaxation time. The results are summarized in Figure 5.32. The relaxation times shrink with cluster size. This is explained as a surface effect. The thermal relaxation rate is determined by electron-electron collisions (Section 3.3.5.3). The electron-electron cross section is determined from a screened Coulomb potential. The screening is maximum in bulk Ag. Surface effects reduce it in two ways. First, the electron density is softened in the surface region (called electron spill-out) which diminishes the polarization, and second, the well bound $d$-core electrons have too little overlap with the surface which curbs their screening effect in the surface zone (an effect which is also responsible for some blue shift of the plasmon, see [TKMBL93] and Section 4.3.4.1). Both effects act in the same direction to reduce screening and thus to enhance the electron-electron collision rates with larger relative surface content (= smaller radii). The theoretical estimate in Figure 5.32 is obtained from the relation for the thermalization rate $\tau_{\text{th}}^{-1} = \gamma_{\text{th}} = \rho_{\text{el}}^{-5/6}\epsilon_d^{-1/2}$ where $\rho_{\text{el}}$ is the density of valence electrons and $\epsilon_d$ the homogenized dielectric function of the $d$-core electrons. This is taken from bulk Ag. The rate is now considered to be such at each point in space inserting the actual quantities there. The $\rho_{\text{el}}(\mathbf{r})$ is taken from a jellium model and the $\epsilon_d$ is distributed like a jellium, but with skipping a surface zone of width $r_s$ [TKMBL93]. The $\gamma_{\text{th}}(\mathbf{r})$ is averaged over the cluster. The inverse is the estimated relaxation time drawn as a solid line in Figure 5.32. The measurements line up very nicely about this simple rate model.

An exploration of very long times scales is shown in Figure 5.33. It deals with excitations of large Ag clusters embedded in glass [SKBG00]. The figure here continues the discussion of Section 5.3.2 in connection with Figure 5.25. The preparation of the sample and the asymptotic modification of optical extinction spectra had been discussed there. The time resolved data shown here shed some light on the underlying mechanisms which take place after heating the cluster with a strong pump pulse. Note that the laser frequency was chosen close to the peak of the unperturbed line (any change can only mean that the peak looses strength). Shown is the relative change of extinction. Large values mean almost total bleaching of the line at 400 nm. The overall pattern is the same in both cases (large as well as moderate fluence): there is a fast initial rise to almost total bleaching with a maximum at around 250 fs. From then on the extinction comes back slowly at several time scales, first a fast one with few ps and finally a very slow relaxation process with a scale of tens of ps. The patterns are explained by a succession of different relaxation processes. The first stage is Landau damping which takes typically 10 fs (see Section 4.3.2 and Figure 5.31 with discussion). This cannot be resolved with the 150 fs laser pulses used here. Next comes the thermalization of the absorbed energy into electronic degrees-of-freedom in the Ag cluster. The typical time scale is here about 250 fs (see also Figure 5.32 and discussion) much in agreement with the time to reach the first peak in

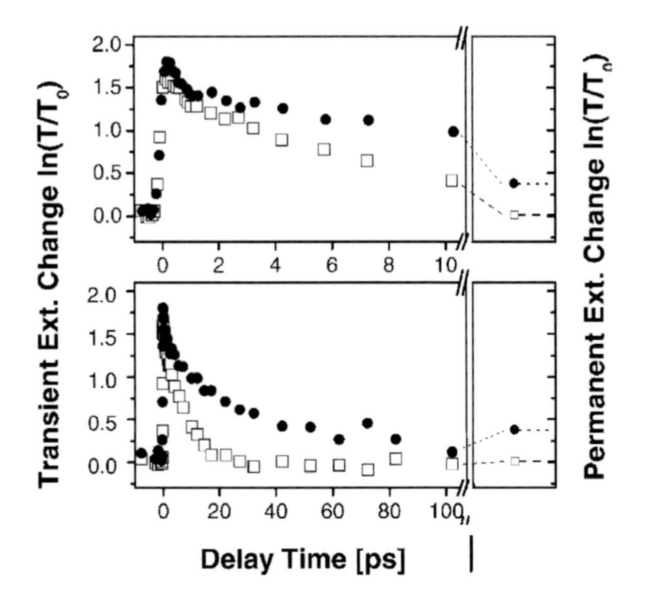


**Figure 5.33:** Pump-and-probe analysis of relaxation processes for Ag clusters (typical diameter 5 nm) in a glass matrix showing the change in optical extinction of the probe pulse (as compared to the sample without pump pulse) versus delay times. The same data are shown on two time scales. The separated right panels indicate the extinction reached asymptotically. The pump and probe laser had wavelength of 400 nm (≡ 3.1 eV) and pulse length of 150 fs. Results are shown for two different laser fluences as indicated: full dots show results for total laser energy of 2.5 J/cm$^2$ and open squares for 0.5 J/cm$^2$. After [SKBG00].

bleaching. The idea is that the strong thermalization leads to a red shift of the plasmon peak which remains still rather narrow. This removes any optical strength from the frequency region of the laser. Ionic thermalization and associated processes such as monomer or fragment evaporation follow at a ps time scale. These make the optical response more diffuse which explains a slow return of some strength at 400 nm. The slowest processes are the diffusion of the thus released Ag through the glass in the vicinity of the original cluster and re-formation of small Ag clusters there. From Figure 5.33 one can deduce relaxation times around 100 ps for this kind of processes. The two different laser strengths yield similar bleaching in the fast initial stages but relax to much different asymptotic states. The more violent case ends up in a significant permanent change of the cluster properties (shown in detail in Figure 5.25) while the case of lower fluence looks like reversible change. It is likely that a major fraction of the produced "Ag dust" re-unites with the original clusters on their diffusion path. In any case, this one time-resolved measurement covers a variety of dynamical processes. This makes the interpretation somewhat risky. Independent measurements are needed to corroborate this picture, some of them are discussed in [SKBG00]. More are possible and desirable. There is thus plenty of space for novel and exciting investigations here.

#### 5.3.4.3 Photonic decay time for clusters in a trap

Experiments with clusters in traps help to overcome several problems which one has with free clusters (see Section 2.2.1.5). One advantage is the long storage time which allows one to follow processes at almost any time scale. This feature has been exploited in [WDD$^+$99] to measure the relaxation time for radiative decay of excited clusters. Before going into the details, let us briefly recall the various competing de-excitation modes. Clusters can get rid of their excitation energy by electron emission, monomer evaporation, fragmentation, or radiative decay. The monomer channel usually dominates in neutral and positively charged

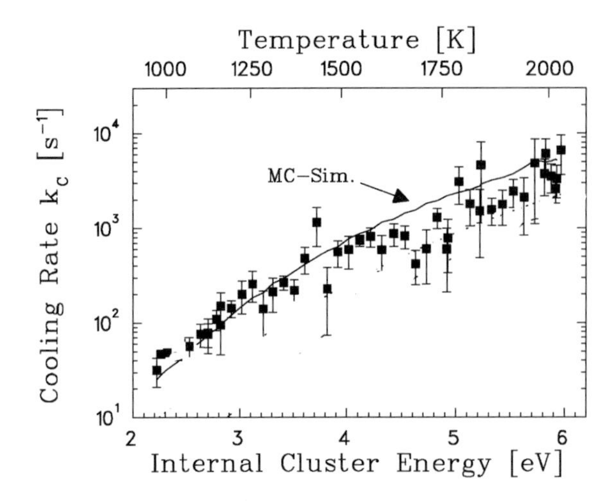


**Figure 5.34:** Rates for radiative cooling of a $V_{13}^{+}$ cluster versus internal excitation energy. Experimental results are drawn as filled squares with error bars. The full line stems from a more detailed Monte-Carlo simulation with fitted dissociation energy of $5.4\,\mathrm{eV}$. From [WDD$^{+}$99].

clusters. The electron channel takes over in cluster anions. The radiative decay time, on the other hand, is in the range of ms and usually overruled by the much faster particle decay channels. However, there are V clusters which are distinguished by particularly strong binding in every respect, i.e. having large IP and large monomer separation energies. These keep excitations long enough to see explicitely radiative decay. The cluster trap allows one to study these slowly decaying objects for sufficiently long time.

Figure 5.34 summarizes the results for studies of radiative decay of the $V_{13}^{+}$ cluster. In preparatory measurements, it was worked out that radiative cooling dominates for internal excitation energies below $E^* \approx 7.5\,\mathrm{eV}$ while cluster fragmentation sets on above that limit. The pump and probe lasers are tuned to have constant sum frequency $\hbar\omega_{\mathrm{pump}} + \hbar\omega_{\mathrm{probe}} = 8.24\,\mathrm{eV}$, i.e. safely in the fragmentation regime. The frequency of the pump pulse is varied in the range 2–6 eV. This deposits a tunable amount of internal excitation energy in the cluster $E^*(t=0) = \hbar\omega_{\mathrm{pump}}$. This energy is always below the fragmentation threshold. The cluster will thus loose its energy slowly through radiative cooling by emitting photons in the far infrared, corresponding to an exponentially decaying $E^*(t)$. No fragmentation will occur at all if the probe pulse comes too late because for then the actual energy $E^*(t) + \hbar\omega_{\mathrm{probe}}$ is already below the fragmentation threshold. The probe has to come sufficiently early to trigger final fragmentation. Evaluation of the branching as a function of delay time allows one to determine the relaxation time for radiative cooling. The summary of many such measurements is the radiative cooling rate (= inverse relaxation time) shown in Figure 5.34. One sees, quite expectedly, a steady decrease of the relaxation time with internal excitation energy. A simple theoretical estimate can be performed on the basis of Weisskopf's rule for the monomer evaporation rate [Wei37] and a similar estimate for the photon decay rate. Assuming only one-photon decay processes and adopting the photon cross-section from Mie theory (see Eq. 4.19) one obtains already a good first guess [WDD$^{+}$99]. A more realistic model allows for sequences of photon decays. This was simulated with Monte-Carlo techniques and the uncertainties in the model were given free for fit, namely the monomer dissociation energy and the photon cross-section. With a slightly enhanced cross-section (by 10%) and the quoted dissociation energy, one obtains the perfect reproduction given by the full line in Figure 5.34.

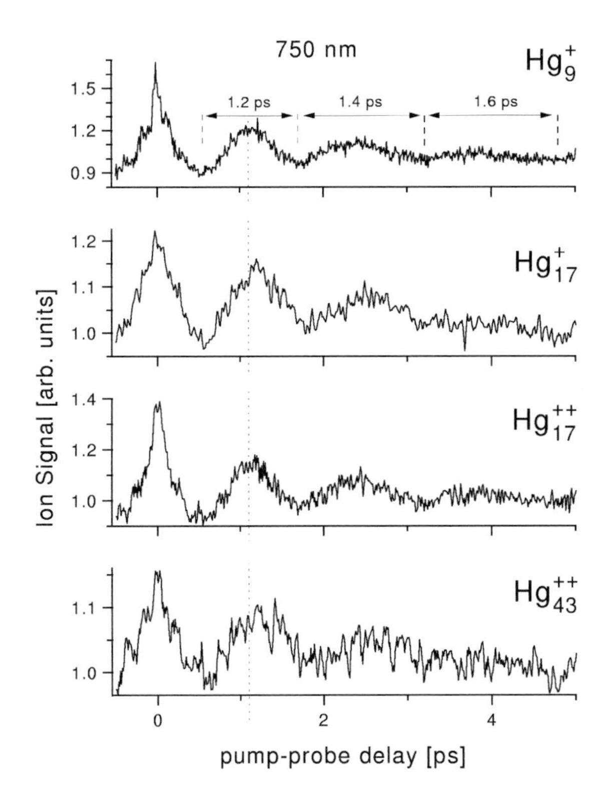


**Figure 5.35:** Yield of ionized Hg clusters (final states as indicated) after pump and probe excitation with a laser of wavelength 750 nm. The laser intensity was tuned such that a single pulse gives only very low yield. From [BLW+99].

As a final remark, note the relaxation times thus measured. They are in the range of ms. Such times are accessible for free clusters only through storage in a trap.

#### 5.3.4.4 Chromophores in free Hg clusters

Pump-and-probe experiments with free clusters are rare because the conditions are less well defined than for deposited, embedded, or trapped clusters. One example is provided by the studies on trimers [LVWW99, HG00] exploring their Born-Oppenheimer surface. An example dealing with larger clusters can be found in [BLW+99]. The test case here was Hg clusters. This material has the interesting feature of changing bonding type with cluster size. The element Hg is two-valued. Two electrons fill the $s$ valence shell leading to a sub-shell closure, so to say a weak rare gas. Small Hg clusters are thus Van-der-Waals bound as rare gases. The enhanced compression in larger clusters induces an overlap of $s$ and $p$ bands and this releases metallic binding, for a model study of this transition see [GPB91]. This transient behavior, with possible coexistence of localized and covalent bonding, will play a role for understanding the following results.

Figure 5.35 shows the ionization yield of Hg clusters following the probe pulse. Pump and probe pulses use here the same frequency which could be varied in a broad range. The pulse length is typically 50 fs. The results shown here have been obtained with a wavelength of 750 nm ($\equiv$ frequency of 1.65 eV). The intensity was chosen just high enough to deliver

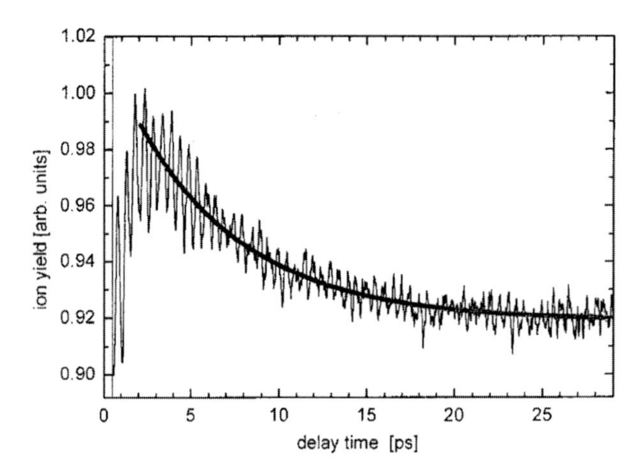


**Figure 5.36:** Photoionization yield after probe pulse for $K_2$ attached to the surface of a large He nano-cluster. The pump pulse lifts $K_2$ in a definite excited state. The probe pulse allows ionization of the $K_2$ out of the excited state. After [SV01].

sufficient counting rates in pump-and-probe but low enough that a single pump pulse produces only insignificant ionization rates. The initial Hg clusters were neutral. Thus the intermediate clusters after pump are excited but still neutral. The probe pulse then delivers the final kick to ionization. The final cluster sizes and charges (determined by charge and mass selection after probe) are indicated in Figure 5.35. Single and double ionization is found. The ionization yield shows marked oscillations with delay time, much similar in all cases but somewhat different in quantitative detail. The same patterns were found for the other laser frequencies. It was also found that the branching between double and single ionization is almost independent of intensity when varying it by a factor of ten in the range $10^{11}$ to $10^{12}\,\mathrm{W/cm^2}$. A tentative interpretation of these results assumes that the same small subunit in all clusters acts as a chromophore, absorbing the light, storing it and finally releasing it to the cluster for various reactions, and amongst them ionization. It is most probably a Hg dimer at the center. The view is supported by theoretical calculations of Hg clusters [KGB95]. The oscillations seen in Figure 5.35 run at an ionic time scale. The chromophore is most probably excited into a neutral Rydberg state. This changes the polarization interaction with the surrounding Hg atoms which, in turn, leads predominantly to an ionic breathing mode. It is determined by the forces in the closest layer and thus rather independent of the initial system size.

#### 5.3.4.5 Exploring cluster dynamics with a marker molecule

A somewhat similar situation to the chromophore case of Section 5.3.4.4 was studied in [SV01, CSS02]. A $K_2$ dimer was attached to a large He cluster. The predominantly metallic K is repelled from the interior and the K dimer rests at the surface of the He droplet. Figure 5.36 shows a result from pump-and-probe measurements with this setup. The pump pulse was chosen to excite specifically the $K_2$ dimer into an excited bound state. The probe pulse adds just enough energy for ionization out of the relaxed state. The final ionization yield is shown as a function of delay time. There are marked oscillations on top of a slow trend. The oscillations are clearly related to vibrations of the $K_2$ along its excited Born-Oppenheimer surface. The slower global trends are due to the interplay of the dimer with the He droplet. The strong increase for the first 2 ps is related to a relaxation of the He droplet adjusting to the new situation

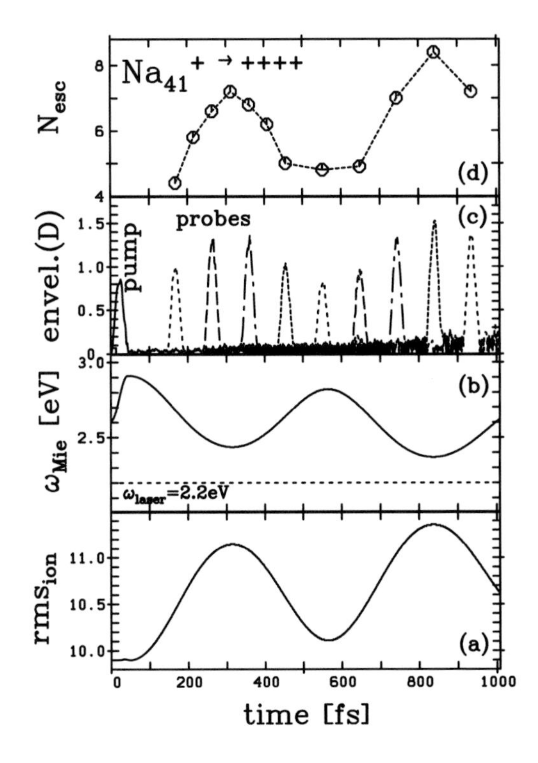


**Figure 5.37:** Illustration of pump and probe spectroscopy of ionic breathing vibrations of $Na_{41}{}^{+}$. Panel (a): time evolution of the ionic r.m.s. radius, $\sqrt{\sum_I \mathbf{R}_I^2}$, after the initial pump pulse. Panel (b): time evolution of the mean plasmon frequency after the pump pulse. The laser frequency is drawn (horizontal dotted line) for comparison. Panel (c): dipole response (envelope of dipole moment) to pump pulse (first 50 fs) and for various probe pulses at different delay times. The pulses have the same photon frequency $\omega = 2.2\,\text{eV}$, intensity ($\equiv 1.1 \times 10^{12}$ $\text{W/cm}^2$), and a $\sin^2$ shape with FWHM= 24 fs. Panel (d): Net ionization after probe pulses with varying delay time. The pump laser was tuned such that initially $N_{\text{esc}} = 3$. The peak time of the probe laser is indicated by the open circles. After [ARS02b].

with an excited $K_2$. This effect has been counterchecked by very detailed calculations for the response of the He droplet including time-dependent correlations [KZ01]. The increase then turns into a slow exponential decay. This is due to the desorption of the molecule from the droplet. The experiment demonstrates a typical setup exploiting a “marker molecule” to get access to the otherwise rather inert material of a He cluster. The experimental preparation is probably more laborious than for the case of homogeneous clusters. As a payoff, the interpretation becomes simpler because the constituents are well controllable (as compared to the previous example of Hg chromophores inside Hg clusters).

#### 5.3.4.6 Collective ionic modes in free clusters

The test case presented in the previous two subsections revealed a very involved dynamical scenario due to the rather complex setup, the particular bond structure of Hg clusters and the combination of a He cluster with a marker molecule. Clusters from simple metals give simpler mechanisms which are (nonetheless or just because of that) very instructive. Figure 5.37 shows the result of a theoretical exploration for the cluster $Na_{41}{}^{+}$ with TDLDA-MD (see Section 3.4.1). The basic mechanism is here the dependence of the Mie plasmon frequency on the cluster radius $\propto R^{-3/2}$, see Eq. (1.18) in Section 1.3.3.3. The test case $Na_{41}{}^{+}$ was chosen because it comes close to a spherical shape which simplifies considerations because the ionic breathing mode will dominate the dynamics. The pump pulse is chosen to ionize the initial cluster into charge state $Na_{41}{}^{4+}$, large enough to produce sizeable Coulomb pressure for a slow expansion with a well developed ionic oscillation. These oscillations of the cluster radius after the pump pulse are shown in the lowest panel of Figure 5.37. The initial cluster was spherically symmetric. Thus the Coulomb pressure is also nearly isotropic which altogether triggers a breathing mode (radial oscillations) of the whole cluster, as shown in panel (a). Quadrupole components and other multipoles were found to remain small in that case. The Mie plasmon resonance depends on charge state and cluster radius. An estimate is shown in panel (b). One sees the fast initial blue-shift which is due to the fast ionization accompanying the laser pump pulse (duration 50 fs). After that, one finds oscillations which perfectly map the radius oscillations with $\omega(t) \propto R(t)^{-3/2}$. The laser frequency for the probe pulse is indicated by an horizontal full line. It was chosen safely below the Mie resonance such that the actual Mie frequency always stays away from resonance with the laser. The electronic response to a probe pulse is small if the Mie frequency is far from the laser and larger if it comes close, see panel (c). The subsequent further ionization is proportional to the response. The net ionization (evaluated 200 fs after the probe pulse, and thus close to the asymptotic value) is shown in panel (d). It behaves precisely as expected: large radii bring the Mie frequency closer to the laser frequency than small radii and thus yield more ionization. This maps perfectly the radius oscillations. Even a detail as the slight drift to increasing average radii is well reproduced. The example shows that properly chosen laser conditions allow a direct mapping of global oscillations for metal clusters. The same setup applied to deformed clusters and exploiting different polarizations for the probe pulse allows one to disentangle also possible ellipsoidal (quadrupole) oscillations of a metal cluster [ARS03]. Such scenarios are yet awaiting experimental tests for the case of free clusters.

However, there are already time-resolved measurements of ionic breathing for large Ag clusters embedded in a glass matrix [PGM$^+$00]. The experimental setup is similar to the one used for producing Figure 5.33. But the dynamics is driven in a more moderate regime to allow studies of the original embedded cluster. The theoretical description is complicated by the large cluster size and the glass environment for which a fully microscopic description is not readily available. The gross features of the experiment could be sufficiently well modeled by the macroscopic dielectric description which is still applicable for moderate excitations.

## 5.4 Excitation by particle impact

Their widespread availability and enormous flexibility leaves lasers in the prime position as tools for analyzing cluster dynamics. Nonetheless, it is important to have an alternative and complementing access by collisions with particles. These probes "see" often different aspects of a cluster and they are in some cases more suited to explore unusual modes. In this section, we will address briefly a few typical examples of excitation by highly charged ions (Section 5.4.2), neutral atoms (5.4.3), and electrons (5.4.4). Before going into these details comes a more general preparatory section dealing with general questions of stability and excitation mechanisms.

### 5.4.1 Stability of clusters

The mechanism of electron extraction is much different for lasers as compared to collisions with highly charged ions. In particular, nanosecond lasers proceed by pumping steadily energy

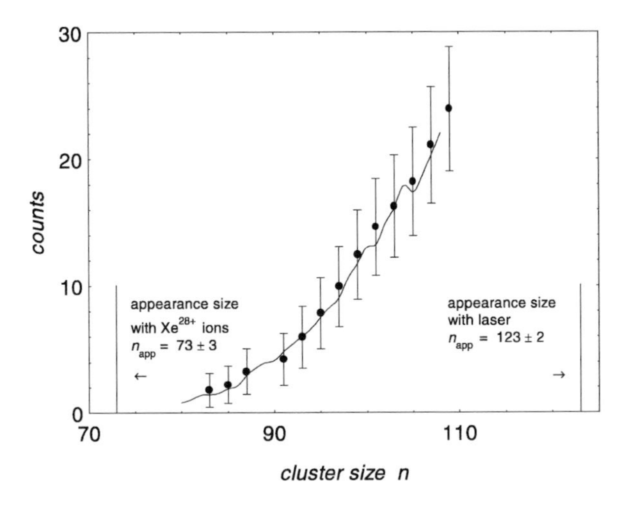


**Figure 5.38:** Yield of stable $Na_n^{4+}$ clusters after collisions of $Xe^{28+}$ with neutral Na clusters. The results as a function of $n$ converge to a limit of stability of $n \approx 73$ which is indicated on the left bottom corner. The stability limit reached with nano-second lasers (see [NBF+97]) is indicated on the right side. From [NHC+02].

into the cluster until energy and oscillation amplitude become large enough to allow emission of one electron. A highly positively charged ion passing by exerts an enormously attractive Coulomb field which gently soaks off a few electrons like a vacuum cleaner. It is thus plausible that laser excitations leave more internal excitation energy in the cluster than ionic collisions. This has consequences for the charge stability of clusters. A key feature is here the appearance size which is the minimum cluster size $n$ down to which a given charge state $q$ (i.e. a cluster $XX_n^{q+}$) remains stable. The systematics of appearance sizes (see Section 2.3.11) is usually explored with ionization through nanosecond lasers, for a summary see e.g. [NBF+97], and unexpectedly large values of $n$ are then found. This means that appearance sizes are not necessarily a direct measurement of the limit of stability. The problem comes from the substantial heating of the cluster during laser ionization. One would like to reduce this heating as much as possible. Ionization by swift collisions with highly charged ions should leave the cluster at a lower temperature which, in turn, should allow one to access larger appearance sizes. This was indeed found. An example is given in Figure 5.38 which shows the yield of cluster sizes for fixed charge state $q = 4$. The appearance size from nanosecond laser experiments lies at $n = 123$ as indicated in the figure. Ionic collisions allow one to produce much smaller sizes for $q = 4$. By extrapolation, one can estimate the appearance size for the collision experiments at $n = 73$ which is as much as 40% lower than for nanosecond lasers.

The above comparison applies for nanosecond lasers with their very slow excitation process. The picture is different again for femtosecond lasers. The laser energy is concentrated on a dramatically short time interval. The thus large actual field strength favors a direct electron emission somehow as the strong Coulomb field of a closely bypassing ion. Moreover, the very short excitation times leave branching into internal excitation less chances. The two processes, ion collision versus excitation by femtosecond lasers, should thus be more comparable. Figure 5.39 shows that this is indeed the case. Those two processes yield similar amounts of internal heating. At second glance, we see interesting differences in the trends: the internal excitation roughly grows $\propto q^1$ for fs lasers but $\propto q^2$ for the ionic collisions. The linear trend for the laser excitation is explained by the fact that both ionization and energy transfer depend

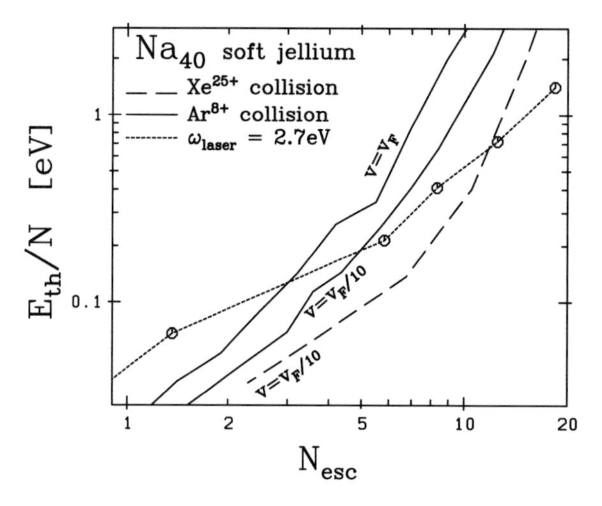


**Figure 5.39:** Internal excitation energy versus achieved ionization for $Na_{40}$ excited with highly charged ions of various sorts and velocities as indicated. The results are compared to the effect of a short-pulse (FWHM= 25 fs) laser with frequency $\hbar\omega_{\rm las} = 2.7$ eV and systematically varied intensity. After [RS00].

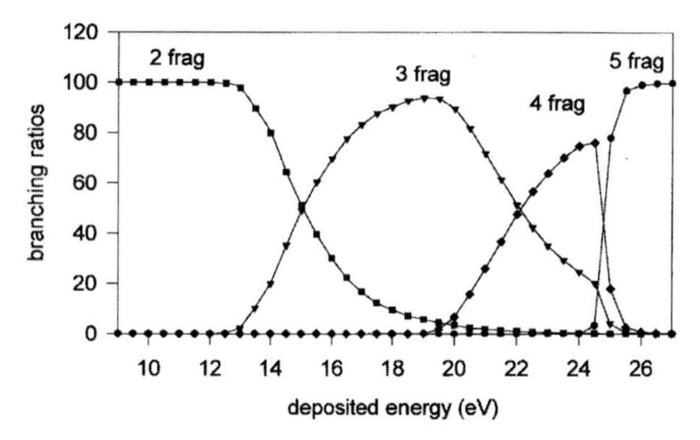


**Figure 5.40:** Relative fragment yields as a function of internal excitation energy for statistical fragmentation of $C_5^+$. Results are obtained by a statistical simulation of a microcanonical ensemble with Coulomb forces properly included. From [WCF+98].

linearly on the intensity. The stronger growth for ionic collisions is made plausible by the fact that the distortion through the Coulomb field grows very fast with decreasing collision distance (which is needed to achieve high ionization). The different trends let us conclude that short-pulse laser excitation is favorable for extremely high ionization while ionic collisions perform better for the moderate charge states which are needed for exploring the appearance sizes.

Gentle ionization is only one aspect of excitation by highly charged ions. Many experiments are more interested in a high energy deposit. One important aspect of collisions is here that they allow one to perform the excitation in a very short time interval. The effect of a high excitation is destruction of the cluster in various reaction channels. The dominant channels change, of course, with excitation energy. Results from an exploratory calculation for a simple and overseeable example are shown in Figure 5.40. The fragment yields as a function of internal excitation for the test case $C_5^+$ were computed here with the statistical equilibrium model from [GH95]. It turns out that the number of emerging fragments is a clear signature for the violence of the process. The average fragment size then naturally decreases with increasing excitation energy. We will see an experimental example in Section 5.4.2.

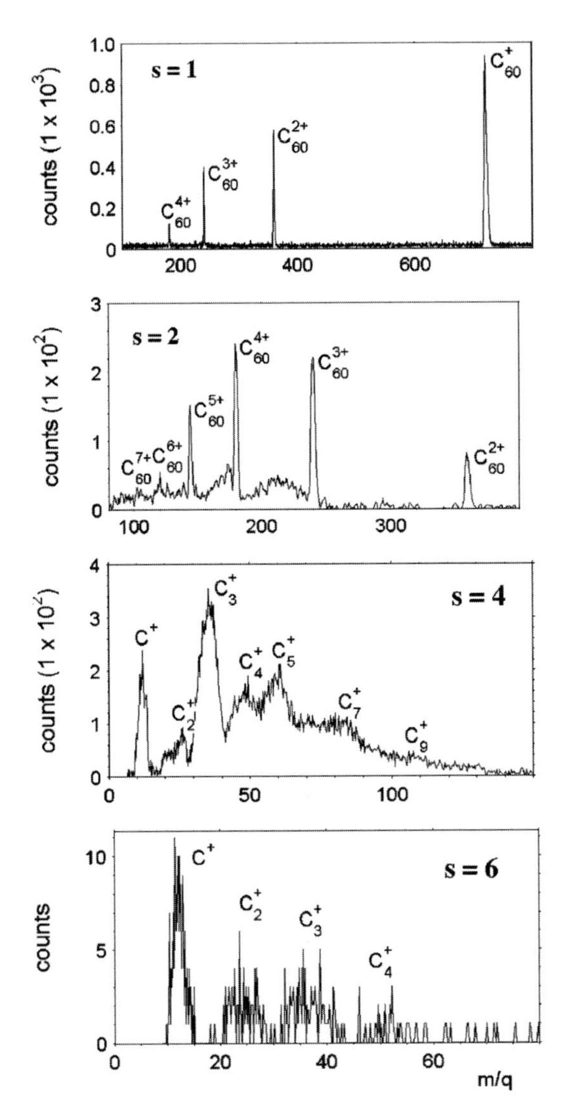


**Figure 5.41:** Fragment distributions of collisions of $Ar^{8+}$ with $C_{60}$ at 16 keV. The results are sorted according to outgoing charge state of the Ar projectile $s = 8 - q_{out}$ (number of transferred electrons from $C_{60}$ to $Ar^{8+}$). The cases $s$ =1, 2, 4 and 6 are shown as indicated. After [CFH+00].

## 5.4.2 Collisions with ions

### 5.4.2.1 From the non-destructive to the destructive regime

We first discuss here the information content in an ion-collision experiment by a typical example using a well collimated and energy resolved beam of $Ar^{8+}$ ion colliding with a swarm of $C_{60}$ clusters. The outgoing Ar beam is analyzed with respect to its charge state, i.e. to the number of electrons absorbed from the target. The reaction products are mass analyzed and the data are stored in coincidence with the charge state of the final Ar ion. Figure 5.41 shows resulting fragment distributions. The peaks in the spectra are labeled by the corresponding fragments. The number $s$ of transferred electrons is a measure for the violence of the reaction. One sees that the $C_{60}$ target remains basically intact for the softest collisions with $s = 1$. Only various ionization stages are produced. The same features hold for $s = 2$. But generally higher ionization is achieved due to the increasing ion impact. The picture changes dramatically for the higher charge states ($s = 4$, $s = 6$). In these cases the $C_{60}$ cluster is broken into small pieces with average fragment sizes decreasing with increasing violence ($s = 4 \rightarrow s = 6$). The interpretation assumes that the transferred charge is a measure of the actual impact parameter (distance of closest approach) for the collision. Low values of $s$ mean large impact parameter (remote collision) while large $s$ means close collisions. It is plausible that the violence of the collision (i.e. the deposited energy) increases with decreasing impact parameter. High energy allows one to break the cluster into more and smaller pieces, in accordance with the findings of Figure 5.41. The situation is very similar to nuclear collisions where the number of produced free neutrons (the smallest fragment here) is taken as a measure for the impact parameter [Pha92]. The elaborate setup of these collision experiments on $C_{60}$ allows one to extract much more information which we cannot report here, for details see [CFH$^+$00].

The complication with collisions is that the cross-sections correspond to an average over all impact parameters. It is not easily possible to conclude from this detailed information about the separate reaction paths. Theoretical support is thus strongly required to shed some light on the underlying dynamical processes. Gentle collisions of neutral compounds can be simulated rather easily with MD using effective atom-atom potential, see e.g. [HIM93]. Charged projectiles are better simulated by taking electronic degrees-of-freedom explicitely into account. This leads naturally to TDLDA-MD, as outlined in Section 3.4.1. Figure 5.42 shows results from TDLDA-MD simulations for collisions of various ions with a $C_{60}$ cluster. The TDLDA method used in these simulations [KJSS00, KS01] is performed on a LCAO basis moving with the atoms [SS96]. This scheme is particularly suited for these collision processes where ionic motion over large distances is to be simulated. The price to be paid is that electron emission is underestimated in the LCAO basis. It would be negligible anyway at low collisional energies. At larger energies, the calculations can still be considered as explorative, giving an idea of the fragmentation process as a function of excitation. Nonetheless, the full TDLDA-MD calculations are expensive and only the initial stages of the collision process can be followed. But that suffices because these are the most crucial moments for the further evolution. The examples in Figure 5.42 nicely illustrate the dramatic dependence of the reaction mechanism on the projectile mass and velocity. The rather light $H^+$ ion is reflected by the $C_{60}$ while leaving very little excitation energy in the $C_{60}$ target. On the other hand, the heavier $C^+$, and even more so $Ar^+$, all at the same velocity, lead to destruction of the $C_{60}$. Higher velocity diminishes the effect as can be seen in the rightmost panels. The heavy projectile $Ar^+$ which destroys the cluster efficiently at the low velocity $v$=0.02 seems just to pass through it at the high velocity of $v$=0.45. Note that $v$=0.45 is close to the electronic Fermi velocity in $C_{60}$ and corresponds to a projectile energy of 240 keV. The projectile at $v$=0.02 has 0.5 keV energy. In order to relate the results of these simulations to the experimental data of Figure 5.41, remember that a low impact parameter of 1 $a_0$ corresponds to a violent collision (nearly "head on") with associated large charge transfer. Moreover, the experiments employ much higher charges ($Ar^{8+}$ ion).

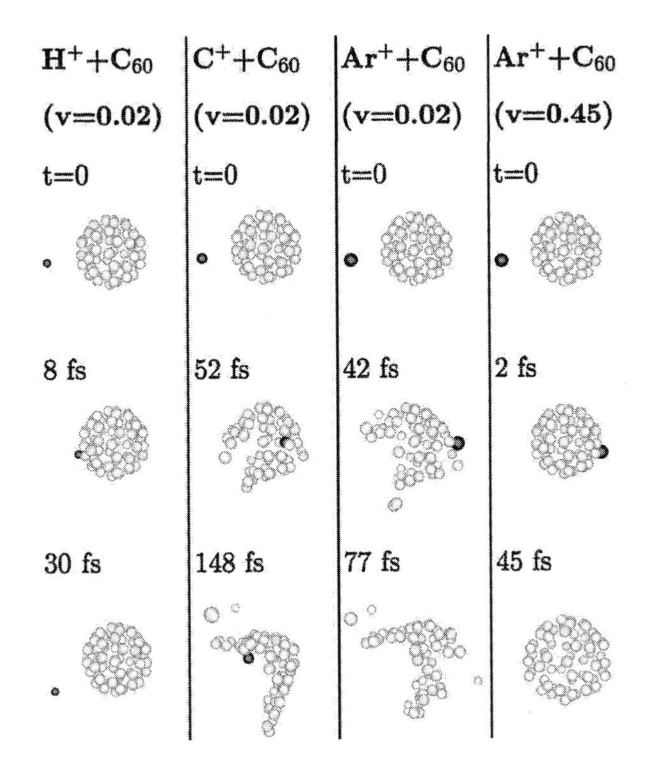


**Figure 5.42:** Snapshots of ion-$C_{60}$ collisions with various projectiles and velocities (in natural units $\hbar\, a_0 Ha$) as indicated, produced with TDLDA-MD simulations. The impact parameter was 1 $a_0$ for each case. The initial direction of the projectile goes horizontally from left to right. The projectile is indicated by a black dot and the C atoms by grey dots. From [KS01].

#### 5.4.2.2 Results in the fragmentation regime

Systematic measurements of fragmentation patterns (with a bias on tracking phase transitions) in finite (small) systems have also recently been performed in the case of hydrogen clusters [GFF+02]. The fragmentation of $H_3^+(H_2)_m$ clusters (with $m$ ranging between 6 and 14) has been studied on an event by event basis in the spirit of techniques developed in nuclear physics [DST00]. Clusters are hit by energetic He atoms with 60 keV/amu kinetic energy. The fragment mass distributions are analyzed as a function of the energy deposited in the cluster. This allows one to plot a caloric curve which exhibits a phase transition in a region around 60 eV excitation energy. This transition can be seen in Figure 5.43 where fragment mass distributions are presented for the various clusters, as a function of normalized fragment size (to allow comparison between the various initial clusters) and for various deposited energy bins (the average deposited energy is indicated in brackets on each plot). The data had been inspected preferably for a signal of a phase transition in connection with the appearance of a negative heat capacity (see also the discussion of Section 5.1.3). But an interest of these experimental data also lies in the way the analyses has been performed. It is indeed inspired from data analysis developed in the search of nuclear multi-fragmentation [DST00]. Fragments are studied on an event by event basis using a multi-detector. Both these demanding steps (event by event analysis and use of a multi-detector) have been identified as key requirements for a proper analysis of such complex situations as multi-fragmentation dynamics. Such anlysis are still relatively rare in cluster physics. They will hopefully be further developed in the future.

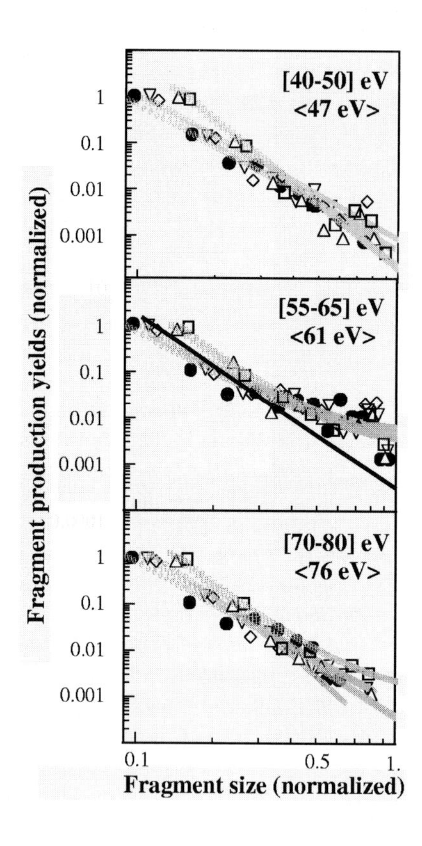


**Figure 5.43:** Fragment distributions from fragmentation of $H_3^+(H_2)_m$ ($6 \leq m \leq 14$) hit by 60 keV/amu He atoms. The fragment size distributions gather results from the various initial cluster masses and are sorted according to various deposited energy bins (40–50, 55–65 and 70–80 eV from top to bottom). Note the overabundance of heavy fragments (as compared to the two other panels) in the medium panel, which is thus associated with a smaller temperature and hence a signal of a negative heat capacity (see also Section 5.1.3). From [GFF+02].

## 5.4.3 Collisions with neutral atoms

Atom–cluster collisions are a generalization of atom–atom or atom–molecule collisions which constitute a standard tool of investigation in atomic physics, see e.g. [Bra83, SNBB02]. Several aspects can be investigated depending on the kinematical conditions. The atomic projectile can be used as a hard sphere for fast excitation of the target (particularly in connection with He projectiles). Less violent reactions explore transfer and thus allow one to disentangle details of the outer electron shells. We will discuss one example from each of these aspects.

### 5.4.3.1 Fragmentation dynamics

The He atom is a special case amongst the neutral projectiles. It is the most inert of the atoms and can be considered as a frozen projectile under many circumstances. Scattering of He has been used, e.g., as a means to probe the structure of surfaces [SBT+97] and it may serve for this purpose in connection with clusters as well. Different, and more dynamical aspects, are revealed when looking at inelastic collisions with substantial energy loss. Figure 5.44 shows an example from collisions of neutral He with free $Ar_n^+$ clusters. The mechanism explored here is energy deposition by the colliding He atom and subsequent evaporation of Ar monomers until the deposited energy is used up. As with all collision experiments, the data collect events from all impact parameters and thus from a wide band of excitation energies. Accordingly,

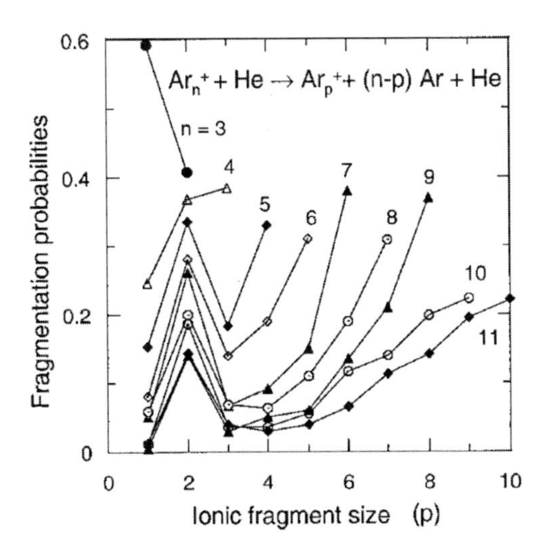


**Figure 5.44:** Relative yield of fragment sizes $p$ after collisions with He atoms having 4.8 keV kinetic energy for various initial $Ar_n^+$ cluster sizes as indicated beneath each curve. From [BBFP02].

one sees a broad distribution of final ion sizes. Plotted results are fragmentation probabilities as a function of ionic fragment size $p$ ($Ar_n^+$ + He $\rightarrow$ $Ar_p^+$ + $(n-p)$Ar + He). Large impact parameters are related to low excitation but have the largest geometrical cross-section. Thus the fragment yields have a maximum at the largest fragment and steadily decrease towards smaller ones, except for the sudden peak at $Ar_2^+$. This is caused by the particular stability of that cluster ion which sets a halting point for further evaporation. An exception is collision with $Ar_3^+$ which has a single $Ar^+$ as preferred fragment. A detailed analysis shows that the mix of reaction channels seems here different from the case of larger clusters in that a larger fraction of electronic excitation is involved [BBFP02]. Such a close inspection requires more elaborate observables as, e.g., double differential cross-sections. Collision experiments (with atoms or highly charged ions) thus need to collect as many observables as possible to reveal their strengths.

#### 5.4.3.2 Electron transfer reactions

Soft collisions are widely used in atomic physics to study transfer processes. There exist meanwhile a few studies on electron transfer in atom–cluster collisions. Figure 5.45 shows results for a transfer reaction from a neutral Cs projectile colliding with a metal cluster. That combination is particularly suited for transfer because Cs has a loosely bound valence electron, ready to be caught by the attractive potential well of the positively charged target cluster. There thus emerge sizeable transfer cross-sections as seen in the figure. It is also plausible to see that the doubly charged cluster $Na_{31}^{2+}$ has a much larger cross-section than $Na_{31}^+$ because its Coulomb attraction is much larger. Such reactions are well suited for treatment by time-dependent mean-field theories. An analysis of metal-atom to metal-cluster transfer processes was performed in [KJSS00] with TDLDA using a LCAO basis. An even more refined analysis of transfer reactions is reported in [RLWL+99]. It considers Na on $C_{60}^+$ collisions where the Na atom was used in two different states, namely in the ground state (basically $3s$ of the valence electron) as well as a laser excited $3p$ state. This hints at the enormous potential

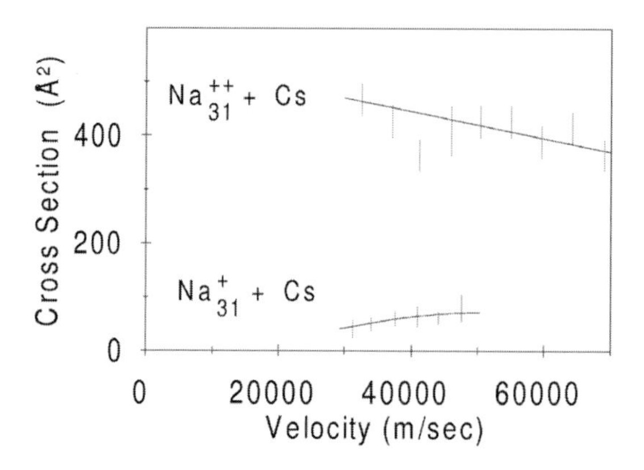


**Figure 5.45:** Transfer cross-sections for the reactions $\mathrm{Na}_{31}^{+} + \mathrm{Cs} \longrightarrow \mathrm{Na}_{31} + \mathrm{Cs}^{+}$ and $\mathrm{Na}_{31}^{2+} + \mathrm{Cs} \longrightarrow \mathrm{Na}_{31}^{+} + \mathrm{Cs}^{+}$ as a function of the collision energy of the Cs atom. From [BCCL01].

of this kind of experiment. The proper interpretation requires detailed theoretical reaction calculations which, however, are very demanding and, to some extent, still in a developing stage.

### 5.4.4 Electron scattering

The lightest projectile is the electron. And electron scattering is a most promising tool for cluster analysis [PEFS02]. One can be inspired by the enormous amount of information gained from electron scattering in nuclear physics: analysis of elastic scattering cross-sections, for example, allows one to conclude on the nuclear charge distribution, see e.g. [FV82], and carefully triggered inelastic processes give access to non-standard collective modes which are masked by the dipole resonance in photon excitation, see e.g. the discovery of shear modes (scissors) in [BRS$^+$84]. Electrons as probes are also widely used in molecular and surface physics. There is the tunneling electron microscope (TEM) which is already employed for the handling of nano-particles on surfaces, see e.g. [MWL$^+$01], or as an injector for probing surfaces by electron scattering [EFSP00]. Surface structures are also regularly analyzed with low-energy electron diffraction (LEED), for a review see [Hei95].

First theoretical estimates show that inelastic electron scattering from clusters can give access to collective modes other than just the Mie surface plasmon [GISG98, GIPS00]. A special case of inelastic electron scattering is electron energy-loss spectroscopy (EELS) which uses predominantly the forward direction [IM82]. Here one can also hope to get access to non-dipole modes of the cluster, preferably the volume modes (see Section 5.2.2.3). From a slightly different perspective, one may also gain useful information from the bremsstrahlung generated by the collision with fast electrons [GIS98]. In a simple "brute force" approach, one can use energetic electrons to deposit large amounts of excitation energy in a cluster, similar to collisions with highly charged ions which were discussed in Section 5.4.2. An example of electron-impact-induced fragmentation of fullerene ions can for example be found in [HAA$^+$00].

Elastic scattering is probably the simplest first step. The electron diffraction pattern will give hints on gross properties such as system radius and structure. Such analysis was done,

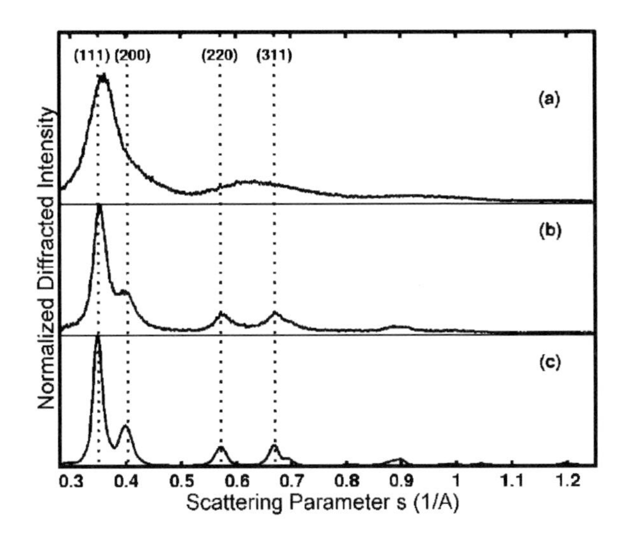


**Figure 5.46:** Diffraction pattern for scattering of 80 keV electrons from a beam of free Pb clusters drawn after Debye functional analysis in terms of scattering parameters. Panel (a): for Pb clusters produced with He as carrier gas. Panel (b): for Pb clusters produced with Ar as carrier gas. Panel (c): theoretical result for fcc bulk Pb. From [HWB$^+$01].

e.g., in [RHUM93, HRU93, HWB$^+$01] for various sorts of clusters. Figure 5.46 shows a result of electron diffraction on Pb clusters produced in a gas aggregation source. The case is similar to Debye scattering from a crystalline powder and it is analyzed with the same methods. Analysis of the mean diameter using the diffraction pattern shows that growth in He yields the smaller clusters while growth in Ar allows one to grow larger clusters. Comparing the experimental results with various structures, one finds a best fit for decahedral geometry, for details see [HWB$^+$01]. The short example here gives a hint on the large amount of structural information contained already in elastic electron scattering. Much more can be recovered when refining the analysis, from the experimental as well as from the theoretical side.

An example of an elaborate electron scattering analysis in the case of water molecules is given in Figure 5.47. The key quantity to describe fully resolved reaction kinematics for collision induced ionization is the five-fold differential cross-section depending on incident energy, projectile scattering angle (in-plane and out-of-plane angle), kinetic energy of ejected electron, and angle of ejected electron (again in-plane and out-of-plane). In Figure 5.47 is shown a cut through two of these dependences with integration over the others. A theoretical description requires a fully fledged distorted-wave Born approximation (DWBA) [CHH02] because electron propagation in the target is everything else than perturbative. Experiments are not simple either, for a similar reason, namely because low energy electrons are easily perturbed by any weak electron-magnetic field around in the set-up. Nonetheless, the figure demonstrates that such experiments are feasible when investing sufficient patience. The case of clusters is again an order of magnitude more demanding than atoms, for example, due to the lower counting rates, particularly for free clusters. But progress in experimental techniques will deliver sooner or later more detailed information on clusters.

## 5.5 Strongly non-linear laser processes

We come now to the last point mentioned in the introduction to this chapter: laser dynamics in the regime of strong excitations. One is facing here completely new phenomena as was

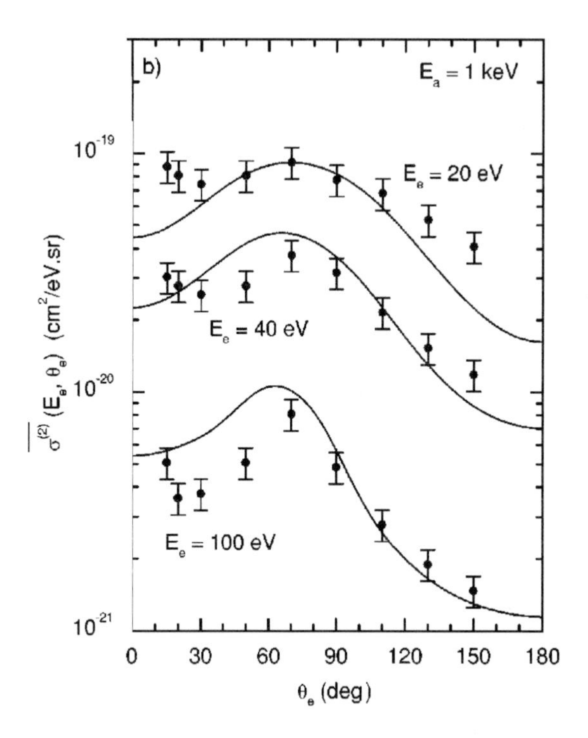


**Figure 5.47:** Double differential cross section depending on the energy and in-plane angle $\theta_e$ of the ejected electron induced by electron impact on a $H_2O$ molecule. The impinging electron had an initial kinetic energy of 1 keV. Theoretical results (full lines) from [CHH02] and experimental data (full circles) from [BR85].

already hinted at by Figure 5.3 showing that energetic radiation is emitted by those highly excited clusters. One such new aspect is that the dynamics in the violent regime is more or less independent of the cluster material. Note the many examples of dynamics with low or moderate intensities brought up in previous chapters and previous sections of this chapter. They span a broad variety of different dynamical scenarios changing from one cluster to another depending on the type of the cluster (metallic, ionic, covalent, or rare gas) and on its composition. Already simple optical response exhibits a rich selection of different spectra and the variety grows huge when proceeding to the semi-linear regime of femtosecond spectroscopy with moderate laser pulses, see Section 5.3. This changes dramatically when stepping to even higher laser intensities. The switching point was discussed in connection with Figure 3.3 where the various forces in a cluster are compared with the external force exerted by the laser. The laser forces break into the domain of core electrons for intensities above $I = 10^{16}\,\mathrm{W/cm}^2$. Once the core electrons are released, all types of clusters look the same. It does not matter any more whether one deals with a rare gas or with a metal. One always finds a large number of free electrons mixed with ions of various charge states and both packed in a rather small spatial region, which reminds one of a typical plasma. Due to the involved sizes it is usually called a nano-plasma. Cluster physics in very intense external fields is thus turning to the physics of small plasma drops, a field of its own, showing its own set of phenomena and requiring its own theoretical modeling.

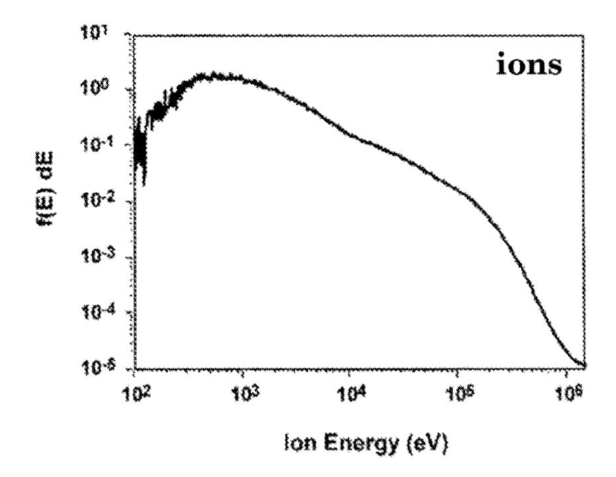


**Figure 5.48:** Spectrum of ion energies from xenon clusters (2500 atoms) and laser pulses with peak intensity of $2 \times 10^{16}$ W/cm$^2$. From [DST$^+$98].

### 5.5.1 Signals from exploding clusters

The investigation of the non-linear dynamics of atomic clusters has livened up with the advent of table-top devices that produce sub-picosecond laser pulses with intensities easily exceeding $10^{17}$ W/cm$^2$ [MLT$^+$94, BSC$^+$96, DTS$^+$97b, LDNS98, KSK$^+$99, SKvIH98]. Due to the nano-plasma dynamics discussed above, there may be a strong electromagnetic coupling to the Mie plasmon frequency that leads to nearly 100% absorption rates [DTS$^+$97a]. Moreover, the clusters are isolated objects such that no surrounding environment will absorb their excitation energy, in contrast to solids. This allows them to reach enormously high charge and energy states. The emission of high-energy electrons in the keV range [LDNS98], highly charged and very energetic ions [DTS$^+$97b] and fragments [KSK$^+$99] as well as X-ray production [MLT$^+$94, DLS$^+$97] (see also Figure 5.3) are then the spectacular manifestations of this laser-cluster interaction. These phenomena have been observed in a similar fashion for many different sorts of clusters, from rare gases to metals and even up to organic species [BSC$^+$96].

Figures 5.48 and 5.49 complement the example of energetic photon emission (from Figure 5.3) by energy spectra from heavy charged fragments. Figure 5.48 presents a result from a strong laser excitation of large Xe clusters. It shows the distribution of kinetic energies of the emitted ions. Most of the ions have kinetic energies of a few keV. Nonetheless the mean kinetic energy is $45 \pm 5$ keV. This is due to the remarkably long tail towards very high ion energies, delivering ions with energies up to 1 MeV. Figure 5.49 shows a much different test case and a different way of analysis. It deals with strong excitations of an heterocyclic compound. The emerging fragments are sorted according to kinetic energies and charge to mass ratio. One sees enormous energies for the highly charged fragments and a quick decrease of energies for lower charges. This hints clearly at the Coulomb force as the main actor. For comparison, estimated kinetic energies from a pure Coulomb explosion model are drawn (filled triangles). The model assumes immediate charging and a start of the Coulomb trajectories from the cluster site. The agreement with the data confirms such a picture for that particular example. Much depends, of course, on the kinetics of the excitation. We will come to that in Section 5.5.2.

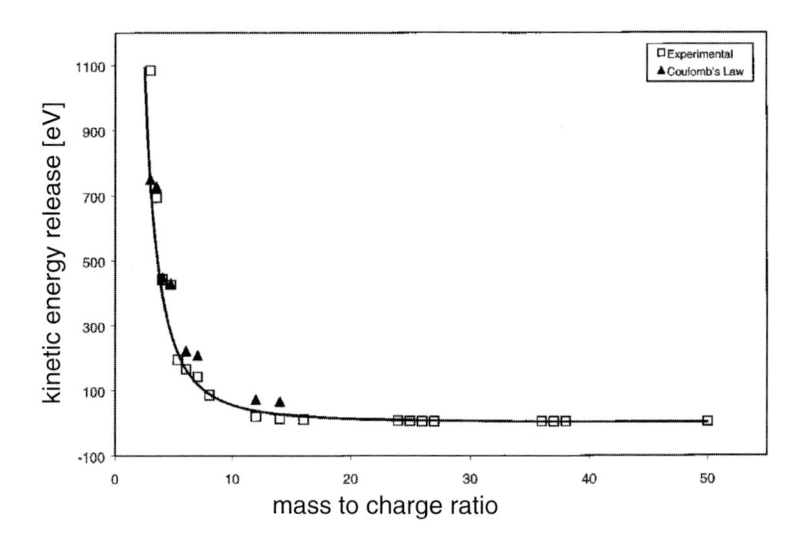


**Figure 5.49:** Kinetic energies versus mass to charge ratio for charged fragments (excluding mere protons) ejected from explosion of a pyridine cluster after strong laser impact. Shown are maximum energies for each ratio. Results are compared with a simple estimate from Coulomb explosion starting at an initial cluster configuration. From [CWFJ02].

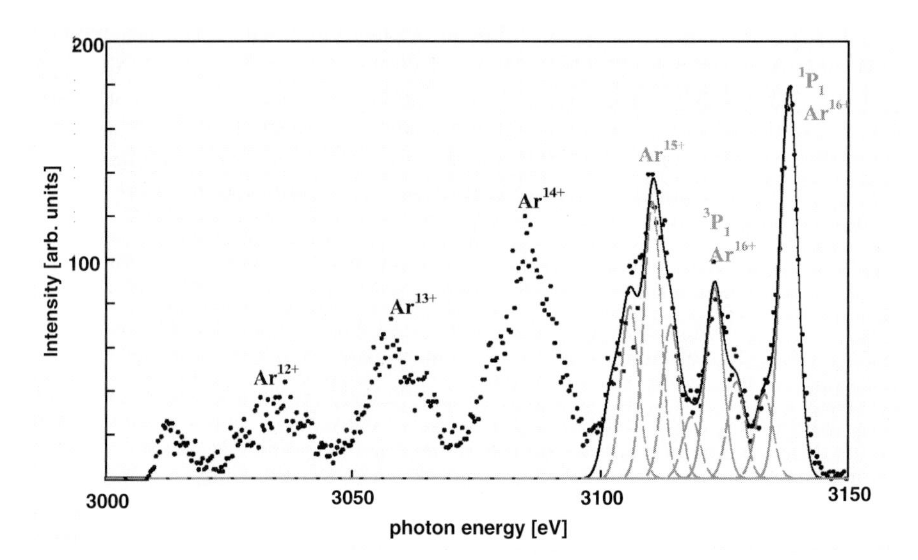


**Figure 5.50:** High resolution X-ray spectrum from irradiation of Ar clusters by a laser pulse with FWHM=80 fs, intensity $6.7 \times 10^{17}$ W/cm$^2$, and wavelength 800 nm. The various peaks originate from K shell ionization of $Ar^{q+}$ in various charge states as indicated. Each peak covers several transitions as indicated by dashed lines for the two highest charge states. After [LDG$^+$01].

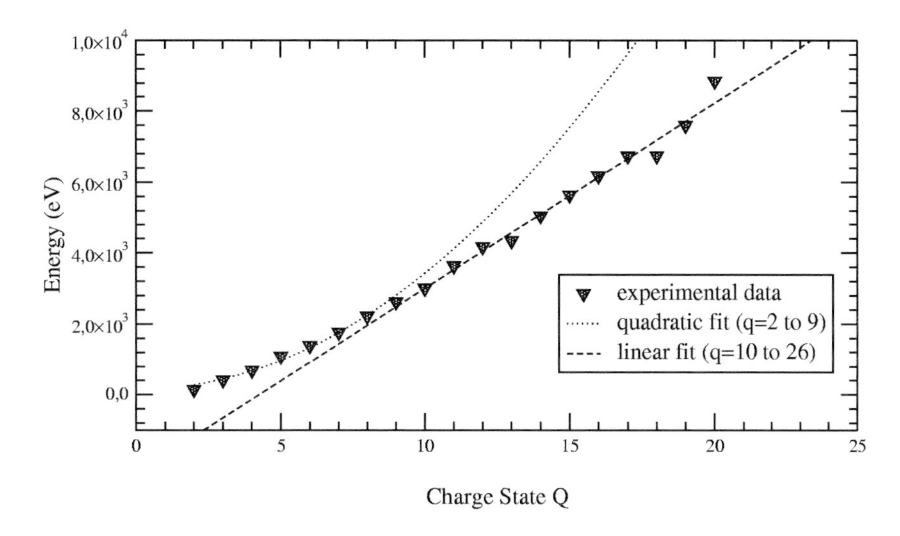


**Figure 5.51:** Average kinetic energies of ions emitted from an exploding Pb cluster versus charge state. Results are an average over 1000 events with laser pulses with FWHM=760 fs, intensity $3{*}10^{15}\ \mathrm{W/cm}^2$, and wavelength 790 nm. Average cluster size is $N = 250$. The dashed line shows a linear trend fitted to high $q$ data and the dotted line a quadratic trend applicable for low $q$ data. From [LVC$^+$02].

A detailed look at the mechanism for X-ray production is taken in Figure 5.50. It shows the X-ray spectrum with high resolution in a narrow window of energies for K shell transitions ($1s \leftrightarrow 2p$). Several well separated peaks appear. Each one can be associated with a different charge state. Several sub-lines contribute within a given ion as indicated by the dashed lines for $Ar^{15+}$ and $Ar^{16+}$. These are, however, not resolved in detail. The spectrum demonstrates that a mix of highly charged ions is present in the excited cluster. It is obvious that the very intense laser pulse has efficiently removed many electrons from the Ar core states. They are floating around throughout the whole cluster forming, together with the ions, a nano-plasma.

More information on emitted ions is shown in Figure 5.51 for the case of a highly excited Pb cluster [LVC$^+$02]. The spectrum of kinetic energies has been recorded for each charge state separately. The figure shows the average kinetic energies versus charge state. Two regimes can be recognized. For low charge states $q$, the kinetic energy depends quadratically on $q$. This can be explained by Coulomb explosion of an initially localized ensemble of Pb ions with charge $q$. The trend turns to a linear dependence on $q$ for high charge states. This is interpreted in [LVC$^+$02] as coming from hydrodynamical flow of the highly charged ions following the expansion of the nano-plasma. Such an interpretation is, of course, not sufficiently supported by the few available experimental data. It needs to be corroborated by model calculations. These will be discussed in the next subsection.

Most experiments in the highly excited domain are done with rare gas clusters. This is for the simple technical reason that rare gases allow more easily the production of large clusters in supersonic beams. There exist also many experiments with metal clusters, see e.g.

[KSK+99, LVC+02] and the example of Figure 5.51. Figure 5.49 has exemplified a much different species, namely organic compounds. The results are all very similar because details of binding become less important in these strong fields. The laser releases quickly many electrons into the cluster volume where they then form a nano-plasma. The properties of this plasma state are more or less independent of the initial material.

## 5.5.2 Modeling exploding clusters

Various models have been developed to describe the cluster dynamics in the violent regime. The complexity of the problem calls for simple schemes. One family of approaches, coming from molecular physics, concentrates on a detailed tracking of the trajectories in the Coulomb explosion. This amounts to a MD calculation with Coulomb forces from an initial configuration with a given ensemble of ions, as e.g. in [PJ98]. This seems to provide a pertinent picture of the energy distributions of ions.

An important ingredient in the excitation dynamics are the free electrons. They build, together with the ions, a nano-plasma. Several models have been developed in that direction [BR94, RPSWB97, IB00, RLB01, LJ00, KR01] which all take into account the role played by these free electrons inside the cluster. The basic mechanisms at work can be summarized as follows. The laser very quickly strips a sizeable number of electrons from their parent atoms. These electrons form a reservoir of quasi free electrons, which can strongly couple to the laser when the electron density matches a critical value characteristic of the dipole eigenfrequency of the system (Mie resonance). The global response of the cluster is then characterized by a heating of the electron cloud and electronic emission, in addition to the possibly enhanced dipole oscillations. The large net charge and the high excitation energy thus acquired by the cluster lead to its final explosion. Again, it should be noted that this violent scenario is, to a large extent, independent of the nature of the irradiated clusters. The general scenario we have just outlined, however, does not precisely specify the relative importance and the time scales of the various mechanisms at work. As a consequence, although the global scheme is generally accepted, the debate on the relative importance of the competing processes (various ionization mechanisms inside the cluster, role of the plasma resonance, of electron temperature, of net charge, etc.) remains largely open and various models are still investigated to explain experimental trends. As one typical (and probably the most prominent) example for modeling in the regime of violent laser excitations of large clusters, we will outline in the following the nano-plasma model initiated in [DDR+96].

Most examples for violent events deal with large clusters and reach up to rather high electronic and ionic velocities. Drastic simplifications are needed to cope with that situation. Therefore, one starts with a macroscopic hydrodynamical picture for the flow of ions and electrons augmented with rate equations for transitions between species. The relevant dynamical variables then reduce to a set of integral quantities (radius, ionization, etc.). Such an oversimplified picture aims at describing just the gross properties of the system. This is appropriate for the case of high intensity irradiation of clusters because the details of the excitation mechanism tend to be washed out to the benefit of a global plasma-like behavior, in fact a nano-plasma due to the finite and small extension. The model has to cover a complete scenario of the various interaction processes, taking into account ionization, heating, polarization fields, electronic emission and cluster expansion simultaneously.

The basic dynamical degrees of freedom in the nano-plasma model are

$$\begin{aligned} N_j &= \text{number of ions in charge state "j"}, \\ N_e &= \text{number of "free" electrons}, \\ E_{\text{int}} &= \text{internal energy of the electron cloud}, \\ R &= \text{radius of the cluster.} \end{aligned}$$

Ions, electrons and energy are assumed to be distributed homogeneously in a sphere of radius $R$. The time evolution of these dynamical parameters is determined by rate equations. We present here a summary of them and refer the reader to Appendix F.4 for a more detailed discussion.

The evolution of ion numbers is determined by the rate equations

$$\frac{dN_j}{dt} = W_j^{\text{tot}} N_{j-1} - W_{j+1}^{\text{tot}} N_j \tag{5.10}$$

where $W_j^{\text{tot}}$ is the rate for ionization of ions in charge state $N_j$. The electron number changes as

$$\frac{dN_e}{dt} = \sum_j j \frac{dN_j}{dt} - \frac{dQ}{dt} \tag{5.11}$$

where $Q$ is the total net charge of the cluster and its change is determined by the integrated net flow through the surface. The change in cluster radius is driven by the total pressure

$$\frac{\partial^2 R}{\partial t^2} = 5 \frac{P_C + P_H}{n_i m_i} \frac{1}{R} \tag{5.12}$$

where $P_C$ accounts for the Coulomb forces and $P_H$ is the thermodynamic pressure of the electron gas. The change of the internal energy of the electron cloud (temperature $T_{\text{e}}$) is determined by heating through absorption of electromagnetic energy ($\dot{E}^{(\text{el-magn})}$), cooling through global expansion ($\partial R/\partial t$ term), ionization processes ($dN_j/dt$ term), and energy loss by electron flow through the cluster surface ($E_{\text{fs}}$), yielding altogether

$$\frac{d}{dt}(E_{\text{int}}) = \dot{E}^{(\text{el-magn})} - \left( \frac{2E_{\text{int}}}{R} \frac{\partial R}{\partial t} + \sum_j I_P^{(j)} dN_j/dt + E_{\text{fs}} \right) \tag{5.13}$$

with $I_P^{(j)}$ the ionization potential of an ion of charge state $j$ and $E_{\text{fs}}$ accounting for energy loss by electron flow through the cluster surface. Some entries require temperature rather than intrinsic energy. But these are related trivially as in the ideal gas by the relation

$$E_{\text{int}} = \frac{3}{2} N_e k T_e \quad . \tag{5.14}$$

The equations (5.10) through (5.14) (with more detailed entries as given in Appendix F.4) constitute the dynamics of the nano-plasma model. After all, the model contains the basic competing processes in the dynamics of a cluster in the nano-plasma state. It is technically simple. But it requires several empirical ingredients, as e.g. the various ionization rates, and

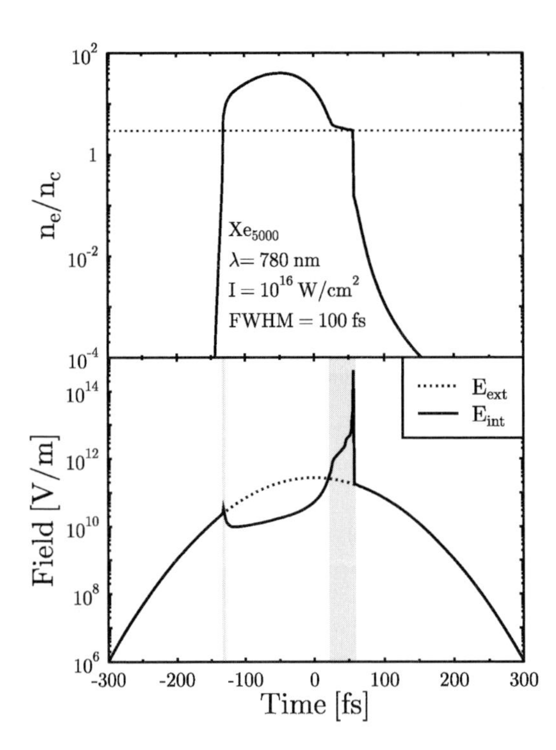


**Figure 5.52:** Result of the nano-plasma model for the time evolution of key observables for an exploding rare gas cluster. Test case is an exploding $Xe_{5000}$ cluster irradiated by a laser pulse with FWHM=100 fs, wavelength of 780 nm, and peak intensity of $10^{16}$ W/cm$^2$. Upper panel: number of free electrons relative to the critical number $n_c$. Lower panel: field strength, from external laser field (dotted line) and net field (full line). After [MBB$^+$03].

involves strong simplifications, e.g. for the emission rate. Thus there remain many open ends which may be questioned and which call for improvement. For example, the assumption of homogeneous distributions is not strictly valid in the presence of different charges and strong Coulomb force. Nonetheless, the nano-plasma model serves as a good starting point to get first insights into the detailed time evolution of an exploding nano-plasma.

A typical result from the nano-plasma model is shown in Figure 5.52. The peak of the laser pulse sets the zero point of the time scale. As soon as the field strength becomes sufficiently large, the electrons are stripped off their parent atoms mainly by optical ionization via tunneling, and the number of free electrons (upper panel) increases rapidly. In fact the plotted quantity is the electronic density scaled to a "critical" density. The concept of a critical density $n_c$ is taken from the volume plasma. It represents the electron density at which the laser frequency is in resonance with the volume plasmon frequency $\omega_p$. The critical point in a nano-plasma lies at the Mie frequency $\omega_{\text{Mie}} = \omega_p/\sqrt{3}$, i.e. at a density of $3n_c$, see Section 1.3.3.1. It is indicated in the upper panel by a horizontal dotted line. As soon as the electron density reaches this critical value, the Mie plasmon response sets on as can be seen from the then growing difference between external and intrinsic fields in the lower panel of Figure 5.52. The electric field is then shielded due to the high electron densities reached and the field drops. The tunnel ionization rate falls off, but electrons are still created through thermal collisions. From times $t \simeq -100$ fs onwards, the electron temperatures are high enough for some electrons to leave the cluster leading to a decreasing density until time $t \simeq 10$ fs when the second, more important resonance takes place. The inner field is then amplified to 1.5 times the external field. This leads to a large heating of the cluster with electronic temper-

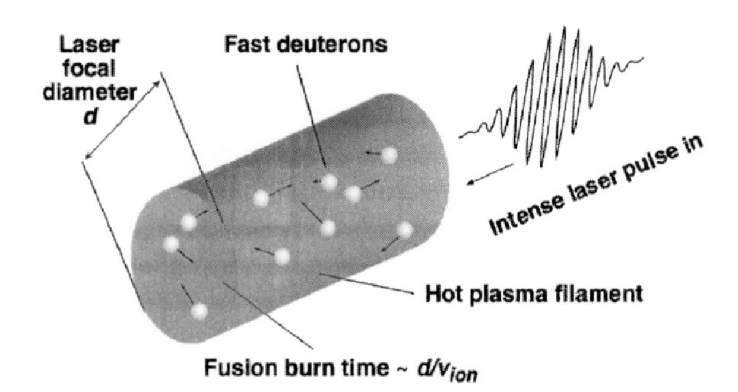


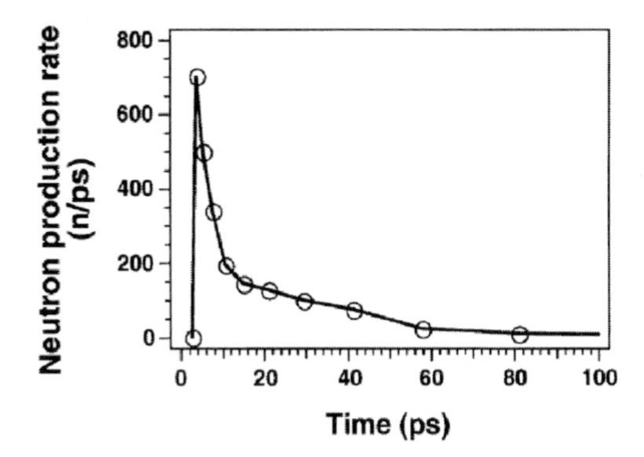


**Figure 5.53:** Neutrons from exploding Dt clusters. Upper panel: schematic view of the experimental setup. Lower panel: neutron production rate as a function of time. After [Zwe$^+$00].

ature reaching the keV range. The rate of electron emission increases consequently giving a highly positively charged cluster which will explode due to huge Coulomb repulsion between ions. Fast emission and explosion then let the electron density drop quickly as can be seen from the figure.

Even though the nano-plasma model provides a pertinent description of the gross features of the interaction of intense lasers with clusters, it cannot keep up with all experimental results observed, especially those far from average values. For example, the model agrees with the average ion charges observed experimentally but fails in describing the far tails of ion charge distributions which are at the origin of X-ray generation. This is also the case concerning ion energies. One needs a description in terms of more detailed distributions to accommodate all regions of phase space up to the outer tails.

### 5.5.3 Nuclear reactions

The dynamics of a highly excited nano-plasma establishes a further link between clusters and nuclear physics. The fast ions which are ejected from an exploding cluster can be used as projectiles for nuclear fusion. The upper part of Figure 5.53 shows a schematic setup for such experiments. An ensemble of deuterium lusters is brought into contact with an intense laser beam. The strong laser field induces Coulomb explosion of the clusters with subsequent emission of energetic deuterons. These deuterons collide with the deuterons from other clusters.

A fraction of them has relative energies above the fusion threshold [HGGE97]. This allows a nuclear reaction which can be identified by the emission of free neutrons with a well defined energy. The lower panel of the figure shows that one finds indeed a large amount of fusion neutrons with a peak very shortly after the laser pulse. Exploding clusters thus provide a new and surprising road to controlled fusion. The energy balance, however, is not sufficient for large scale production but applications, for example in terms of table top neutron beams, are conceivable. Nonetheless, this final example demonstrates nicely the enormous richness of phenomena which is offered by cluster dynamics.

## 5.6 Conclusion

In this chapter we have tried to explore new frontiers in cluster dynamics. The title of the chapter was ambitious and in a sense reducing, to the extent that we precisely tried to go beyond the actual boundaries of the field, if one can decide where they lie. Still, even if such concepts of frontiers are by nature time dependent, it is clear that cluster dynamics is a field of physics in full development. We have tried here to review several emerging directions, which we consider as particularly promising in terms of cluster dynamics. This of course does not imply an exhaustive review of all that is going on in the field and there for sure exist further research axes which we have overlooked. But this is the very nature of such an exercise.

To sort material in the field of cluster dynamics is not an easy task and one needs a guiding line to organize the presentation. We have, quite naturally, sorted the presented results as a function of the violence of the interaction process with the probe. This has led us to basically distinguish three classes of phenomena corresponding to three dynamical domains. The linear response domain, in particular around the optical response, has already been explored, and discussed at length, for many years. There however remain several interesting investigations to be pursued in the domain or in connex fields. The linear domain corresponds to a reasonably well defined theoretical frame. On the other extreme lies the strongly non-linear domain, which, in turn, raises specific theoretical problems, in addition to the experimental challenges. We have mostly discussed this domain in the context of the interactions with intense laser beams, a widely used device both for fundamental purposes and for potential applications. In the strongly non-linear domain one is led to explore physical situations in which the cluster, which is violently destroyed in the interaction process, may provide a laboratory for other investigations rather than be the object of investigation itself.

In between these two extreme dynamical domains lies what we call the semi-linear domain in which the cluster experiences a sizable excitation but remains stable or at least metastable during a sufficiently long time to allow dedicated investigations. This is so to say the domain of choice for truly investigating cluster dynamics. We have thus devoted a large fraction of the chapter to examples of applications in this dynamical domain. Typical examples concern here photoelectron spectroscopy and pump and probe dynamics. Here again lasers play a central role and we have mostly considered them as probes in the presented examples. Still many investigations are also starting to be performed with charged or neutral projectiles hitting clusters. We have presented a few examples of applications. It should however be realized that the share of the collisional process is hardly comparable to the one of lasers. Finally we ought to mention that, although the chapter was focused on dynamical problems, we have also briefly discussed a few promising developments in the field of cluster structure. Still the examples presented here were, to some extent, related to dynamical problems.

The field of cluster dynamics is obviously a quickly and strongly developing subfield of cluster physics. We have presented in this chapter a few exciting new directions of research which we hope will lead to even more new developments.

# 6 Concluding remarks

Cluster physics is a young and fast developing field of science, with important issues both in terms of fundamental science and potential technological applications. Throughout this book, we have tried to illustrate this "simple" statement with particular emphasis on cluster dynamics. In fact, although "simple", this covers already several aspects. Let us briefly comment on these various points. Cluster science, as such, started only in the last three decades of the twentieth century. A key barrier before that time had been the production of free clusters. This restricted the field to embedded or deposited clusters which stay in strong interaction with an external environment. However, practical applications of clusters had been around for centuries, long before cluster science started. The reader is reminded of the use of these objects in colored glasses since ancient times or, more recently, in photography.

In spite of its short history, cluster science has become in a few decades a major domain of science with its own and genuine questions both experimentally and theoretically. The discovery of fullerenes, as specific carbon clusters, has led to the development of a new field of carbon compounds and nanotubes (although still related to its "mother" domain of cluster science). But cluster science is much more than the mere technological developments around fullerenes. Several other types of clusters may possess interesting properties. And as such they deserve dedicated studies. These experimental investigations actually call for equally dedicated theoretical developments, accounting for the special features of the cluster many-body problem. We have seen this for several examples throughout this book.

Inside the huge field of cluster science, we have focused in this book on dynamical questions. Usually, the first identification of a physical system requires "static" investigations in order to characterize its equilibrium properties. Once the structure has been understood, one starts to explore dynamical aspects. These depend also on the actual excitation mechanism and thus provide a broader field of investigations than mere structure studies. Clusters followed this "natural" evolution and dynamical investigations started soon after structure studies. Clusters are bound by electromagnetic interactions and are thus explored mainly by electromagnetic probes. Lasers provide here a particularly versatile tool with a variety of tunable parameters: frequency, intensity, and pulse profile. The remarkable development of laser technology during the past two decades has also pushed studies in cluster dynamics. The importance of lasers for cluster dynamics is obvious from many examples in this book. To be complete, of course, one should also mention the alternative means of investigation provided by collisions with charged particles, as actually illustrated on the cover page of this book. Although such investigations are still in their infancy, they should bring essential information complementing laser-based investigations. As a retrospection, and at the same time a look forward to future developments, we try now to summarize salient aspects of the discussions carried through in this book.

*Introduction to cluster dynamics.* Paul-Gerhard Reinhard, Eric Suraud

ISBN: 3-527-40345-0

**Clusters are specific scalable objects** Although clusters might naively be viewed as big molecules or small pieces of bulk, experience shows that they are more. By their very nature, clusters interpolate between atoms/molecules and the bulk, but in between these two extremes they exhibit specific properties, in particular as a function of their size. This scalability of clusters is probably one of their most striking properties. It should be mentioned again that these size effects are by no means trivial. They depend essentially on the nature of the observable. A typical generic effect is the appearance of electronic shell effects which are related to the quantum nature of cluster electrons and to the finite size. These effects are particularly pronounced in metal clusters and depend only in small details on the material. More specific of the bonding type of the cluster materials are the atomic shell effects. They are common to all types of clusters and represent stages in the progressive growth of clusters towards bulk. The actual sequence of atomic shell closures is usually related to the asymptotic crystalline structure of the bulk material. In a dynamical context, the optical response of clusters exhibits marked finite size quantum effects. The optical response of a cluster (in particular for metals) does strongly depend on its size. This was actually already exploited by Roman craftworkers to obtain, by cluster inclusions, colored glasses. The optical response has been discussed at many places in this book, both from the experimental and theoretical points of view. It is indeed an essential dynamical quantity, which provides useful information on cluster structure, in particular in the case of metal clusters. The weight given to optical response in this book reflects the fact that it constitutes the doorway to cluster dynamics. Metal clusters exhibit here the clearest signals of scalability and we have indeed devoted a large part of the examples to this specific class of clusters. It is also related to the numerous studies performed on metal clusters and hence the important share these clusters have taken in research. Their remarkable capability to couple to light (and hence to lasers) makes them especially suited for investigations of dynamical properties.

**Cluster production or the key to physics** Cluster production techniques constitute the key to experiments on free clusters. Cluster sources usually produce first a poorly defined ensemble particularly in terms of size and temperature which constitutes a strong hindrance to accurate measurements. In order to perform any sensible measurement, one needs to perform at least a mass selection of produced clusters. The production chain (source + possible mass analysis) thus constitutes an essential part of any experimental setup. This is at variance with other fields of physics where the studied species delivered by an external source or an accelerator is well defined (e.g., in atomic or nuclear physics). This difficulty in the production mechanism is a major constraint for cluster experiments. A second aspect concerns the type of observables. A simple and basic tool is mass analysis, e.g., by a time-of-flight measurement. The new generation of experiments try to address new quantities such as the characteristics of the electrons emitted by the excited clusters. These are electron kinetic energy distributions or angular distributions, ideally measured both together. In the longer term, one may measure correlations between emitted electrons, as already done with atoms. The variety of the ongoing experiments exemplified in this book raises optimism for future experimental achievements. The ever increasing capabilities of lasers, as already outlined above, foster that optimism.

**The finite many-body problem in clusters** Clusters may be small but not too small and large but not too large. They are, by nature, finite and they can cover a very broad range of sizes. They constitute an archetype of a finite many-body problem where many standard approximations fail because they are tuned either to small or to infinite systems. Seeing that positively, clusters offer an ideal testing ground for theoretical approaches to the many-body problem of finite systems. This is particularly true for density functional theory (DFT) methods, which play a prominent role in this book because of their usefulness in dynamical problems. Time-dependent DFT is still in a developing stage and applications in cluster physics provide a useful testing ground. Although DFT provides a robust basis for a description of cluster dynamics, it has to be augmented by dynamical correlations in the case of high excitation energies as, e.g., induced by intense laser beams. The inclusion of dissipative mechanisms in an effective mean-field theory such as DFT constitutes an important task for future development.

**The semi-linear domain, realm of cluster dynamics** We have considered numerous observables for describing the dynamics of clusters which can be associated to various degrees of non-linearity in terms of the response of the system. The linear domain deals with situations in which the cluster remains very close to its ground state. It thus mostly provides information on cluster structure. The optical response belongs to this linear domain. At the same time, the spectral distribution of optical absorption (=dipole) strength is basic information for the design of experiments beyond the linear domain. The true realm of cluster dynamics resides in what we call the semi-linear domain. This domain corresponds to situations in which the investigated cluster is significantly perturbed, for example by sizable energy deposit and/or non negligible electron emission, as for example in the course of irradiations by moderately intense lasers. But the cluster remains intact as a whole for a sufficiently long time, which allows one to study the dynamics of the system in the course of its de-excitation. Many recent, ongoing and forthcoming investigations (both from the theoretical and experimental sides) belong to the semi-linear domain. Let us cite as prominent examples the measurement of photoelectron spectra and the time-resolved dynamical studies in pump and probe setups. This latter technique promises to become one of the most powerful means of dynamical investigation, may be even in the femtosecond domain but for sure well below the ionic time scale. We have seen that such time-resolved techniques have numerous applications, from the analysis side as well as for applications. Modern high-intensity lasers allow one to go beyond the semi-linear domain. Strongly non-linear response is expected and observed. In such cases, though, the cluster is sooner or later fully disintegrated. The actual explosion allows interesting investigations, for example, of the physics of fragmentation. But the dynamical evolution soon becomes independent of the initial cluster material. Clusters then serve as a laboratory for a finite piece of plasma, a nano-plasma.

As a final word we would like to insist upon the incompleteness of our enterprise, which, beyond mere accidental omissions of our own, reflects the fact that many new investigations appear in brief. This is, in our opinion, a sign of dynamism of a young and promising field of physics to which this introductory book is devoted, namely "cluster dynamics".

# A Conventions of notations, symbols, units, acronyms

## A.1 Units

We list here basic physical constants and units (data are taken from [Sue63]):

electron mass: $m_e c^2 = 510.9\,\mathrm{keV} = 37.57\times10^3\,\mathrm{Ry} = 18.78\times10^3\,\mathrm{Ha}$

$m_e = 0.0156\,\mathrm{eV\,fs^2 a_0^{-2}} = 0.5\,\mathrm{Ry^{-1}a_0^{-2}} = 1\,\mathrm{Ha^{-1}a_0^{-2}}$

light velocity: $c = 5670\,\mathrm{a_0\,fs^{-1}} = 274.12\,\mathrm{Ry\,a_0} = 137.06\,\mathrm{a_0\,Ha^{-1}}$

fine-structure constant: $\alpha = \dfrac{e^2}{\hbar c} = 0.007297 = \dfrac{1}{137.03}$

charge: $e^2 = 1\,\mathrm{Ha\,a_0} = 2\,\mathrm{Ry\,a_0} = 14.40\,\mathrm{eV\,Å}$

dielectric constant: $\epsilon_0 = \dfrac{1}{4\pi} \equiv$ Gaussian system of units

Bohr energy: $E_B = \dfrac{e^4 m_e}{2\hbar^2} = \dfrac{\alpha^2 m_e c^2}{2} = 13.604\,\mathrm{eV} = 1\,\mathrm{Ry} = \dfrac{1}{2}\,\mathrm{Ha}$

Bohr radius: $\mathrm{a_0} = \dfrac{\hbar^2}{m_e c^2} = 0.5291\,\mathrm{Å} = 0.05291\,\mathrm{nm} = 0.5291\times10^{-10}\,\mathrm{m}$

Bohr magneton: $\mu_\mathrm{B} = \dfrac{\hbar e}{2m_e} = 5.788\mathrm{eV\,T^{-1}}$

Boltzmann constant: $k = 8.6174\,10^{-5}\,\mathrm{eV\,K^{-1}}$

energy scales: $1\,\mathrm{Ha} = 2\,\mathrm{Ry} = 27.2\,\mathrm{eV}$

$10^{-6}\,\mathrm{eV} = 0.2418\,h\,\mathrm{GHz} = 8.066\times10^{-3}\dfrac{hc}{\mathrm{cm}}$

$1\,h\,\mathrm{GHz} = 4.136\times10^{-6}\,\mathrm{eV}\ ;\ 1\dfrac{hc}{\mathrm{cm}} = 0.1240\times10^{-3}\,\mathrm{eV}$

time scales: $1\,\mathrm{fs} = 10^{-15}\,\mathrm{s} = 1.519\dfrac{\hbar}{\mathrm{eV}} = 20.66\dfrac{\hbar}{\mathrm{Ry}} = 41.32\dfrac{\hbar}{\mathrm{Ha}}$

$1\dfrac{\hbar}{\mathrm{Ha}} = 0.5\,\dfrac{\hbar}{\mathrm{Ry}} = 0.0242\,\mathrm{fs}$

laser intensity: $I = \dfrac{c}{8\pi}|E_0|^2\ ;\ \dfrac{I}{\mathrm{W\,cm^{-2}}} = 27.8\left|\dfrac{E_0}{\mathrm{V\,cm^{-1}}}\right|^2$

scale factors: $\hbar c = 1.9731\times10^{-7}\,\mathrm{eV\,m} = 1973.1\,\mathrm{eVÅ} = 274.12\,\mathrm{Ry\,a_0}$

$\dfrac{\hbar^2}{m_e} = 1\,\mathrm{Ha\,a_0^2} = 2\,\mathrm{Ry\,a_0^2} = 7.617\,\mathrm{eV\,Å^2}$

Note that we are using the Gaussian system of units for electromagnetic properties. In dynamical situations one also needs to treat simultaneously energy, distance and time scales and one has to consider proper combinations of these three quantities, taken from the values above. There are thus some standard "working" packages such as: eV, $\mathrm{a_0}$ and fs; Ry, $\mathrm{a_0}$, $\hbar\,\mathrm{Ry^{-1}}$ ($1\,\hbar\,\mathrm{Ry^{-1}} = 0.0484$ fs); and Ha, $\mathrm{a_0}$, $\hbar\,\mathrm{Ha^{-1}}$ ($1\,\hbar\,\mathrm{Ha^{-1}} = 0.0242$ fs, called the atomic unit).

*Introduction to cluster dynamics.* Paul-Gerhard Reinhard, Eric Suraud

ISBN: 3-527-40345-0

## A.2 Notations

Throughout this book we use the following general notations:

- vectors are denoted with bold characters (e.g., the position $\mathbf{r}$)
- operators are denoted by a hat (e.g., the many-body Hamiltonian operator $\hat{H}$)
- we use Dirac notation for states $|\Psi>$;
  the corresponding wavefunction reads $\Psi(\mathbf{r}) = <\mathbf{r}|\Psi>$
- we use standard atomic labeling $1s$, $2s$, $2p$, ...
- elements are denoted by their standard symbol: H, He, Li, ...

Frequently used symbols are explained in the following tables.

### General quantities

| | |
|---|---|
| $\mathbf{r}$, $r$ | position |
| $x, y, z$ | components of $\mathbf{r}$ |
| $d\mathbf{r}, d^3r$ | volume element for 3D integration |
| $k$, $\mathbf{k}$ | wave vector, wave momentum |
| $\mathbf{p}$ | momentum conjugate to $\mathbf{r}$ |
| $\nabla$ | gradient operator |
| $\Delta$ | Laplacian operator |
| $t$ | time |
| $v$, $\mathbf{v}$ | particle velocity |
| $T$ | temperature |
| $F, \mathbf{F}$ | force |
| $E$ | energy, mostly for total systems or parts of total system |
| $E_{\text{int}}$ | internal energy in a thermodynamic sense |
| $\epsilon$ | dielectric constant of a material |
| $V$ | potential in general |
| $C_{\text{heat}}$ | heat capacity |
| $\mu$ | magnetic moment |

### Cluster components: atoms and electrons

| | |
|---|---|
| $i, j$ | labels of electrons |
| $m_e$ | electron mass |
| $\mathbf{r}_i$ | coordinate of electron number $i$ |
| $I, J$ | labels of ions |
| $Z_I$ | valence of ion $I$ |
| $Z$ | charge of atom or ion |
| $Q_I$ | net charge of ion at site $I$ |
| $M_{\text{I}}$ | ionic mass |

| | |
|---|---|
| $\mathbf{R}$ | ionic coordinate |
| $\mathbf{P}$ | ionic momentum |
| | |
| $H$ | classical Hamiltonian |
| $\hat{H}$ | quantal Hamiltonian |
| $U_{\rm back}$ | one-electron potential from ionic background |
| $U$ | potential field for one electron |
| $U_{\rm C}$ | mean Coulomb potential |
| $\hat{U}_{\rm ex}$ | one-body operator for exchange in HF |
| $h$ | classical one-electron Hamiltonian |
| $\hat{h}$ | quantal one-electron Hamiltonian |
| $\hat{V}_{\rm PsP}$ | pseudo-potential, possibly non-local |
| $\epsilon_{\rm xc}(\rho)$ | exchange correlation energy density |
| | |
| $\alpha, \beta$ | labels for one-electron states |
| $\sigma$ | spin label |
| $\varphi, \varphi_\alpha$ | single electron wavefunction, single electron state $\alpha$ |
| $\phi$ | general single electron wavefunction, expansion basis |
| $\varepsilon$ | energy of one-electron state |
| $\rho(\mathbf{r}, \mathbf{r}')$ | one-electron density matrix |
| $\rho(\mathbf{r})$ | local one-electron density distribution |
| $\Phi$ | $N$ electron wavefunction in independent particle form |
| $\Psi$ | general $N$ electron wavefunction |
| $\psi$ | single electron state in discussing pseudo-potentials |
| $\mathcal{A}$ | anti-symmetrization operator |

**Global cluster characteristics**

| | |
|---|---|
| $N$ | particle number, atoms, ions or electrons |
| $R$ | cluster radius |
| $r_s$ | Wigner-Seitz radius |
| $\varepsilon_{\rm F}$ | Fermi energy |
| $k_{\rm F}$ | Fermi momentum |
| $v_{\rm F}$ | Fermi velocity |
| $p_{\rm F} = \hbar k_{\rm F}$ | Fermi momentum |
| $\rho_0$ | bulk equilibrium density of electrons |
| $f(\mathbf{r}, \mathbf{p})$ | electronic phase-space distribution |
| | |
| $r_{\rm rms}$ | rms radius |
| $\beta_2, \gamma$ | deformation parameters in Hill-Wheeler form |
| $\omega_x, \omega_y, \omega_z$ | parameters of oscillator potential |
| $I_{ii}$ | moment of inertia |
| $R_{\rm jel}$ | cluster radius of jellium background |
| $\alpha_{lm}$ | parameters of multipole deformation |

| | |
|---|---|
| $\sigma_{\rm jel}$ | width of jellium profile |
| $\sigma_{\rm WS}$ | width of Woods-Saxon profile |

**External fields and response**

| | |
|---|---|
| $I$ | laser intensity, sometimes ion label |
| $\mathbf{E}$, $\mathcal{E}$ | electrical field |
| $f(t)$, $F(t)$ | laser profile in time, integral thereof |
| $T_{\rm pulse}$ | laser pulse duration |
| $p_{\rm las}$ | photon momentum |
| $B$, $\mathbf{B}$ | magnetic field |
| $\mathbf{A}$ | electromagnetic vector potential |
| $\omega_{\rm Mie}$ | frequency of Mie surface plasmon |
| $\omega_{\rm spl}$ | frequency of surface plasmon, sum rule |
| $\omega_{\rm p}$ | volume plasmon frequency |
| $\mathbf{D}(t)$ | dipole moment in time domain |
| $\tilde{D}_i(\omega)$ | dipole moment in frequency domain |
| $S_{D_i}$ | dipole strength distribution |
| $\mathcal{P}_{D_i}$ | dipole power spectrum |
| $\varepsilon_{\rm kin}$ | kinetic energy of emitted electrons |
| $n_{\rm phot}$ | number of photons |
| $n_{\rm plas}$ | number of plasmon states |

## A.3 Acronyms

| | |
|---|---|
| 1D, 2D, 3D | One-, two-, three-dimensional |
| ADSIC | Average-Density SIC |
| AE | All Electrons |
| ALDA | Adiabatic LDA |
| amu | atomic mass unit |
| ATI | Above Threshold Ionization |
| BO | Born-Oppenheimer |
| BOA | Born-Oppenheimer Approximation |
| BOS | Born-Oppenheimer Surface |
| CAPS | Cylindrically Averaged Pseudo-potential Scheme |
| CI | Configuration Interaction |
| DDTB | Distance Dependent TB |
| DFT | Density Functional Theory |
| ETF | Extended TF |
| EELS | Electron Energy Loss Spectroscopy |
| FWHM | Full Width at Half Maximum |
| GAM | Global Averaged Method (approximation to SIC) |
| GCM | Generator Coordinate Method |
| GGA | Generalized Gradient Approximation |
| HF | Hartree-Fock |
| HOMO | Highest Occupied Molecular Orbital |
| ICR | Ion Cyclotron Resonance |
| IP | Ionization Potential |
| IR | Infra-Red |
| KLI | Krieger-Li-Iafrate (approximation to SIC) |
| KS | Kohn-Sham |
| LCAO | Linear Combination of Atomic Orbitals |
| LCGO | Linear Combination of Gaussian Orbitals |
| LDA | Local Density Approximation |
| LEED | Low-Energy Electron Diffraction |
| LFC | Local Field Correction |
| LSDA | Local Spin-Density Approximation |
| LUMO | Lowest Unoccupied Molecular Orbital |
| MC | Monte-Carlo |
| MCHF | Multi-Configurational HF |
| MD | Molecular Dynamics |
| OPM | Optimized Potential Method |
| PES | Photo-Electron Spectra (spectroscopy) |
| PsP | Pseudo-Potential |
| R2PI | Resonant two Photon Ionization |
| RPA | Random-Phase Approximation |
| SAJM | Structure Averaged Jellium Model |

| | |
|---|---|
| SAPS | Spherically Averaged Pseudo-potential Scheme |
| SEM | Scanning Electron Microscope |
| SIC | Self-Interaction Correction |
| SH | Second Harmonics |
| SHG | Second Harmonic Generation |
| TB | Tight Binding |
| TDDFT | Time-Dependent DFT |
| TDHF | Time-Dependent HF |
| TDLDA | Time-Dependent LDA |
| TDLDA-MD | TDLDA Molecular Dynamics |
| TDTF | Time-Dependent TF |
| TEM | Tunneling Electron Microscope |
| TF | Thomas-Fermi |
| TOF | Time Of Flight |
| UV | Ultra-Violet |
| Vlasov-MD | Vlasov Molecular Dynamics |
| VUU | Vlasov-Ühling-Uhlenbeck |
| VUU-MD | VUU Molecular Dynamics |
| X-ray | photons in Röntgen regime (several keV) |
| ZEKE | Zero Electron Kinetic Energy |

## A.4 A few reference books on cluster physics and related domains

We list here a few textbooks used in the preparation of this manuscript, as well the proceedings of a few major conferences on cluster physics. This list is by no means exhaustive.

- **General physics**
  - J D Jackson, *Classical Electrodynamics*, Wiley, New York, 1962
  - A Messiah, *Quantum Mechanics Vol. I*, Wiley, New York, 1970
- **General many body theory**
  - R M Dreizler and E K U Gross, *Density Functional Theory: An Approach to the Quantum Many-Body Problem*, Springer, Berlin, 1990
  - A L Fetter and J D Walecka, *Quantum theory of Many-Particle Systems*, McGraw-Hill, New York, 1971
  - R G Parr and W Yang, *Density-Functional Theory of Atoms and Molecules*, Oxford University Press, Oxford, 1989
  - D Pines and P Nozières, *The theory of Quantum Liquids*, Benjamin, New York, 1966
  - G D Mahan, *Many Particle Physics*, Plenum, New York, 1993
  - E Krotscheck and J Navarro, *Microscopic Approaches to Quantum Liquids in Confined Geometries*, World Scientific, Singapore, 2002
- **Solid state physics**
  - N W Ashcroft and N D Mermin, *Solid State Physics*, Saunders College, Philadelphia, 1976
  - J Callaway, *Quantum Theory of the Solid State*, Academic, London, 1991
  - S Elliott, *The Physics and Chemistry of Solids*, Wiley, New York, 1998
- **Atomic and molecular physics - Chemistry**
  - R H Brandsen, *Atomic Collision Theory*, Benjamin, Reading, 1983,
  - L Szasz, *Pseudopotential Theory of Atoms and Molecules*, Wiley, New York, 1985
  - F H M Faisal, *Theory of Multiphoton Processes*, Plenum Press, New York, 1987
  - Y Yamaguchi and Y Osamure and J D Goddard and H E Schaefer III, *A New Dimension to Quantum Chemistry*, Oxford University Press, Oxford, 1994,
  - A H Zewail, *Femtochemistry, Vol. I & II*, World Scientific, Singapore, 1994,
  - D Pettifor, *Bonding and Structure of Molecules and Solids*, Clarendon, Oxford, 1995
  - B H Bransden and J Joachain, *Physics of Atoms and Molecules*, Longman, London, 1997
  - H Pauly, *Atom, Molecule, and Cluster Beams I & II*, Springer, Berlin, 2000

- **Cluster physics**

  - S Sugano, *Microclusters* Springer, Berlin, 1987
  - H Haberland, *Clusters of Atoms and Molecules 1 – Theory, Experiment, and Clusters of Atoms*, Springer Series in Chemical Physics, vol. 52, Springer, Berlin, 1994
  - H Haberland, *Clusters of Atoms and Molecules 2 – Solvation and Chemistry of Free Clusters, and Embedded, Supported and Compressed Clusters*, Springer Series in Chemical Physics, vol. 56, Springer, Berlin, 1994
  - W Ekardt, *Metal Clusters*, Wiley, New York, 1999
  - J Jellinek, *Theory of Atomic and Molecular Clusters*, Springer, Berlin, 1999
  - *Proceedings of the ISSPIC conferences of the past decade:*

    ISSPIC 6: Z. Phys. D **26** (1993)
    ISSPIC 7: Surf. Rev. Lett. **3** (1996)
    ISSPIC 8: Z. Phys. D **40** (1997)
    ISSPIC 9: Eur. Phys. J. D **9** (1999)
    ISSPIC 10: Eur. Phys. J. D **16** (2001)
    ISSPIC 11: Eur. Phys. J. D **24** (2003)

# B Gross properties of atoms and solids

Some basic properties and characteristics of atoms and bulk have been discussed in a general manner in Chapter 1. The aim of this appendix is to provide quantitative information complementing the more qualitative discussions of Chapter 1.

## B.1 The periodic table of elements

The Mendeleev table of elements provides a classification of elements on the basis of their properties, in particular at the chemical level. It is shown on the next page. The layout follows a standard pattern. The atomic number $Z$ (top left) is the number of protons in the nucleus. The atomic mass (bottom) is weighted by isotopic abundances in the Earth's surface, relative to the mass of the $^{12}$C isotope, defined to be exactly 12 unified atomic mass units (amu). Relative isotopic abundances often vary considerably, both in natural and commercial samples. A number in parentheses is the mass of the longest lived isotope of that element when no stable isotope exists. For elements 110 to 112, the atomic numbers of known isotopes are given. These data are from [AW93, oAWA96]. The names given below for elements 104 to 109 are those recommended by the International Union of Pure and Applied Chemistry in late 1997.

Elements belonging to a vertical column form a *group*, while elements along a horizontal line constitute a *period*. The elements of each group have consistently high or low values of certain physical and chemical properties. As is well known the small energies involved in chemistry imply that only the least bound electrons do play a role in the interactions between atoms, and hence in molecules. A key factor is thus the occupation of the last (valence) electronic shell. As a consequence shell closure as observed in rare gases (group 18, He, Ne, Ar, Kr, Xe ...) makes these atoms rather inert. In turn, groups 1 (alkalines) and 17 (halogens) are chemically particularly active, because of their "closeness" to rare gases. Atoms of group 1 have a single weakly bound $s$ electron, they are strongly metallic, usually soft and very reactive with oxygen or water, in particular. Halogens (group 17) possess a $p$ subshell lacking one electron for shell closure and they thus develop a strong tendency to gain an electron from outside. They can in particular form especially stable arrangements with alkalines. In between groups 1 and 17 tendencies interpolate between the ones observed in alkalines or halogens and they are globally less marked. Groups 3 to 12 correspond to the so called transition metals in which the energetic position of the $d$ and $f$ orbitals relative to $s$ and $p$ shells is fluctuating from element to element. This inhibits a prediction of the properties of these elements by simple systematics.

*Introduction to cluster dynamics.* Paul-Gerhard Reinhard, Eric Suraud

ISBN: 3-527-40345-0

## PERIODIC TABLE OF THE ELEMENTS

| 1<br>IA | 2<br>IIA | 3<br>IIIB | 4<br>IVB | 5<br>VB | 6<br>VIB | 7<br>VIIB | 8<br>VIII | 9<br>VIII | 10<br>VIII | 11<br>IB | 12<br>IIB | 13<br>IIIA | 14<br>IVA | 15<br>VA | 16<br>VIA | 17<br>VIIA | 18<br>VIIIA |
|---|---|---|---|---|---|---|---|---|---|---|---|---|---|---|---|---|---|
| 1 H<br>Hydrogen<br>1.00794 | | | | | | | | | | | | | | | | | 2 He<br>Helium<br>4.002602 |
| 3 Li<br>Lithium<br>6.941 | 4 Be<br>Beryllium<br>9.012182 | | | | | | | | | | | 5 B<br>Boron<br>10.811 | 6 C<br>Carbon<br>12.0107 | 7 N<br>Nitrogen<br>14.00674 | 8 O<br>Oxygen<br>15.9994 | 9 F<br>Fluorine<br>18.9984032 | 10 Ne<br>Neon<br>20.1797 |
| 11 Na<br>Sodium<br>22.989770 | 12 Mg<br>Magnesium<br>24.3050 | | | | | | | | | | | 13 Al<br>Aluminum<br>26.981538 | 14 Si<br>Silicon<br>28.0855 | 15 P<br>Phosph.<br>30.973761 | 16 S<br>Sulfur<br>32.066 | 17 Cl<br>Chlorine<br>35.4527 | 18 Ar<br>Argon<br>39.948 |
| 19 K<br>Potassium<br>39.0983 | 20 Ca<br>Calcium<br>40.078 | 21 Sc<br>Scandium<br>44.955910 | 22 Ti<br>Titanium<br>47.867 | 23 V<br>Vanadium<br>50.9415 | 24 Cr<br>Chromium<br>51.9961 | 25 Mn<br>Manganese<br>54.938049 | 26 Fe<br>Iron<br>55.845 | 27 Co<br>Cobalt<br>58.933200 | 28 Ni<br>Nickel<br>58.6934 | 29 Cu<br>Copper<br>63.546 | 30 Zn<br>Zinc<br>65.39 | 31 Ga<br>Gallium<br>69.723 | 32 Ge<br>German.<br>72.61 | 33 As<br>Arsenic<br>74.92160 | 34 Se<br>Selenium<br>78.96 | 35 Br<br>Bromine<br>79.904 | 36 Kr<br>Krypton<br>83.80 |
| 37 Rb<br>Rubidium<br>85.4678 | 38 Sr<br>Strontium<br>87.62 | 39 Y<br>Yttrium<br>88.90585 | 40 Zr<br>Zirconium<br>91.224 | 41 Nb<br>Niobium<br>92.90638 | 42 Mo<br>Molybd.<br>95.94 | 43 Tc<br>Technet.<br>(97.907215) | 44 Ru<br>Ruthen.<br>101.07 | 45 Rh<br>Rhodium<br>102.90550 | 46 Pd<br>Palladium<br>106.42 | 47 Ag<br>Silver<br>107.8682 | 48 Cd<br>Cadmium<br>112.411 | 49 In<br>Indium<br>114.818 | 50 Sn<br>Tin<br>118.710 | 51 Sb<br>Antimony<br>121.760 | 52 Te<br>Tellurium<br>127.60 | 53 I<br>Iodine<br>126.90447 | 54 Xe<br>Xenon<br>131.29 |
| 55 Cs<br>Cesium<br>132.90545 | 56 Ba<br>Barium<br>137.327 | 57–71<br>Lantha-<br>nides | 72 Hf<br>Hafnium<br>178.49 | 73 Ta<br>Tantalum<br>180.9479 | 74 W<br>Tungsten<br>183.84 | 75 Re<br>Rhenium<br>186.207 | 76 Os<br>Osmium<br>190.23 | 77 Ir<br>Iridium<br>192.217 | 78 Pt<br>Platinum<br>195.078 | 79 Au<br>Gold<br>196.96655 | 80 Hg<br>Mercury<br>200.59 | 81 Tl<br>Thallium<br>204.3833 | 82 Pb<br>Lead<br>207.2 | 83 Bi<br>Bismuth<br>208.98038 | 84 Po<br>Polonium<br>(208.982415) | 85 At<br>Astatine<br>(209.987131) | 86 Rn<br>Radon<br>(222.017570) |
| 87 Fr<br>Francium<br>(223.019731) | 88 Ra<br>Radium<br>(226.025402) | 89–103<br>Actinides | 104 Rf<br>Rutherford.<br>(261.1089) | 105 Db<br>Dubnium<br>(262.1144) | 106 Sg<br>Seaborg.<br>(263.1186) | 107 Bh<br>Bohrium<br>(262.1231) | 108 Hs<br>Hassium<br>(265.1306) | 109 Mt<br>Meitner.<br>(266.1378) | 110<br>(269, 273) | 111<br>(272) | 112<br>(277) | | | | | | |

|  | | | | | | | | | | | | | | | |
|---|---|---|---|---|---|---|---|---|---|---|---|---|---|---|---|
| Lanthanide series | 57 La<br>Lanthan.<br>138.9055 | 58 Ce<br>Cerium<br>140.116 | 59 Pr<br>Praseodym.<br>140.90765 | 60 Nd<br>Neodym.<br>144.24 | 61 Pm<br>Prometh.<br>(144.912745) | 62 Sm<br>Samarium<br>150.36 | 63 Eu<br>Europium<br>151.964 | 64 Gd<br>Gadolin.<br>157.25 | 65 Tb<br>Terbium<br>158.92534 | 66 Dy<br>Dyspros.<br>162.50 | 67 Ho<br>Holmium<br>164.93032 | 68 Er<br>Erbium<br>167.26 | 69 Tm<br>Thulium<br>168.93421 | 70 Yb<br>Ytterbium<br>173.04 | 71 Lu<br>Lutetium<br>174.967 |
| Actinide series | 89 Ac<br>Actinium<br>(227.027747) | 90 Th<br>Thorium<br>232.0381 | 91 Pa<br>Protactin.<br>231.03588 | 92 U<br>Uranium<br>238.0289 | 93 Np<br>Neptunium<br>(237.048166) | 94 Pu<br>Plutonium<br>(244.064197) | 95 Am<br>Americ.<br>(243.061372) | 96 Cm<br>Curium<br>(247.070346) | 97 Bk<br>Berkelium<br>(247.070298) | 98 Cf<br>Californ.<br>(251.079579) | 99 Es<br>Einstein.<br>(252.08297) | 100 Fm<br>Fermium<br>(257.095096) | 101 Md<br>Mendelev.<br>(258.098427) | 102 No<br>Nobelium<br>(259.1011) | 103 Lr<br>Lawrenc.<br>(262.1098) |

## B.2 Atomic trends

The evolution of gross atomic properties with atomic number helps one to understand qualitatively the behavior of an element inside a compound. Particularly important is how atomic properties vary within a given group or period. A basic aspect is electron binding as a function of the atomic number $Z$ and thus the way atomic radii and ionization potentials (IP) evolve. Intuitively, when $Z$ increases (along a period or down a group) one might expect that electrons become more strongly bound. But the larger the $Z$, the larger the number of bound electrons and the larger the effect of the Pauli principle to make the least bound electrons farther and farther away from the nucleus, and thus the electrons are less and less strongly bound. Both effects of course act against each other, but altogether it is the shielding effect which dominates. The distance factor thus wins and electrons are globally less and less strongly bound when $Z$ increases. This means an increase of atomic radius and a decrease of ionization potential. These trends are represented in Figure B.1 showing schematically the evolution of atomic gross properties along groups (vertical) and periods (horizontal).

It is also interesting to consider, in this respect of size and IP, the anion and cation corresponding to a given atom. Starting from the neutral species X with radius $R$ and ionization potential $\varepsilon_0$, the radius and IP of $X^-$ are smaller than $R$ and $\varepsilon_0$ respectively, while the radius and IP of $X^+$ are larger.

The trends shown in Figure B.1 are global. There are of course fine details, in particular due to quantum effects, which cannot be seen from this figure. As a complement we thus display in Figure B.2 the explicit (experimental) values of two key atomic observables, radii and ionization potentials. We recover, of course, the general trends outlined in Figure B.1, along periods or groups, but now with fine quantum details, which nevertheless do not alter the general scheme outlined above.

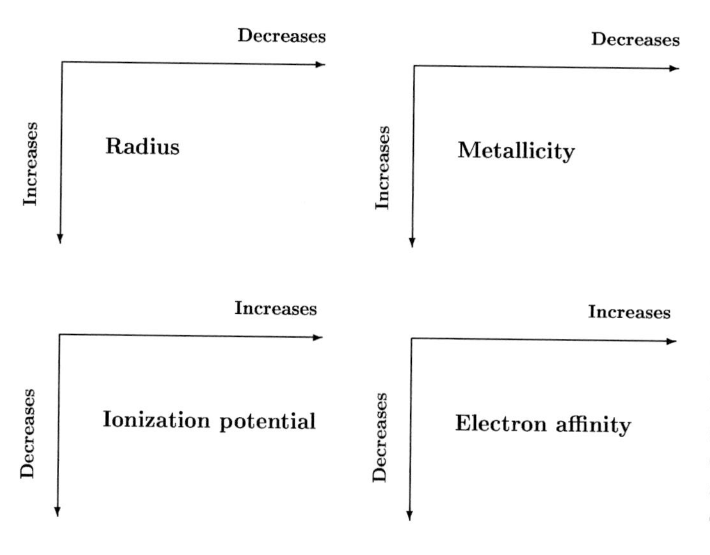


**Figure B.1:** Trends for evolution of radius, metalicity, ionization potential and electron affinity with changing element. The directions have the same meaning as in the periodic table: horizontal direction stands for changes within one period and vertical direction within one group.

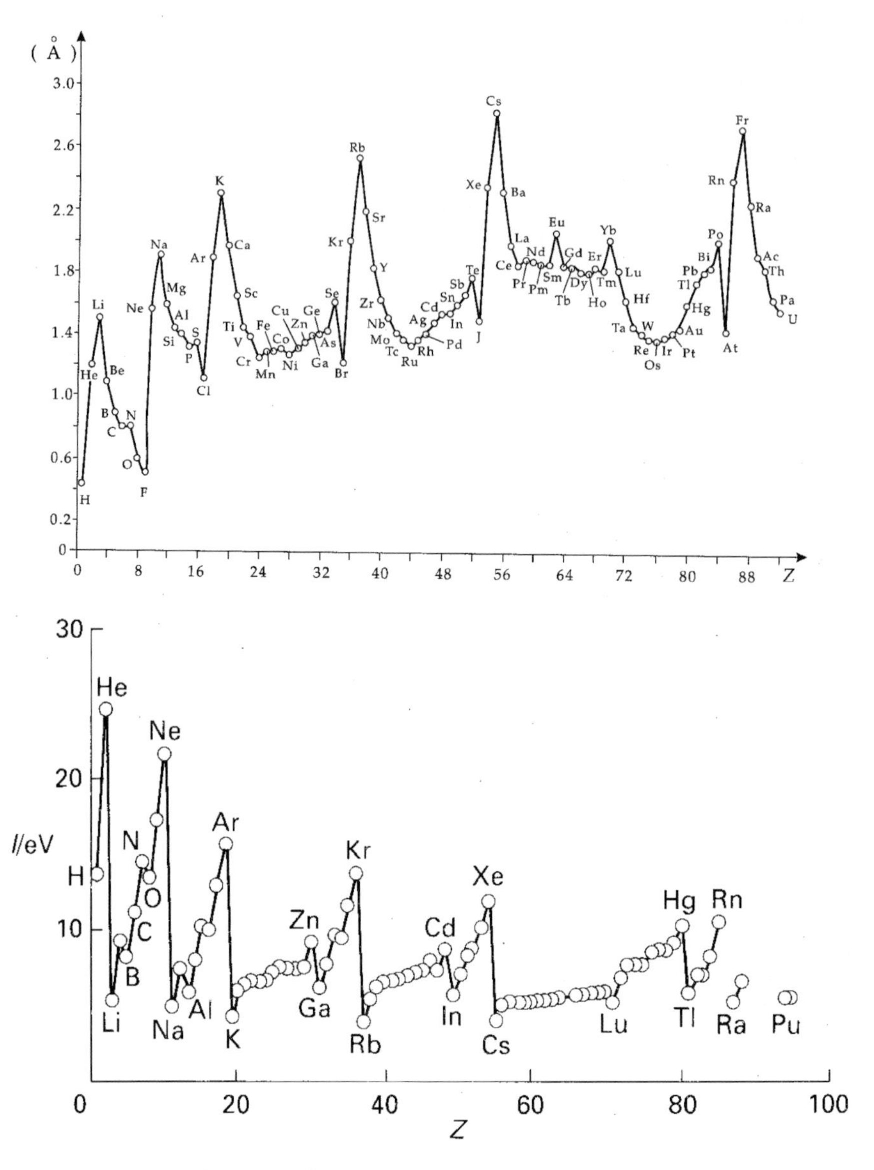


**Figure B.2:** Upper: Atomic radii (in Å) as a function of atomic number $Z$ up to $Z = 92$. The overall evolution is a soft increase of radii as a function of $Z$. This evolution is strongly affected by shell effects which lead to a marked oscillatory pattern. Halogens have consistently small radii while alkalines with their weakly bound $s$ valence electron have consistently large radii. Lower: Atomic IP's (in eV) as a function of atomic number $Z$. The general trend is a soft decrease of IPs as a function of $Z$. As radii, IPs are strongly affected by shell effects, which leads to marked oscillations on top of the average evolution. Rare gas atoms consistently exhibit the largest IP reflecting their stability. In turn alkalines exhibit consistently small IPs reflecting the fact that their valence $s$ electron is weakly bound.

## B.3 Electronic structure of atoms

The next step is to consider the explicit electronic structure of atoms. As discussed in Section 1.1.1, the explicit electronic structure of atoms is involved. Nevertheless, and particularly in light atoms, Hund's rules provide a reliable scheme for predicting the electronic structures (see Section 1.1.1.1). These electronic structures are reported in Table B.1. The tables provide the atomic number, name and electronic structure for most elements in standard notation.

**Table B.1**: Table of electronic structures. The electronic configurations are taken from [MW95]. Standard notation is used. This means, e.g., for the element Si: starting point is a Ne electronic core (see electronic structure of Ne), two $3s$ and two $3p$ electrons are added.

| $Z$ | Element | | Configuration |
|---|---|---|---|
| 1 | H | Hydrogen | $1s$ |
| 2 | He | Helium | $1s^2$ |
| 3 | Li | Lithium | (He)$2s$ |
| 4 | Be | Beryllium | (He)$2s^2$ |
| 5 | B | Boron | (He)$2s^2 2p$ |
| 6 | C | Carbon | (He)$2s^2 2p^2$ |
| 7 | N | Nitrogen | (He)$2s^2 2p^3$ |
| 8 | O | Oxygen | (He)$2s^2 2p^4$ |
| 9 | F | Fluorine | (He)$2s^2 2p^5$ |
| 10 | Ne | Neon | (He)$2s^2 2p^6$ |
| 11 | Na | Sodium | (Ne)$3s$ |
| 12 | Mg | Magnesium | (Ne)$3s^2$ |
| 13 | Al | Aluminum | (Ne)$3s^2 3p$ |
| 14 | Si | Silicon | (Ne)$3s^2 3p^2$ |
| 15 | P | Phosphorus | (Ne)$3s^2 3p^3$ |
| 16 | S | Sulfur | (Ne)$3s^2 3p^4$ |
| 17 | Cl | Chlorine | (Ne)$3s^2 3p^5$ |
| 18 | Ar | Argon | (Ne)$3s^2 3p^6$ |
| 19 | K | Potassium | (Ar)$4s$ |
| 20 | Ca | Calcium | (Ar)$4s^2$ |
| 21 | Sc | Scandium | (Ar)$3d4s^2$ |
| 22 | Ti | Titanium | (Ar)$3d^2 4s^2$ |
| 23 | V | Vanadium | (Ar)$3d^3 4s^2$ |
| 24 | Cr | Chromium | (Ar)$3d^5 4s$ |
| 25 | Mn | Manganese | (Ar)$3d^5 4s^2$ |
| 26 | Fe | Iron | (Ar)$3d^6 4s^2$ |
| 27 | Co | Cobalt | (Ar)$3d^7 4s^2$ |
| 28 | Ni | Nickel | (Ar)$3d^8 4s^2$ |
| 29 | Cu | Copper | (Ar)$3d^{10} 4s$ |
| 30 | Zn | Zinc | (Ar)$3d^{10} 4s^2$ |

| $Z$ | Element | | Configuration |
|---|---|---|---|
| 31 | Ga | Gallium | (Ar)$3d^{10}4s^2 4p$ |
| 32 | Ge | Germanium | (Ar)$3d^{10}4s^2 4p^2$ |
| 33 | As | Arsenic | (Ar)$3d^{10}4s^2 4p^3$ |
| 34 | Se | Selenium | (Ar)$3d^{10}4s^2 4p^4$ |
| 35 | Br | Bromine | (Ar)$3d^{10}4s^2 4p^5$ |
| 36 | Kr | Krypton | (Ar)$3d^{10}4s^2 4p^6$ |
| 37 | Rb | Rubidium | (Kr)$5s$ |
| 38 | Sr | Strontium | (Kr)$5s^2$ |
| 39 | Y | Yttrium | (Kr)$4d5s^2$ |
| 40 | Zr | Zirconium | (Kr)$4d^2 5s^2$ |
| 41 | Nb | Niobium | (Kr)$4d^4 5s$ |
| 42 | Mo | Molybdenum | (Kr)$4d^5 5s$ |
| 43 | Tc | Technetium | (Kr)$4d^5 5s^2$ |
| 44 | Ru | Ruthenium | (Kr)$4d^7 5s$ |
| 45 | Rh | Rhodium | (Kr)$4d^8 5s$ |
| 46 | Pd | Palladium | (Kr)$4d^{10}$ |
| 47 | Ag | Silver | (Kr)$4d^{10}5s^2$ |
| 48 | Cd | Cadmium | (Kr)$4d^{10}5s^2$ |
| 49 | In | Indium | (Kr)$4d^{10}5s^2 5p$ |
| 50 | Sn | Tin | (Kr)$4d^{10}5s^2 5p^2$ |
| 51 | Sb | Antimony | (Kr)$4d^{10}5s^2 5p^3$ |
| 52 | Te | Tellurium | (Kr)$4d^{10}5s^2 5p^4$ |
| 53 | I | Iodine | (Kr)$4d^{10}5s^2 5p^5$ |
| 54 | Xe | Xenon | (Kr)$4d^{10}5s^2 5p^6$ |
| 55 | Cs | Cesium | (Xe)$6s$ |
| 56 | Ba | Barium | (Xe)$6s^2$ |
| 57 | La | Lanthanum | (Xe)$5d6s^2$ |
| 58 | Ce | Cerium | (Xe)$4f5d6s^2$ |
| 59 | Pr | Praseodymium | (Xe)$4f^3 6s^2$ |
| 60 | Nd | Neodymium | (Xe)$4f^4 6s^2$ |
| 61 | Pm | Promethium | (Xe)$4f^5 6s^2$ |
| 62 | Sm | Samarium | (Xe)$4f^6 6s^2$ |
| 63 | Eu | Europium | (Xe)$4f^7 6s^2$ |
| 64 | Gd | Gadolinium | (Xe)$4f^7 5d6s^2$ |
| 65 | Tb | Terbium | (Xe)$4f^9 6s^2$ |
| 66 | Dy | Dysprosium | (Xe)$4f^{10}6s^2$ |
| 67 | Ho | Holmium | (Xe)$4f^{11}6s^2$ |
| 68 | Er | Erbium | (Xe)$4f^{12}6s^2$ |
| 69 | Tm | Thulium | (Xe)$4f^{13}6s^2$ |
| 70 | Yb | Ytterbium | (Xe)$4f^{14}6s^2$ |
| 71 | Lu | Lutetium | (Xe)$4f^{14}5d6s^2$ |
| 72 | Hf | Hafnium | (Xe)$4f^{14}5d^2 6s^2$ |

| $Z$ | Element | | Configuration |
|---|---|---|---|
| 73 | Ta | Tantalum | (Xe)$4f^{14}5d^3 6s^2$ |
| 74 | W | Tungsten | (Xe)$4f^{14}5d^4 6s^2$ |
| 75 | Re | Rhenium | (Xe)$4f^{14}5d^5 6s^2$ |
| 76 | Os | Osmium | (Xe)$4f^{14}5d^6 6s^2$ |
| 77 | Ir | Iridium | (Xe)$4f^{14}5d^7 6s^2$ |
| 78 | Pt | Platinum | (Xe)$4f^{14}5d^9 6s$ |
| 79 | Au | Gold | (Xe)$4f^{14}5d^{10} 6s$ |
| 80 | Hg | Mercury | (Xe)$4f^{14}5d^{10} 6s^2$ |
| 81 | Tl | Thallium | (Xe)$4f^{14}5d^{10} 6s^2 6p$ |
| 82 | Pb | Lead | (Xe)$4f^{14}5d^{10} 6s^2 6p^2$ |
| 83 | Bi | Bismuth | (Xe)$4f^{14}5d^{10} 6s^2 6p^3$ |
| 84 | Po | Polonium | (Xe)$4f^{14}5d^{10} 6s^2 6p^4$ |
| 85 | At | Astatine | (Xe)$4f^{14}5d^{10} 6s^2 6p^5$ |
| 86 | Rn | Radon | (Xe)$4f^{14}5d^{10} 6s^2 6p^6$ |
| 87 | Fr | Francium | (Rn)$7s$ |
| 88 | Ra | Radium | (Rn)$7s^2$ |
| 89 | Ac | Actinium | (Rn)$6d 7s^2$ |
| 90 | Th | Thorium | (Rn)$6d^2 7s^2$ |
| 91 | Pa | Protactinium | (Rn)$5f^2 6d 7s^2$ |
| 92 | U | Uranium | (Rn)$5f^3 6d 7s^2$ |
| 93 | Np | Neptunium | (Rn)$5f^4 6d 7s^2$ |
| 94 | Pu | Plutonium | (Rn)$5f^6 7s^2$ |
| 95 | Am | Americium | (Rn)$5f^7 7s^2$ |
| 96 | Cm | Curium | (Rn)$5f^7 6d 7s^2$ |
| 97 | Bk | Berkelium | (Rn)$5f^9 7s^2$ |
| 98 | Cf | Californium | (Rn)$5f^{10} 7s^2$ |
| 99 | Es | Einsteinium | (Rn)$5f^{11} 7s^2$ |
| 100 | Fm | Fermium | (Rn)$5f^{12} 7s^2$ |
| 101 | Md | Mendelevium | (Rn)$5f^{13} 7s^2$ |
| 102 | No | Nobelium | (Rn)$5f^{14} 7s^2$ |

## B.4 Properties of bulk material

We provide here some data on bulk materials. Most of the discussions in this book have been based on pictures stemming from atoms rather than from the bulk. Condensed matter techniques provide valuable tools for very large systems and they have also been often used in cluster physics. We thus give here in table form a few properties of solids. Solid state physics is a huge field. We restrict ourselves here to the few basic numbers used at some places in the book: the Wigner Seitz radius and the work function. The Wigner Seitz radius $r_s$ is defined as the radius "occupied" by one atom in a sample. In other words, if $\rho$ is the material density, $r_s = (3/(4\pi\rho))^{1/3}$. The Wigner Seitz radius thus provides information on the compactness of a material. The work function $W$ is just the bulk analogue of the ionization potential in an atom or a cluster. It provides a measure of the global electronic binding of the material. Table B.2 shows $r_s$ and $W$ for a selection of simple and noble metals, most of them being extensively used in examples throughout the book. As expected, one can see that the values of $W$ are relatively small, even as compared to atomic IP. This reflects the particular softness of metals among other materials.

**Table B.2:** Measured gross properties of simple and noble metals. We list the Wigner Seitz radius $r_s$ of the material (second column) and the work function $W$ (third column).

| Element | $r_s$ (a$_0$) | $W$ (eV) |
|---|---|---|
| Li | 3.25 | 2.38 |
| Na | 3.96 | 2.35 |
| K | 4.86 | 2.22 |
| Rb | 5.20 | 2.16 |
| Cs | 5.62 | 1.81 |
| Cu | 2.67 | 4.4 |
| Ag | 3.02 | 4.3 |
| Au | 3.01 | 4.3 |
| Be | 1.87 | 3.92 |
| Mg | 2.66 | 3.64 |
| Ca | 3.27 | 2.80 |
| Sr | 3.57 | 2.35 |
| Ba | 3.71 | 2.49 |
| Nb | 3.07 | 3.99 |

| Element | $r_s$ (a$_0$) | $W$ (eV) |
|---|---|---|
| Fe | 2.12 | 4.31 |
| Mn | 2.14 | 3.83 |
| Zn | 2.30 | 4.24 |
| Cd | 2.59 | 4.1 |
| Hg | 2.65 | 4.52 |
| Al | 2.07 | 4.25 |
| Ga | 2.19 | 3.96 |
| In | 2.41 | 3.8 |
| Tl | 2.48 | 3.7 |
| Sn | 2.22 | 4.38 |
| Pb | 2.30 | 4.0 |
| Bi | 2.25 | 4.4 |
| Sb | 2.14 | 4.08 |

# C Some details on basic techniques from molecular physics and quantum chemistry

## C.1 The Born-Oppenheimer approximation

The starting point is a fully coupled quantum mechanical description of ions and electrons with the total wavefunction $\Psi(\mathbf{r}_1, \ldots, \mathbf{r}_{N_{\rm el}}; \mathbf{R}_1, \ldots, \mathbf{R}_{N_{\rm ion}})$. We abbreviate that in the following as $\Psi(\mathbf{r}; \mathbf{R})$ where $\mathbf{r}$ stands for the entity of all electronic coordinates and $\mathbf{R}$ for all ionic ones. The full Schrödinger equation (3.3) reads

$$\left( \frac{\hat{P}^2}{2M} + V_{\rm ion}(\mathbf{R}) + \hat{H}_{\rm el} + \hat{H}_{\rm coupl}(\mathbf{r}, \mathbf{R}) \right) \Psi = E\Psi \quad .$$

We make now the ansatz (3.2) for adiabatic separation, i.e. $\Psi(\mathbf{r}; \mathbf{R}) = \psi_{\rm ion}(\mathbf{R})\phi_{\mathbf{R}}(\mathbf{r})$. The electronic wavefunction $\phi_{\mathbf{R}}(\mathbf{r})$ thus depends parametrically on the ionic coordinates $\mathbf{R}$. The key assumption is that this dependence is weak (in case of exact separation, there would be no dependence at all). We thus approximate

$$\begin{aligned} \hat{P}^2\Psi(\mathbf{r}; \mathbf{R}) &= \left(\hat{P}^2\psi_{\rm ion}(\mathbf{R})\right) \phi_{\mathbf{R}}(\mathbf{r}) \\ &\quad \underbrace{+2\left(\hat{\mathbf{P}}\psi_{\rm ion}(\mathbf{R})\right)(-\imath\hbar\nabla_{\mathbf{R}})\phi_{\mathbf{R}}(\mathbf{r}) - \hbar^2\psi_{\rm ion}(\mathbf{R})\Delta_{\mathbf{R}}\phi_{\mathbf{R}}(\mathbf{r})}_{\approx 0} \\ &\approx \left(\hat{P}^2\psi_{\rm ion}(\mathbf{R})\right) \phi_{\mathbf{R}}(\mathbf{r}) \quad . \end{aligned} \tag{C.1}$$

Inserting this into the full Schrödinger equation (3.3) yields

$$\left( \frac{\hat{P}^2}{2M} + V_{\rm ion} \right) \psi_{\rm ion} + \frac{(\hat{H}_{\rm el} + \hat{H}_{\rm coupl})\phi_{\mathbf{R}}}{\phi_{\mathbf{R}}} \psi_{\rm ion} = E\psi_{\rm ion} \quad .$$

The term containing $\phi_{\mathbf{R}}$ is the only one carrying electronic coordinates. That cannot be compensated by any other term in the equation. Thus one has to require that the following electronic eigenvalue problem is fulfilled

$$(\hat{H}_{\rm el} + \hat{H}_{\rm coupl})\phi_{\mathbf{R}} = E_{\rm el}\phi_{\mathbf{R}} \quad . \tag{C.2}$$

Note that $H_{\rm coupl}$ and $\phi_{\mathbf{R}}$ in this equation do still depend on the ionic configuration $\mathbf{R}$. The eigenvalues $E_{\rm el}$ thus also depend on $\mathbf{R}$. They serve as an additional potential $E_{\rm el} = \langle\phi_{\mathbf{R}}|\hat{H}_{\rm el}+$

*Introduction to cluster dynamics.* Paul-Gerhard Reinhard, Eric Suraud

ISBN: 3-527-40345-0

$\hat{H}_{\mathrm{coupl}}|\phi_{\mathbf{R}}\rangle$ in the Born-Oppenheimer potential Eq. (3.4)

$$V_{\mathrm{BO}}(\mathbf{R}) = V_{\mathrm{ion}}(\mathbf{R}) + \langle\phi_{\mathbf{R}}|\hat{H}_{\mathrm{el}} + \hat{H}_{\mathrm{coupl}}|\phi_{\mathrm{R}}\rangle$$

entering the ionic Schrödinger equation

$$\left(\frac{\hat{P}^2}{2M} + V_{\mathrm{BO}}(\mathbf{R})\right)\psi_{\mathrm{ion}}(\mathbf{R}) = E\psi_{\mathrm{ion}}(\mathbf{R}) \quad . \tag{C.3}$$

The fully coupled problem (3.3) has thus been separated into two simpler steps: first, the electronic eigenvalue problem (C.2), and second, the subsequent ionic Schrödinger equation (C.3).

The Born-Oppenheimer separation as outlined above contains already the seed for going beyond. The mixed derivative $2\left(\hat{\mathbf{P}}\psi_{\mathrm{ion}}(\mathbf{R})\right)(-\imath\hbar\nabla_{\mathbf{R}})\phi_{\mathbf{R}}(\mathbf{r})$ neglected in the approximate step (C.1) couples to excited states of the electronic spectrum at given $\mathbf{R}$. This gives rise to diabatic effects, namely transitions between electronic configurations which are driven by that term and which are proportional to the ionic momentum $\psi^*_{\mathrm{ion}}\hat{\mathbf{P}}\psi_{\mathrm{ion}}$. It is reasonable to assume that the electronic transitions proceed on a fast scale as compared to the ionic motion. This allows one to derive a jump probability from the diabatic term and to follow a stochastic dynamics with occasional trajectory hopping and Born-Oppenheimer propagation in between, as proposed e.g. in [Tul90] and applied to cluster fragmentation in [BGS+00]. The hopping is switched on by probabilities and those can only be resolved by stochastic methods. The scheme then generates an ensemble of Born-Oppenheimer trajectories. It is interesting to note that these trajectories can develop, in the course of propagation, very different electronic mean fields. That is a wanted feature in situations where one wants to describe complex reactions with several very different asymptotic channels (e.g. fragmentation versus fission versus monomer evaporation). Somewhat similar equations generating ensembles with large fluctuations of mean fields have been discussed also in the context of energetic nuclear reaction, see e.g. [RS92, AARS96].

## C.2 Ab initio methods for the electronic problem

Ab initio approaches are described in Section 3.3.2 where the basic methods of quantum chemistry are briefly reviewed. We discuss here a few more details, focusing in particular on the question of the basis on which wavefunctions are actually represented. This is an important issue in quantum chemistry and has given rise to numerous developments, all carrying their dedicated acronyms. We try here to keep track of this scenario.

### C.2.1 Hartree Fock as a starting point

In the Hartree-Fock picture one is bound to solve a set of one electron self consistent equations:

$$\hat{h}_{\mathrm{HF}}\varphi_\alpha = \varepsilon_\alpha\varphi_\alpha \tag{C.4}$$

with ingredients as described in Section 3.3.2.1. The many body wavefunction is assumed to be a single Slater determinant $\Phi$ built from the $\varphi_\alpha$. A key point is the representation of the single-electron orbitals $\varphi_\alpha$. In standard quantum chemistry, the usual way of approaching this question is to expand these orbitals on a set of well chosen (given) orbitals. This is the essence of the so called Linear Combination of Atomic Orbitals (LCAO) mentioned at a few places along this book, see for example Section 1.1.2.3. One then writes

$$\varphi_\alpha = \sum_\mu C_\alpha^\mu \chi_\mu \tag{C.5}$$

where the $\chi_\mu$ are the given atomic orbitals. There remains here a choice in the nature of them. The most frequently used basis sets are the so called Slater type orbitals (STO) [Sla29] and Gaussian type orbitals (GTO) [Boy50]. In these bases, the angular part is described by powers of cartesian coordinates, while the radial part is either an exponential of the distance to the atomic center (STO) or a gaussian (GTO). Both these bases have been extensively used and optimized over the years giving raise to numerous "standard" basis sets.

With a basis set at hand, the Hartree-Fock equations turn to a secular equation (Section 1.1.2.3) in which the expansion coefficients $C_\alpha^\mu$ become the quantities to be evaluated. Of course, the final results of the calculation depend on the actual characteristics of the basis, in particular its size. Such an approach thus provides at best (as any numerical scheme) an approximate solution to the Hartree-Fock problem. Within a finite basis set the best approximation to the Hartree-Fock solution is the so-called self consistent field (SCF) wavefunction. The SCF wavefunction thus results from the minimization of the total electronic energy with respect to changes of the molecular orbital coefficients $C_\alpha^\mu$ under the constraint of orthonormality.

### C.2.2 Beyond HF: CI and MCHF/MCSCF

The next step beyond the HF approach is to expand the Hilbert space for the many body electronic wavefunctions $\Phi$ by expressing it as a linear combination of a finite number of Slater determinants (called configurations) or configuration state functions (CSF). These CSF functions are constructed as linear combinations of determinants to be eigenfunctions of the square of the total electronic spin angular momentum operator. In any case, the expansion reads

$$\Psi = \sum_n \Phi_n c_n \tag{C.6}$$

where the $c_n$ are the CI expansion coefficients and $\Phi_n$ the CSFs. Each configuration $\Phi_n$ is constructed using a set of molecular orbitals from a given wavefunction $\Phi_0$, called the reference wavefunction. In most cases the reference wavefunction (or configuration) is either an SCF or a multiconfiguration SCF wavefunction (see below) and the configurations are obtained as $ph$ excitations built on $\Phi_0$. In the standard configuration interaction (CI) method the variational coefficients are the $c_n$ which are obtained as solutions of the secular equation (3.22b) while keeping the configuration wavefunctions $\Phi_n$ fixed.

When all possible configurations, namely with all possible electronic excitations accounted for, are included in the CI expansion, one says that one is performing a "full-CI" calculation. This represents the best variational calculation for a given basis set, but this is practically achievable only for systems with typically not more than about 10 electrons. When not all configurations are taken into account one speaks of "limited CI", for which the most frequently used case is to account only for $1ph$ and $2ph$ excitations. Among standard and efficient approximations used in CI calculations let us furthermore cite the so called "frozen core" and "deleted virtual" approximation. In the frozen core approximation, a limited number of low lying occupied orbitals are kept frozen, namely do not enter possible $ph$ excitations. Correspondingly in the deleted virtual approximation a limited number of high-lying (virtual) unoccupied levels are eliminated from possible $ph$ excitations, i.e excitations into such levels are not permitted. Both these approximations amount to reduce the number of possible $ph$ excitations to be computed in the CI set of configurations and thus to reduce the computational cost.

In the construction of the CI wavefunctions, only the expansion coefficients are taken as variational parameters, while configurations are given. The next step is to relax the latter condition and consider both expansion coefficients and configurations as variational quantities. In the so called Multi-Configurational (MC) HF (or MCSCF) method, the electronic energy is thus minimized with respect to both the molecular orbitals and the CI coefficients. Among MCSCF methods the most frequently used technique is the so called "complete active space" (CAS) method (CASSCF). The CASSCF yields MCSCF wavefunctions corresponding to a full CI but within a limited configuration space (the so-called active space) and a limited number of electrons. This methods turns out to provide a very powerful starting point for the study of a variety of chemical systems. Still, because the number of configurations increases quickly with the number of electrons in a full CI calculation, not very large systems can be attacked in practice. For example, in a case without any spatial symmetry, a CASSCF with 12 electrons and 12 active orbital generates 226512 singlet CSFs.

# D More on pseudo-potentials

In Section 3.2.2, we discussed briefly the various types of pseudo-potentials used in cluster physics. We want here to address this question again, giving some details on the actual construction of modern pseudo-potentials. Indeed the simple approach developed in Section 3.2.2.1 needs improvements to achieve a quantitative level. Since these early steps in pseudo-potential theory, more robust construction methods have been developed leading in particular to the so called norm conserving pseudo-potentials. We outline here some basic features of such pseudo-potentials.

## D.1 Construction of norm conserving pseudo-potentials

As pointed out in Section 3.2.2.1 the simple manipulation on core and pseudo-valence wavefunctions as done in Eq. (3.7), namely

$$|\varphi_v) = |\psi) - \sum_{c=1}^{N_{\rm c}} |\varphi_c)(\varphi_c|\psi) \quad ,$$

does not allow a proper normalization of the pseudo-wavefunction. This question is essential and has given rise to specific developments in terms of what is called norm conserving pseudo-potentials, which take as a starting point a somewhat different point of view. We follow here the general prescription given by Troullier and Martins [TM91].

Pseudo-potentials are primarily constructed from atoms in which electronic structure calculations have been performed including all electrons (AE). Many such all electrons calculations have been performed in density functional theory and we shall follow that line here as well. Starting from an atom in free space spherical symmetry can be assumed and the one electron wavefunctions of the problem can be developed in radial and angular parts:

$$\phi^{\rm AE}(\mathbf{r}) = \sum_{l,m} R_l^{\rm AE}(r) Y_{lm}(\Omega) \tag{D.1}$$

with obvious notations. The radial Kohn-Sham equations then read

$$\left( -\frac{\hbar^2}{2m_{\rm e}}\frac{d^2}{dr^2} + \frac{\hbar^2 l(l+1)}{2m_{\rm e} r^2} + V^{\rm AE}(r) \right) r R_l^{\rm AE}(r) = \varepsilon_l^{\rm AE} R_l^{\rm AE}(r) \tag{D.2}$$

where the self consistent one electron potential reads

$$V^{\rm AE}(r) = -\frac{Ze^2}{r} + V_{\rm H}(r) + V_{\rm xc}^{\rm AE}(r) \tag{D.3}$$

*Introduction to cluster dynamics.* Paul-Gerhard Reinhard, Eric Suraud

ISBN: 3-527-40345-0

including the chosen DFT exchange correlation potential $V_{\rm xc}^{\rm AE}(r)$ and where $Z$ is the atomic number and $V_{\rm H}$ the Hartree potential. The solution of this set of equations provides a set of AE single electron radial wavefunctions $R_l^{\rm AE}(r)$ and eigenenergies $\varepsilon_l^{\rm AE}$ which will serve to construct the pseudo-potential and pseudo-wavefunctions.

The radial pseudo-potential wavefunctions $R_l^{\rm PP}(r)$ are then constructed according to the following general conditions:

- The radial wavefunction $R_l^{\rm PP}(r)$, which is solution of the pseudo-atom Kohn-Sham equation

$$\left(-\frac{\hbar^2}{2m_{\rm e}}\frac{d^2}{dr^2}+\frac{\hbar^2 l(l+1)}{2m_{\rm e}r^2}+V_l^{\rm PP}(r)\right)rR_l^{\rm PP}(r)=\varepsilon_l^{\rm PP}R_l^{\rm PP}(r) \tag{D.4}$$

  must be node-less in order to be smooth. It is to be noted here that the, still undefined, pseudo-potential $V_l^{\rm PP}(r)$ is a priori dependent on the angular momentum $l$.

- Beyond a certain cutoff radius $r_c(l)$, characteristic of the core radius, the pseudo and AE wavefunctions should be equal (or almost equal) ($R_l^{\rm PP}(r)=R_l^{\rm AE}(r)$ for $r>r_c(l)$).

- The charge enclosed within $r_c(l)$ must be the same for the pseudo and AE wavefunctions

$$\int_0^{r_c(l)} dr|R_l^{\rm PP}(r)|^2r^2=\int_0^{r_c(l)} dr|R_l^{\rm AE}(r)|^2r^2 \quad . \tag{D.5}$$

- The AE and PP single energies should be equal ($\varepsilon_l^{\rm PP}=\varepsilon_l^{\rm AE}$).

Pseudo-potentials which fulfill these four conditions are known as norm conserving pseudo-potentials. It is obvious that the conditions listed here solve the above mentioned normalization problem of the pseudo-wavefunctions. These pseudo-potentials usually exhibit relatively good transferability and can thus be widely used in molecular, cluster or bulk problems. The transferability of the pseudopotential is strongly correlated to the cutoff radius $r_c(l)$. A large value of $r_c(l)$ leads to especially smooth pseudo-potentials but to a poor transferability due the loss of physically relevant information, and vice versa. In any case $r_c(l)$ should be larger than the radius of the outermost node of all AE wavefunctions. Note that, generally speaking, a good criterion for transferability is given by the equality of logarithmic derivatives of AE and PP wavefunctions.

It should finally be noted that the above conditions do not explicitely give access to the pseudo-potential itself. Of course, one can invert Eq. (D.4) to express $V_l^{\rm PP}(r)$ as a function of $R_l^{\rm PP}(r)$. But the above general rules leave in fact a lot of freedom for the actual construction of pseudo-potentials. They allow in particular the construction of especially soft pseudo-potentials which are well suited for calculations on coordinate-space grids or on a basis of plane waves (the number of necessary plane waves can then be limited to a tractable amount by a low cutoff energy in reciprocal space).

## D.2 Ultra soft pseudo-potentials

The remaining freedom for constructing norm conserving pseudo-potentials can be used to build smooth pseudo-potentials with a relatively large cutoff radius $r_c(l)$, while still allowing proper transferability. We outline here again the classic proposal of Troullier and Martins [TM91]. The radial PP wavefunction is modeled inside the core by a simple form:

$$R_l^{\mathrm{PP}}(r) = r^l \exp\left(p_l(r^2)\right) \tag{D.6}$$

where $p_l$ is a polynomial of sixth order in $r^2$ fulfilling the following conditions

- Norm conservation within the core radius $r_c(l)$.
- Continuity of the PP wavefunctions and their first 4 derivatives at cutoff radius $r_c(l)$.
- Zero curvature of the pseudo-potential at the origin, + which turns out to be an important issue to produce smooth pseudo-potentials.

There is no general rule which proves the validity of the above conditions. Still, it turns out that these characteristics deliver excellent results in particular in terms of transferability and convergence as a function of the basis set.

## D.3 Examples of simple local pseudo-potentials

Although general pseudo-potentials are by nature non-local, it turns out that one can construct in some cases simple, yet efficient and robust, local pseudo-potentials. This is typically the case in simple metals. We provide here a few details on practical local pseudo-potentials, which can be typically used in alkaline metals. We consider again the particularly soft parameterization provided by a combination of two error functions [KBR99] (Eq. (3.10) in Section 3.2.2.2)

$$V_{\mathrm{PsP}}(\mathbf{r}) = \sum_{i=1}^{2} c_i \frac{\mathrm{erf}(\mathbf{r}/\sigma_i)}{|(\mathbf{r})|} \qquad \text{with} \quad \mathrm{erf}(x) = \sqrt{\frac{2}{\pi}} \int_0^x e^{-y^2}\, dy \quad . \tag{D.7}$$

The $\sigma_i$ are widths and the strength parameters $c_i$ have to line up to the total charge of the ionic core $c_1 + c_2 = Z_{\mathrm{ion}}$.

Table D.1 shows examples of working pseudo-potentials for the simple most local form Eq. (3.10) repeated just above. The alkalines, except for Li [BG95], are the simplest elements in that respect. They can be very well described with a local pseudo-potential which, at the same time, is still fairly soft and convenient for numerical purposes.

One may wonder why hydrogen appears as an element with pseudo-potential while the hydrogen atom is perfectly described with a pure Coulomb potential. But the Coulomb singularity is numerically hard to handle. The pseudo-potential gives the chance to reproduce the

| | $c_1$ | $c_2$ | $\sigma_1[a_0]$ | $\sigma_2[a_0]$ |
|---|---|---|---|---|
| H | 2.200 | –1.200 | 0.255 | 0.382 |
| Na | 2.292 | –3.292 | 0.681 | 1.163 |
| K | 1.318 | –2.318 | 0.847 | 1.695 |
| Cs | 0.853 | –1.853 | 0.951 | 2.168 |

**Table D.1:** Two local PsP in the form Eq. (3.10), for H see [RS01b] and for Na see [KBR99].

exact Coulomb spectrum with a soft potential which is much better suited for numerical handling. It can be used, e.g., in connection with finite meshes in coordinate space. For example, the above pseudo-potential yields acceptable hydrogen spectra in connection with a mesh size of 0.3 $a_0$.

# E More on density functional theory

## E.1 Kohn-Sham equations with spin densities

We present here a more complete version of the Kohn-Sham equations. They were outlined in Section 3.3.3.2 for spin-saturated systems. We repeat here the basic relations for the more involved case with spin densities associated with the local spin-density approximation (LSDA). The total energy Eq. (3.23) in full detail taking care of the electronic spin degree of freedom reads

$$\begin{aligned}
E_{\text{total}} &= E_{\text{kin}}(\{\varphi_\alpha\}) + E_{\text{C}}(\rho) + E_{\text{xc}}(\rho) + E_{\text{ext}}(\rho, t) \\
&\quad + E_{\text{el,ion}}(\rho, \{\mathbf{R}_I\}) + E_{\text{kin,ion}}(\{\mathbf{P}_I\}) + V_{\text{ion}}(\{\mathbf{R}_I\}) \quad , \qquad \text{(E.1a)}
\end{aligned}$$

$$E_{\text{kin}} = \sum_{\alpha=1}^{N_{\text{el}}} \int d^3r \varphi_\alpha^+ \frac{\hat{p}^2}{2m} \varphi_\alpha \quad , \qquad \text{(E.1b)}$$

$$E_{\text{C}} = \frac{e^2}{2} \int d^3r d^3r' \, \frac{\rho(\mathbf{r})\rho(\mathbf{r}')}{|\mathbf{r}' - \mathbf{r}|} \quad , \qquad \text{(E.1c)}$$

$$E_{\text{xc}} = \int d^3r \rho(\mathbf{r}) \epsilon_{\text{xc}} \left(\rho_\uparrow(\mathbf{r}), \rho_\downarrow(\mathbf{r})\right) \quad , \qquad \text{(E.1d)}$$

$$E_{\text{ext}} = \int d^3 \, \rho(\mathbf{r}) U_{\text{ext}}(\mathbf{r}, t) \quad , \qquad \text{(E.1e)}$$

$$E_{\text{el,ion}} = \int d^3 \, \rho(\mathbf{r}) U_{\text{back}}(\mathbf{r}) \quad , \qquad \text{(E.1f)}$$

$$\rho_\sigma(\mathbf{r}) = \sum_\alpha \varphi_\alpha^+(\mathbf{r}) \hat{\Pi}_\sigma \varphi_\alpha(\mathbf{r}) \quad , \quad \sigma \in \{\uparrow, \downarrow\} \quad , \qquad \text{(E.1g)}$$

$$\rho(\mathbf{r}) = \rho_\uparrow(\mathbf{r}) + \rho_\downarrow(\mathbf{r}) \quad , \qquad \text{(E.1h)}$$

$$E_{\text{kin,ion}} = \sum_I \frac{\mathbf{P}_I^2}{2M_I} \quad , \qquad \text{(E.1i)}$$

$$V_{\text{ion}} = \sum_{I<J} \frac{e^2 Z_I Z_J}{|\mathbf{R}_I - \mathbf{R}_J|} \quad , \qquad \text{(E.1j)}$$

where $\hat{\Pi}_\sigma$ is the projector on spin $\sigma$ and the potential $U_{\text{back}}$ in Eq. (E.1f) is to be inserted according to the chosen approximation for the ionic background, see Section 3.2. Note that $U_{\text{back}}$ may become a non-local functional if non-local pseudo-potentials are used. This is ignored in the above notation. The exchange-correlation energy density $\epsilon_{\text{xc}}$ is taken from one

*Introduction to cluster dynamics.* Paul-Gerhard Reinhard, Eric Suraud

ISBN: 3-527-40345-0

of the elsewhere provided functionals, see e.g. [PW92] for a widely used LSDA functional. Note that the matrix elements (in the kinetic energy) employ the hermitian conjugate rather than the complex conjugate because the wavefunction $\varphi_\alpha$ carries a Pauli spinor.

The Kohn-Sham potential which is to be inserted into the Kohn-Sham (KS) equations (3.24) or (3.24d) reads in detail

$$\hat{h}_{\mathrm{KS},\sigma} = \frac{\hat{p}^2}{2m} + \underbrace{U_{\mathrm{C}}(\mathbf{r}) + \hat{U}_{\mathrm{xc},\sigma} + U_{\mathrm{back}}(\mathbf{r}) + U_{\mathrm{ext}}(\mathbf{r},t)}_{U_{\mathrm{KS}}} \quad , \tag{E.2a}$$

$$U_{\mathrm{C}}(\mathbf{r}) = \int d^3r' \frac{e^2}{|\mathbf{r}-\mathbf{r}'|}\rho(\mathbf{r}') \quad , \tag{E.2b}$$

$$\hat{U}_{\mathrm{xc},\sigma}(\mathbf{r}) = \epsilon_{\mathrm{xc}}\left(\rho_\uparrow(\mathbf{r},\rho_\downarrow(\mathbf{r})\right) + \rho(\mathbf{r})\frac{\partial\epsilon_{\mathrm{xc}}\left(\rho_\uparrow(\mathbf{r},\rho_\downarrow(\mathbf{r})\right)}{\partial\rho_\sigma} \quad , \tag{E.2c}$$

where $\sigma_\alpha$ is the spin polarization of the state $\varphi_\alpha$. The term $\partial\epsilon_{\mathrm{xc}}/\partial\rho_\sigma$ now represents the formal derivative with respect to the spin-density $\rho_\sigma$ considered as an independent variable. The above form applies to systems with trivial spin polarization where each single electron state still has a pure spin direction. Systems with a strong spin-orbit force, such as heavy atoms or the magnetic materials Mn, Fe, or Cr, require mixed single-particle states and a more elaborate formulation [KHSW88], for an example with Cr clusters see [KB99].

## E.2 Gradient corrections in LDA

The local electron density in atoms and molecules is grossly inhomogeneous. This makes LDA somewhat inprecise in these environments. The inhomogeneity of the local density can be characterized by the local gradients $\nabla\rho$. It is an obvious next step to improve LDA energy functionals by augmenting them with a dependence on the gradient as well. Such an idea applied to the kinetic energy makes the step from the Thomas-Fermi model to extended Thomas-Fermi (ETF) which has proven to be a very successful approach, see Section 3.3.5.4 and [BB97]. This concerns the semi-classical treatment of the kinetic energy. In a similar spirit, but for the purpose of full KS calculations, one has also developed, with great care, gradient corrected functionals for the exchange-correlation energy [Bec88]. The result is often called the generalized gradient approximation (GGA) [PBE96], for an example of its effect on $\mathrm{Na}_n$ clusters see [GCKS95]. The gradient terms reduce the self-interaction error and bend the asymptotics of the KS potential into the right direction. It yields a significant improvement in the computation of atomic and molecular binding, lifting, e.g., the description of dissociation energies up to a quantitative level. The functional form of GGA is very involved. Fortunately, there exist ready-to-use subroutines which one can download from the net, see [GGA] which also contains the functional of [PW92] as subset.

The importance of gradient corrections in clusters depends very much on the material and observable of interest. Clusters made from simple metals are the most forgiving systems in this respect and can be treated very well without GGA. The density distribution of their valence electrons is pretty homogeneous and has a soft surface zone. Simple LDA works here surprisingly well for most purposes [Rei92]. Noble metals, in which core or semi-core electrons usually play an important role, are often treated with GGA [MPC98], because they involve more bound (atomic like) electrons. Covalent clusters are more demanding and are also better off with GGA for computing structural details. Remind that the validity of LDA also depends on the observables. It was actually argued in [GCKS95] that gradient corrections make a difference for polarizabilities even in simple Na clusters. And quantities related to single electron energies will require self-interaction correction as outlined in Section E.3.

Corrections to LDA are still a field of intense research. The present GGA is certainly not the last word in that respect. There are for example attempts to achieve further improvements by introducing extra kinetic terms in the functional [PKZB99]. One also tries to use exact exchange and to confine LDA to the correlation part. This is not as simple as it sounds. The success of LDA relies, to some extent, on a lucky counterbalance of the LDA errors in exchange versus those in correlations. One thus needs substantial improvements in the correlation functional when treating exchange exactly. And these schemes surely become much more expensive, computationally speaking. Clusters, particularly large ones, are probably not the most efficient testing ground in that respect.

## E.3 Self interaction correction

Although LDA can work acceptably well for the global binding properties and for optical response of clusters, its KS potential has the wrong asymptotics. For example, an electron escaping a neutral cluster should see a potential $\propto e^2/r$ but the KS potential decays exponentially, see Figure 3.10. This is due to the self-interaction error in the KS scheme. The direct term (Hartree term) employs the total density summed over all states. Thus the KS potential acting on a given state $\varphi_\alpha$ contains also this state $\varphi_\alpha$ which generates a self-interaction. The exact exchange in HF eliminates this self-interaction exactly, but the LDA exchange does not. This has consequences on ionization potentials (IPs). The difference of binding energies $E(N_{\rm el}) - E(N_{\rm el} - 1)$ yields an approximately correct value while the last bound single particle energy $\varepsilon_{\rm HOMO}$ differs. This violates Koopmanns' theorem which states the similarity of these two quantities [Wei78].

A solution which tries to maintain the simplicity of the KS scheme is the self-interaction correction (SIC) [PZ81]. The idea is as simple as in the original formulation of the Hartree theory [Har27]: the contribution from state $\varphi_\alpha$ is subtracted in the KS Hamiltonian used in the KS equation for $\varphi_\alpha$. This ad hoc correction can be formulated variationally, i.e. at the level of the energy functional. The modified Kohn-Sham equations are then derived by variation with respect to the single-electron wavefunctions $\varphi_\alpha^+$. This yields the SIC corrected Kohn-Sham potential ($\rho_\alpha = |\varphi_\alpha|^2$)

$$U_\alpha^{\rm SIC} = U_{\rm KS} - [U_{\rm C}(\rho_\alpha) + U_{\rm xc}(\rho_\alpha)] \quad . \tag{E.2d}$$

The friendly feature of SIC is that $U^{\rm SIC}$ is still a local potential. But there is also a complication involved. The modified KS potential depends on the state $\alpha$ on which it acts. The ortho-normality of the stationary solutions $\{\varphi_\alpha\}$ is thus not guaranteed. It needs to be added as a separate constraint. Moreover, the SIC functional is not invariant under unitary transformations amongst the occupied single particle states $\varphi_\alpha$. In spite of these drawbacks, the simplicity of a still local approach is tempting and the practical success of SIC is for many purposes (e.g. IP) very satisfying. Thus, there exist several applications also in clusters physics, see e.g. [FPM93] for ground state calculations and [PE92] for a time-dependent application.

Recent developments in DFT have cleared the path to have state-independent, local potentials on the grounds of a more complex functional. This is the optimized potential method (OPM) which provides a systematic prescription to derive the best suited local equivalent potential for any given mean field potential [TS76, NK84]. This very general scheme has been applied specifically to SIC and a series of efficient approximations to the full SIC scheme (E.2d) have been developed. We sketch them here briefly.

The optimized potential method (OPM) allows one to deduce an equivalent local and state-independent SIC potential [TS76, NK84]. Still, even OPM is a bit involved as it implies the resolution of an integral equation. It was soon observed that the whole machinery of OPM can be reduced by a few more approximation steps to extremely simple, and yet reliable, local SIC potentials: the KLI method [KLI92], the Slater potential [SH53] as an approximation to SIC-KLI, the globally averaged method (GAM) [PHH$^+$98, URS00] and finally the simplest average density SIC (ADSIC) [LSR02]. These are all approximations directly on the SIC potential (E.2d). They can be summarized as follows (omitting here the "intermediate" GAM approximation):

$$U^{\mathrm{KLI}}(\mathbf{r}) = \sum_\alpha \frac{\rho_\alpha(\mathbf{r})}{\rho(\mathbf{r})} \left\{ U_\alpha^{\mathrm{SIC}}(\mathbf{r}) - \int d^3r' \rho_\alpha(\mathbf{r}') \left[ U^{\mathrm{KLI}}(\mathbf{r}') - U_\alpha^{\mathrm{SIC}}(\mathbf{r}') \right] \right\}, \quad \text{(E.2e)}$$

$$U^{\mathrm{Slater}}(\mathbf{r}) = \sum_\alpha \frac{\rho_\alpha(\mathbf{r})}{\rho(\mathbf{r})} U_\alpha^{\mathrm{SIC}}(\mathbf{r}) \quad , \quad \text{(E.2f)}$$

$$U^{\mathrm{ADSIC}} = U_{\mathrm{KS}} - \left[ U_{\mathrm{C}}\left(\frac{\rho}{N}\right) + U_{\mathrm{xc}}\left(\frac{\rho}{N}\right) \right] \quad . \quad \text{(E.2g)}$$

The expense of the methods decreases from KLI to ADSIC. The final step in the aproximation chain is to replace the single particle densities by an average which leads to ADSIC (E.2g). It is interesting to note that ADSIC can again be formulated variationally at the level of an energy functional while other SIC schemes cannot.

The performance of SIC and the various approaches in connection with metal clusters has been discussed in [LSR02]. Without entering into the details, to be found in [LSR02], it is interesting to note that ADSIC provides an extremely simple, although formally well founded, cheap and robust scheme. There are two limitations to its applicability though: its requires that all electron wavefunctions occupy a similar region in space and energy (which works perfectly well in metals); and because it treats the SIC correction globally it cannot be used in dynamical situations involving dissociation. The OPM approaches, KLI and Slater potential, provide a completely satisfying description in these respects, although they suffer from formal difficulties (unitary robustness (important for dynamics), variational formulation). But note that there is a high extra expense in these two methods as compared to ADSIC, as one needs to compute the Coulomb potential for each single-particle density $\rho_\alpha$ separately. There are also subtle differences concerning the smoothness of the corrective potential, which may lead to observable effects, for example in photoelectron spectra [MCR01]. Note finally that semi-classical approaches, as discussed in Section 3.3.5, cannot distinguish single electron states. Only the total density is known. This disables automatically all versions of SIC which explicitely refer to $\varphi_\alpha$. But ADSIC remains applicable. It ought to be mentioned that AD-

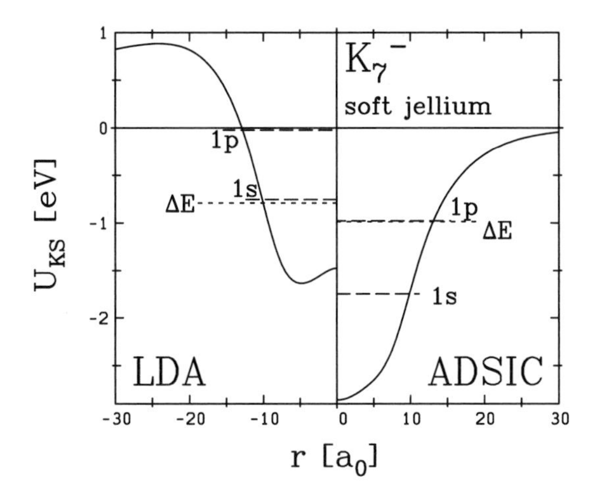


**Figure E.1:** Comparison between LDA and ADSIC for $K_7^-$ with soft jellium background ($r_s = 5.0\,a_0$, $\sigma_{jel} = 1.37\,a_0$). Compared are: the net KS potentials (full lines), the single electron energies (dashed lines), and the one-electron separation energy $\Delta E = E(K_7^-) - E(K_7)$ (dotted lines). The left panel shows results from LDA and the right panel from ADSIC.

SIC was proposed very long ago in [FA34]. This naive approach works surprisingly well for clusters of simple metals. It should be treated with care in other situations [LSR02].

Figure E.1 demonstrates the typical changes caused by the step from LDA to SIC, here for the example of ADSIC. The test case is the metal cluster $K_7^-$ computed with soft jellium background. One sees that ADSIC corrects the asymptotics of the KS potential, from repulsive Coulomb in LDA, to the correct exponential convergence with SIC. This causes a global downshift of all single electron energies in the right direction. After the shift one finds Koopmanns' theorem reestablished, namely the fact that the energy of the HOMO level coincides with the one-electron separation energy $\Delta E$ [Wei78]. Moreover, the theoretical HOMO level ($1p$) comes fairly close to the experimental separation energy of about 1.1 eV while pure LDA is very wrong in that respect.

## E.4 Time-dependent LDA and beyond

As mentioned briefly in Section 3.3.3.5, the TDLDA is an approximation in two directions. As LDA, it is a local approximation assuming that the system can be considered as being composed piecewise from a homogeneous electron gas. Additionally, there is an approximation on the time dependence. Indeed TDLDA employs also an instantaneous approximation ("adiabatic" LDA). It is supposed that exchange and correlations readjust to internal equilibrium much faster than the actual dynamics such that one can use the quasi-static LDA functional. The quest for validity of DFT approximations is even more at stake in the case of TDLDA. For example, we are aiming at describing the Mie plasmon oscillations in finite clusters. But the volume plasmon contributes to the correlations and its frequency is only a factor $\sqrt{3}$ above the Mie plasmon. The adiabatic assumption is thus questionable. On the other hand, TDLDA describes in practice the dynamical response of clusters very well. The reason is that response properties are dominated by the direct Coulomb term while exchange and correlations contribute only little. Rough approximations at that place are acceptable, see also Figure 3.11.

Nonetheless, the case calls for critical inspection and improvement. It is not surprising then that appropriate extensions for dynamical functionals are presently a much discussed topic. The field is still developing, for reviews and first ideas to introduce a true frequency dependence of the functionals see e.g. [GUG95, GDP96, vL01]. One line of development is to involve the electron currents as truly dynamical quantities in the functional [CG97, VUC97, vFBvL$^+$02]. It is also to be noted that there is a similarity between LDA and the local field correction (LFC) widely used in plasma physics, for classical plasmas see [STLS68, IIT87] and for recent development in a quantum framework including dynamical effects see [SB93]. Most of these dynamical extensions, including the LFC, are yet confined to small amplitude dynamics. A tractable and reliable extension, also covering large amplitudes, is not yet readily available. Practitioners of cluster dynamics are presently bound to TDLDA, but have to watch out for upcoming developments.

# F Fermi gas model and semi-classics

## F.1 The Fermi gas

The Fermi gas model primarily allows one to treat a system with translational invariance, such as for example an infinite electron gas or infinite nuclear matter (Section 4.4.1). The translational symmetry of the homogeneous system determines the form of the single-particle wavefunctions as

$$\varphi_{\mathbf{k}\sigma} = \frac{1}{(2\pi)^{3/2}\sqrt{V}} e^{\imath\mathbf{k}\mathbf{r}} \chi_\sigma \quad . \tag{F.1}$$

These are so to say plane waves "normalized" to the finite volume $V$ of the considered system. The $\chi_\sigma$ is a Pauli spinor with $\sigma = \pm 1$. The momenta $k_i$ can run, in principle, over all real numbers. In the basic Fermi gas model, no interaction is considered between the particles and the single particle energies $\varepsilon_k$ are thus purely kinetic ($\varepsilon_k = \hbar^2 k^2/2m$). The occupied states thus fill the lowest $|\mathbf{k}|$ up to the Fermi momentum $k_\mathrm{F}$. The latter is adjusted such that a desired density $\rho$ is reached. The density is

$$\rho = \sum_\sigma \int_{|\mathbf{k}|\leq k_\mathrm{F}} d^3k \, |\varphi_{\mathbf{k}\sigma}|^2 = \frac{2}{(2\pi)^3} \int_{|\mathbf{k}|\leq k_\mathrm{F}} d^3k = \frac{k_\mathrm{F}^3}{3\pi^2} \quad \Longrightarrow \quad k_\mathrm{F} = (3\pi^2\rho)^{1/3} \quad . \tag{F.2}$$

The Fermi gas model has found widespread applications in various domains of physics, ranging from nuclear to electronic problems, and even in the physics of compact stars. It is a quantal model with simplified single particle wavefunctions. This allows the computation of many-body observables with the help of tractable many-body wavefunctions, built from the above single particle wavefunctions (F.1). An example, concerning the energy of a simple metal, is detailed in Section F.2. Of course, when used under such circumstances only the simplified single particle wavefunctions Eq. (F.1) are kept for performing actual calculations. The initial assumption of a non-interacting system is usually released.

## F.2 Infinite electron gas at Hartree-Fock level

It is interesting to apply the Fermi gas model to the valence electrons of a (structureless) simple metal. The homogeneous electron gas actually also constitutes a basic system for developing density functionals in DFT. The ionic background is assumed to be smeared to

*Introduction to cluster dynamics.* Paul-Gerhard Reinhard, Eric Suraud

ISBN: 3-527-40345-0

a homogeneous jellium background. The positive background charge exactly compensates the negative homogeneous charge of the electrons. The direct Coulomb term thus vanishes. There remain kinetic energy, Coulomb exchange, and correlations. In a first step, one neglects correlations and one computes the Hartree-Fock energy

$$E = \sum_{\alpha} (\varphi_\alpha | \frac{\hat{p}^2}{2m} | \varphi_\alpha) - \frac{1}{2} \sum_{\alpha\beta} (\varphi_\alpha \varphi_\beta | V_{\text{Coul}} | \varphi_\beta \varphi_\alpha) \tag{F.3}$$

where $\alpha, \beta$ runs over the occupied states, Eq. (F.1). Inserting the plane wave electronic states into the HF energy (F.3) yields simple integrals, but yet an infinite net result, because of the infinite number of electrons involved. A physically sensible finite result is obtained when considering the energy per unit volume $E/V$. This reads

$$\frac{E}{V} = \sum_{\sigma} \int_{|\mathbf{k}| \leq k_{\text{F}}} d^3k \, \frac{\hbar k^2}{2m(2\pi)^3} - \int d(\mathbf{r} - \mathbf{r}') \, \frac{e^2}{2} \sum_{\sigma} \int_{|\mathbf{k}|, |\mathbf{k}'| \leq k_{\text{F}}} d^3k \, d^3k' \, \frac{e^{\imath(\mathbf{k}-\mathbf{k}')(\mathbf{r}-\mathbf{r}')}}{|\mathbf{r} - \mathbf{r}'|} \quad .$$

The energy per particle is related to that as $\rho E/N = E/V$. These integrals can be evaluated analytically with some algebra [FW71]. The result is

$$\frac{E}{N} = \frac{3}{5} \frac{\hbar^2 k_{\text{F}}^2}{2m} - \frac{3e^2}{4\pi} k_{\text{F}} = \underbrace{\frac{3(3\pi^2)^{2/3}}{10m} \rho^{2/3}}_{\epsilon_{\text{kin}}(\rho)} - \underbrace{\frac{3e^2(3\pi)^2}{4\pi} \rho^{1/3}}_{\epsilon_{\text{x}}(\rho)} \tag{F.4}$$

which provides a fair result in comparison to experimental data in the particularly simple case of bulk sodium, see Section 3.3.1. This energy for the homogeneous electron gas can be used as input for the energy functional in LDA. We recognize in the second term the exchange functional Eq. (3.25).

## F.3 The Thomas-Fermi approach

The standard way to obtain the Thomas-Fermi equation is to take the kinetic energy functional $\epsilon_{\text{kin}}$ derived above in Eq. (F.4). The Thomas-Fermi equation can be obtained by variation with respect to the local density starting from an LDA energy functional like (3.23) but now with the kinetic energy replaced by the kinetic energy functional (3.32). The result is then the Thomas-Fermi equation (3.33).

It is interesting to note that the Thomas-Fermi approach can also be derived and viewed in a semi-classical phase space picture. The goal is here to search for a semi-classical ground state distribution $f_0(\mathbf{r}, \mathbf{p})$, which is a stationary state of the Vlasov equation (3.29a), i.e. $\partial_t f = -\{f, h\}$. Stationarity requires $\partial_t f = 0$ and thus

$$\{f_0, h\} = 0 \tag{F.5}$$

which is fulfilled for $f_0$ being any functional of the semi-classical one body Hamiltonian $h$ (Section 3.3.5.2), i.e. $f_0 = f_0(h)$. The ground state additionally minimizes the energy while obeying the Pauli principle $f_0 \leq 2$. This leads to

$$f_0(\mathbf{r}, \mathbf{p}) = 2\theta \left( \varepsilon_{\text{F}} - h(\mathbf{r}, \mathbf{p}) \right) \quad .$$

Note that it poses again a self-consistent problem. The distribution $f_0$ defines a local density $\rho_0$ via Eq. (3.29b), the $\rho_0$ yields the mean field $U_{\rm KS}$, and that, in turn, determines $f_0$ via Eq. (3.33). The momentum dependence is characterized by the local Fermi sphere as given in Eq. (F.2). Integration over momentum space yields the density and with it the Thomas-Fermi equation (3.33).

## F.4 Details of the nano-plasma model

The nano-plasma model provides a description of an irradiated cluster in terms of global quantities, such as numbers of particles or charges. These quantities evolve in time according to rate equations. The model thus constitutes, to some extent, the last conceivable step of simplification, when starting from a fully quantal many-body picture and going down the ladder to more and more macroscopic quantities as dynamical variables. The basic dynamical degrees of freedom in the nano-plasma model are

$$\begin{array}{lcl} N_j & = & \text{number of ions in charge state "}j\text{"}, \\ N_e & = & \text{number of electrons inside the cluster}, \\ E_{\rm int} & = & \text{internal energy of the electron cloud}, \\ R & = & \text{radius of the cluster.} \end{array}$$

It is assumed that ions, electrons and energy are distributed homogeneously inside a sphere of radius $R$. One can thus write down the ion and electron densities as

$$n_j = \frac{3N_j}{4\pi R^3} \quad , \quad n_e = \frac{3N_e}{4\pi R^3} \quad . \tag{F.6}$$

Note that because of the homogeneity of the system, both forms ($n_j$ and $N_j$) carry the same information. They thus will be used side-by-side whatever form yields a simpler expression.

The evolution of ion numbers is determined by rate equations of the form

$$\frac{dN_j}{dt} = W_j^{\rm tot} N_{j-1} - W_{j+1}^{\rm tot} N_j \tag{F.7a}$$

with

$$W_j^{\rm tot} = W_j^{\rm tun} + W_j^{\rm las} + W_j^{\rm th} \tag{F.7b}$$

$$W_j^{\rm tun} = C_{n^*l^*m} \frac{I_P^{(j)}}{\hbar} (3\xi)^{2n^*-m-3/2} e^{-\xi} \quad , \quad \xi = \sqrt{\frac{2m_e}{\hbar^2} \frac{4I_P^{(j)3/2}}{3e\mathcal{E}_{\rm int}}} \quad , \tag{F.7c}$$

$$W_j^{\rm las} = n_e < \sigma_j v >_{\rm las} \quad , \tag{F.7d}$$

$$W_j^{\rm th} = n_e < \sigma_j v >_{\rm th} \quad . \tag{F.7e}$$

In the above expressions $I_P^{(j)}$ is the ionization potential of an ion of charge state $j$, $\sigma_j$ is the scattering cross-section of one electron from the same ion and the prefactor $C_{n^*l^*m}$ in Eq. (F.7c) depends on the quantum numbers (actually effective quantum numbers, whence the stars) of the charge state $j$. The field $\mathcal{E}_{\rm int}$ labels the effective electrical field. The averages in Eqs. (F.7d) and (F.7e) respectively label averages over one optical cycle and over electron

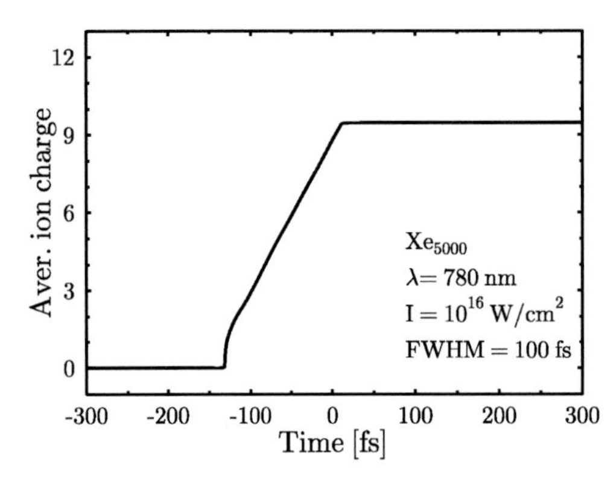


**Figure F.1:** Time evolution of the average charge of an atom in a nano-plasma of a $Xe_{5000}$ cluster excited by a strong laser field. The laser parameters are indicated in the plot. The peak of the laser pulse sets the zero point of the time scale.

thermal velocities. The tunneling rate $W_j^{\rm tun}$ describes the ionization of atoms through tunneling in the effective laser field [ADK86]. The rate $W_j^{\rm las}$ describes ionization which occurs from inelastic collisions of laser driven (free) electrons on atoms/ions. And $W_j^{\rm th}$ stands for ionization through inelastic collisions with thermal electrons. The tunneling rate dominates initially and the two other processes take over with increasing supply of free electrons.A typical situation is sketched in Figure F.1. The initially neutral atoms are ionized by the laser field. The small initial ionization proceeds very fast and turns to a more steady growth as long as the laser stays switched on. The charge finally levels off into an equilibrium where ionization and recombination processes compensate each other.

The electron number changes as

$$\frac{dN_e}{dt} = \sum_j j \frac{dN_j}{dt} - \frac{dQ}{dt} \tag{F.8a}$$

with

$$\frac{dQ}{dt} = \int_{v_{\rm esc}}^{\infty} dv\, v^2 \rho(v) \int_S d\vec{S}\cdot\vec{v} \quad , \tag{F.8b}$$

$$v_{\rm esc} = \sqrt{2K_{\rm esc}/m_e} \quad , \quad K_{\rm esc} = \frac{(Q+1)e^2}{2R} \quad , \tag{F.8c}$$

$$\rho(v) = n_e \frac{1}{4\pi} \left(\frac{2k_B T}{m_e \pi}\right)^{3/2} \exp\left(-\frac{m_e v^2}{2k_B T_{\rm e}}\right) \quad , \tag{F.8d}$$

where $Q$ is the total net charge of the cluster and its change is determined by the integrated net flow $\rho\vec{v}$ through the surface $S$ which is considered as a spherical boundary layer with thickness of the electron mean free path. The velocity integration runs over the thermal velocity distribution of the electrons (temperature $T_{\rm e}$) starting from the minimum energy $K_{\rm esc}$ necessary for electron emission.

The change in cluster radius is driven by the total pressure which is a sum of two components, the hydrodynamic pressure $P_H$ of the hot electron gas which expands outwards, driving the ions and a Coulomb part $P_C$ coming from the space charge build up resulting in a Coulomb repulsion between the ions inside the cluster. It reads

$$\frac{\partial^2 R}{\partial t^2} = 5\frac{P_C + P_H}{n_i m_i}\frac{1}{R} \tag{F.9a}$$

$$P_C = \frac{3}{8\pi}\frac{Q^2 e^2}{R^4} \quad , \quad P_H = n_e k T_e \quad . \tag{F.9b}$$

The change of the internal energy of the electron cloud is mainly driven by heating through absorption of electromagnetic energy, cooling through global expansion, ionization processes, and energy loss by electron flow through the cluster surface:

$$\frac{d}{dt}(E_{\text{int}}) = \dot{E}^{(\text{el-magn})} - \left(\frac{2E_{\text{int}}}{R}\frac{\partial R}{\partial t} + \sum_j I_p^{(j)} dN_j/dt + E_{\text{fs}}\right) \tag{F.10a}$$

with

$$E_{fs} = \int_{v_{\text{esc}}}^{\infty} dv\, v^2\, \rho(v) \int_S d\vec{S}\cdot\vec{v}(m_e v^2/2) \tag{F.10b}$$

$$\dot{E}^{(\text{el-magn})} = -\frac{V}{4\pi}\mathcal{E}_{\text{int}}\cdot\frac{\partial(\epsilon(\omega_{\text{las}})\mathcal{E}_{\text{int}})}{\partial t} \quad , \tag{F.10c}$$

(F.10d)

where $V$ is the volume of the system. The dielectric constant therein is given by the Drude model

$$\epsilon(\omega) = 1 - \omega_p^2/\omega(\omega + i\nu) \quad , \quad \omega_p^2 = 4\pi\frac{n_e}{m_e}e^2 \quad , \quad \nu = \nu_{ei} + A\frac{v}{R} \quad . \tag{F.11a}$$

The $\omega_p$ is the (volume) plasma frequency, see Eq. (1.24) in Section 1.3.3.2. The $\nu_{ei}$ describes damping through electron–ion collisions. The second term in $\nu$ accounts for Landau damping in a homogeneous sphere of radius $R$, $v$ being the average electron velocity, and $A$ an appropriate constant close to unity. The effective intrinsic field $\mathcal{E}_{\text{int}}$ is given by the external field $\mathcal{E}_{\text{ext}}$ and the relation (1.20) for a dielectric sphere. Note that several expressions above need the temperature as input, rather than the intrinsic energy. The temperature can be obtained by the simple (perfect gas) relation

$$E_{\text{int}} = \frac{3}{2}N_e k T_e \quad . \tag{F.12}$$

# G Linearized TDLDA and related approaches

The excitation spectrum of a cluster can be obtained from TDLDA combined with spectral analysis as outlined in Section 4.1.7. TDLDA is driven here with very small amplitudes to ensure that one stays in the linear regime of small deviations from the ground state. This can be exploited to derive from TDLDA, by linearization, a closed set of equations which determine directly the (stationary) eigenmodes of the excitation spectrum. These are the RPA equations (RPA = random phase approximation) which are widely used in all areas of physics. Accordingly, there are many different ways to formulate the RPA equations depending on the underlying means of description (Green's functions, density matrices, Slater states, etc.). To have at least one detailed example, we will present here the linearized TDLDA ($\equiv$ RPA) with the equations-of-motion technique as worked out in [Row70]. Alternative derivations can be found, e.g., in [FW71, RS80, BB94]. The presentation here is based on the formulation given in [RG92] with specialization to metal clusters in [RGB96b]. For simplicity, we will derive the RPA equations under the assumption that a microscopic many-body Hamiltonian $\hat{H}$ is given. The translation to the case of energy-density functionals is sketched at the end.

## G.1 The linearized equations

A mean-field state is described mainly in terms of a set of occupied single-particle wavefunctions $\{\varphi_\alpha(\mathbf{r},t)\}$. It is advantageous for the present, more formal, purposes to switch to a description in terms of the associated Slater state $|\Phi\rangle$ which is the anti-symmetrized product of the single-particle states, i.e. $\langle\mathbf{r}_1\ldots\mathbf{r}_N|\Phi(t)\rangle = \mathcal{A}\left\{\prod\varphi(\mathbf{r}_i,t)\right\}$. For linearization purpose the actual mean-field state $|\Phi(t)\rangle$ is supposed to move close to the ground state $|\Phi_0\rangle$. We thus take the ground state as reference point and represent the time-dependent state by means of the Thouless theorem [Tho60, RS80]

$$|\Phi(t)\rangle = \exp\left(\imath\eta\hat{G}(t)\right)|\Phi_0\rangle \quad , \quad \hat{G}^+ = \hat{G} \quad , \quad \hat{G}\in\{1ph\} \quad . \tag{G.1}$$

We use it here in the form with a hermitian generator $\hat{G}$ which automatically guarantees normalization of $|\Phi(t)\rangle$. The Thouless theorem requires that $\hat{G}$ is a one-particle-one-hole ($1ph$) operator and it asserts that $|\Phi(t)\rangle$ is a Slater state if the reference state $|\Phi_0\rangle$ was also a Slater state. The $\eta$ is some small number which serves as an expansion parameter for the handling of the small amplitudes. The time evolution is determined by the time-dependent variational principle which we write here in the form $\delta(\int dt\langle\Phi(t)|\imath\partial_t - \hat{H}|\Phi(t)\rangle) = 0$. Inserting the expression G.1 for $|\Phi(t)\rangle$ into the variational principle we expand up to second order in $\eta$ which

*Introduction to cluster dynamics.* Paul-Gerhard Reinhard, Eric Suraud

ISBN: 3-527-40345-0

yields

$$\delta \int_0^T dt \left[ \langle \Phi_0 | [\hat{G}, \imath \partial_t \hat{G}] | \Phi_0 \rangle - \langle \Phi_0 | [\hat{G}, [\hat{H}, \hat{G}]] | \Phi_0 \rangle \right] = 0 \quad . \tag{G.2}$$

The variational degree of freedom is here the $1ph$ operator $\hat{G}(t)$. The variation is then also to be taken from the space of $1ph$ operators,

$$\delta \hat{G} \in \left\{ \hat{a}_p^+ \hat{a}_h, \hat{a}_h^+ \hat{a}_p \right\} \equiv \hat{a^+ a} \tag{G.3}$$

where $\hat{a}_\alpha^+$ and $\hat{a}_\alpha$ are the usual Fermion operators, generating and annihilating a Fermion in a single particle state $\alpha$. The $\hat{a^+ a}$ serves as a shorthand notation for both $ph$ and $hp$ operators. Variation then yields

$$\langle \Phi_0 | [\hat{a^+ a}, \imath \partial_t \hat{G}] | \Phi_0 \rangle = \langle \Phi_0 | [\hat{a^+ a}, [\hat{H}, \hat{G}]] | \Phi_0 \rangle \quad .$$

We exploit once again the fact that we consider only small amplitudes. For then the motion decomposes into harmonic oscillations in the various (harmonic) eigenmodes $n$ of the system, i.e.

$$\hat{G} = \sum_n \left[ \hat{C}_n^+ e^{-\imath \omega_n t} + \hat{C}_n e^{+\imath \omega_n t} \right] \tag{G.4}$$

where $\hat{C}_n^+$ is a $1ph$ operator generating the eigenmode $n$ and $\hat{C}_n$ annihilates it. We assume for simplicity that all modes have different frequencies. The oscillation factors $\exp(\pm \imath \omega_n t)$ are then all linearly independent. This yields a set of equations

$$\omega_n \langle \Phi_0 | [\hat{a^+ a}, \hat{C}_n^+] | \Phi_0 \rangle = \langle \Phi_0 | [\hat{a^+ a}, [\hat{H}, \hat{C}_n^+]] | \Phi_0 \rangle \tag{G.5a}$$

$$-\omega_n \langle \Phi_0 | [\hat{a^+ a}, \hat{C}_n] | \Phi_0 \rangle = \langle \Phi_0 | [\hat{a^+ a}, [\hat{H}, \hat{C}_n]] | \Phi_0 \rangle \tag{G.5b}$$

which determine the $\hat{C}_n^+$ and the eigenfrequencies $\omega_n$ uniquely. This form can be evaluated directly by handling operators in terms of wavefunctions as done in [RG92, RGB96b]. A more explicit representation is obtained by expanding the eigen-generators in terms of $1ph$ operators

$$\hat{C}_n^+ = \sum_{ph} \left[ x_{ph}^{(n)} \hat{a}_p^+ \hat{a}_h - y_{ph}^{(n)} \hat{a}_h^+ \hat{a}_p \right] \tag{G.6}$$

where $p > N$ stands for (unoccupied) particle states and $h \leq N$ for (occupied) hole states. This yields the RPA equations in their widely used standard form [RS80]

$$\omega_n \begin{pmatrix} \mathbf{x}^{(n)} \\ \mathbf{y}^{(n)} \end{pmatrix} = \begin{pmatrix} A & B \\ -B^* & -A^* \end{pmatrix} \begin{pmatrix} \mathbf{x}^{(n)} \\ \mathbf{y}^{(n)} \end{pmatrix} \quad , \tag{G.7a}$$

$$A_{ph,p'h'} = \langle \Phi_0 | [\hat{a}_h^+ \hat{a}_p, [\hat{H}, \hat{a}_{p'}^+ \hat{a}_{h'}]] | \Phi_0 \rangle \quad , \tag{G.7b}$$

$$B_{ph,p'h'} = \langle \Phi_0 | [\hat{a}_h^+ \hat{a}_p, [\hat{H}, \hat{a}_{h'}^+ \hat{a}_{p'}]] | \Phi_0 \rangle \quad , \tag{G.7c}$$

where $\left(\mathbf{x}, \mathbf{y}^{(n)}\right)_{ph}$ label the vectors of amplitudes $x, y_{ph}^{(n)}$, respectively. It is a (non-hermitian) eigen-value problem for the expansion coefficients $\mathbf{x}^{(n)}$ and $\mathbf{y}^{(n)}$ and eigen-frequency $\omega_n$.

The equation covers several interesting features. For example, continuous symmetries of the underlying Hamiltonian $\hat{H}$ lead to eigenstate with zero frequency. For extensive discussions of RPA, its solution and deeper formal properties see e.g. [Row70, RS80].

Up to here, we have formulated RPA in terms of a given microscopic Hamiltonian. That is not the situation when dealing with TDLDA. The starting energy functional Eq. (3.23) treats the kinetic energy $\hat{T}$ still microscopically. But the potential energy is given in terms of a local energy functional which we summarize here as $E_{\text{pot}}(\rho)$. The mean-field potential is obtained from the first functional derivative, see Eq. (3.24c). Expansion into small amplitude oscillations then naturally generates the second functional derivative. With successive application of the Thouless theorem and small amplitude expansions, one can identify (for details see [RG92, RGB96b])

$$\begin{aligned}\langle\Phi_0|[\hat{a}_h^+\hat{a}_p,[\hat{H},\hat{a}_{p'}^+\hat{a}_{h'}]]|\Phi_0\rangle &= \langle\Phi_0|[\hat{a}_h^+\hat{a}_p,[\hat{h}_0,\hat{a}_{p'}^+\hat{a}_{h'}]]|\Phi_0\rangle \\ &+\int d\mathbf{r}\,d\mathbf{r}'\,\varphi_p^*(\mathbf{r}')\varphi_h(\mathbf{r}')\frac{\partial^2 E_{\text{pot}}}{\partial\rho(\mathbf{r})\partial\rho(\mathbf{r}')}\varphi_{h'}^*(\mathbf{r}')\varphi_{p'}(\mathbf{r}')\end{aligned}$$

where $\hat{h}_0$ is the Kohn-Sham Hamiltonian (3.24) associated with the ground state $|\Phi_0\rangle$. Similar relations hold for the other combinations of $ph$ and $hp$ operators.

## G.2 Sum rule approximation

The sum rule approximation relies on an intuitive guess for the generators and annihilators $\hat{C}_c^+$ and $\hat{C}_c$ of a dominant collective mode $c$. To see that, one has first to reformulate the above RPA in terms of coordinate- and momentum-like operators. The $\hat{C}_c^+$ and $\hat{C}_c$ can be viewed as basic ladder operators of a harmonic oscillator algebra. These are related to coordinate and momenta as [Mes70] $\hat{Q}_c \propto (\hat{C}_c^+ + \hat{C}_c)$ and $\hat{P}_c \propto (\imath\left(\hat{C}_c^+ - \hat{C}_c\right))$. The above RPA algebra can then be re-mapped to

$$\hat{P}_c = \frac{\imath[\hat{H},\hat{Q}_c]_{ph}}{\langle[\hat{Q}_c[[\hat{H},\hat{Q}_c]]\rangle}, \tag{G.8a}$$

$$\omega_c^2 = \frac{\langle[\hat{P}_c,[\hat{H},\hat{P}_c]]\rangle}{\langle[\hat{Q}_c,[\hat{H},\hat{Q}_c]]\rangle} = \frac{\langle[[\hat{Q}_c,\hat{H}],[\hat{H},[\hat{H},\hat{Q}_c]]\rangle}{\langle[\hat{Q}_c[[\hat{H},\hat{Q}_c]]\rangle}, \tag{G.8b}$$

provided $c$ is an eigenmode. The index $ph$ therein stands for the $1ph$ part of the operator as filtered by the commutator with $a^+a$ in Eq. (G.5). (As an aside, the expression (G.8b) for the excitation energy can also be used as a starting point for deriving RPA from a stationary variational principle [Bra89, RBG90].)

The relations (G.8) require first the knowledge of the eigenmode and its associated "coordinate" operator $\hat{Q}_c$. The sum rule approximation employs an educated guess for that. The starting point is the Thomas-Reiche-Kuhn sum rule [RS80]. With a few formal manipulations, one can deduce a closed expression for the energy weighted dipole strength:

$$\sum_n \hbar\omega_n|\langle\Psi_n|\hat{D}|\Psi_0\rangle|^2 = \frac{\hbar^2}{2m}e^2 N \tag{G.9}$$

where $\hat{D}$ is the dipole operator, $|\Psi_0\rangle$ is the ground state and $\{|\Psi_n\rangle\}$ runs over the RPA excitation spectrum with excitation energies $\hbar\omega_n$. The Mie plasmon resonance exhausts this sum rule completely. Thus it pretends to be the only state around carrying dipole strength. This suggests that the operator $\hat{D}$ plays a crucial role in generating the Mie resonance. The construction $\exp(\imath\hat{D})|\Psi_0\rangle$ describes indeed the velocity field associated with the oscillating electron cloud. The sum rule approximation now consists in identifying $\hat{Q}_c = \hat{D}$ and to insert that into Eq. (G.8). The conjugate momentum turns out to be $\hat{P}_c \propto \sum_{i=1}^{N} \hat{p}_i$, i.e. the momentum of the electronic center-of-mass operator. It is then straightforward to evaluate numerator and denominator in relation (G.8b). The result is precisely Eq. (4.17). This is not surprising because the Mie plasmon exhausts the dipole sum rule.

Such sum rule estimates for resonance frequencies have been developed as versatile and efficient tools in nuclear physics [BLM79] and successfully transplanted to metal clusters in [LS91]. They are the starting point for more detailed and yet simple models in terms of collective flow (fluid dynamics) [Bra89, RGB96a, KB01].

# H Numerical considerations

## H.1 Representation of electron wavefunctions and densities

The various theoretical schemes addressed in Chapter 3 require very different numerical methods. The majority deals with single-particle wavefunctions $\varphi_\alpha(\mathbf{r})$ and local fields as densities or potentials $U(\mathbf{r})$. There are several ways to represent numerically a local function $f(\mathbf{r})$. The most general form is a basis expansion

$$f(\mathbf{r}) = \sum_b \phi_b(\mathbf{r}) c_b \tag{H.1}$$

where $\{\phi_b(\mathbf{r})\}$ is an appropriate set of functions. The set of expansion coefficients $\{c_b\}$ is then the numerical equivalent of $f(\mathbf{r})$. In molecules and clusters one often chooses a basis of atomic wavefunctions (mostly analytical approximations to it) around each atomic center. This is the much celebrated linear combination of atomic orbitals (LCAO). Such a basis is well adapted in cases where one wants to include some core electrons in the calculations. It is, however, not orthogonal, $(\phi_b|\phi_{b'}) \neq \delta_{bb'}$. The single-electron Schrödinger equation then becomes a matrix equation with a norm matrix,

$$(\phi_b|\hat{h}|\phi_{b'})c_{b'} = \varepsilon(\phi_b|\phi_{b'}) \quad . \tag{H.2}$$

Time dependent cases need to take into account additionally a possible time dependence of the basis states $\phi_b$ through motion of the ionic centers [SS96]. But all that is only a minor complication as long as the norm matrix avoids singularities. The hard part is the computation of the matrix elements. Much depends here on appropriate analytical forms for the basis states $\phi_b(\mathbf{r})$. A most versatile and efficient choice are here Gaussian functions $\exp\left(-\lambda(\mathbf{r}-\mathbf{R}_I)^2\right)$ times low order polynomials in $\mathbf{r}$. This is the linear combination of Gaussian orbitals (LCGO). It is widely used in quantum chemistry, for HF and CI calculations [YOGI94] as well as for DFT [Pop99]. The computation of matrix elements can then take advantage of powerful group theoretical recurrence relations [Won70].

LCAO or LCGO are practically unavoidable if a calculation has to cover very different length scales as it happens, e.g., when core electrons are involved. Calculations in clusters which consider only valence electrons or even a smooth jellium background (see Section 3.2.3) or a simple shell model potential, e.g. Eq. (3.27), have to deal with one typical length scale of order $r_s$. In such cases it is advantageous to consider a direct representation of wavefunctions and fields on a grid in coordinate space

$$f(\mathbf{r}) \quad \longleftrightarrow \quad f(\mathbf{r_n}) \quad , \quad \mathbf{r_n} \in \{(n_x\Delta x, n_y\Delta y, n_z\Delta z)\} \quad , \tag{H.3}$$

*Introduction to cluster dynamics.* Paul-Gerhard Reinhard, Eric Suraud

ISBN: 3-527-40345-0

or symmetry restricted versions thereof in cylindrical or spherical coordinates. This is particularly useful in (TD)LDA because a central object there are the local density $\rho(\mathbf{r})$ and the local Kohn-Sham potentials, e.g. in Eq. (3.24). It remains the evaluation of the kinetic energy $\propto \hat{\Delta}$. This can be done by finite differences or Fourier techniques. An extensive comparison of the gridding schemes and their performance can be found in [BLMR92]. As a rule of thumb, Fourier techniques are advantageous in three-dimensional Cartesian grids with not too low a number of grid points (at and above 32 mesh points in each direction) while finite differences perform better in restricted symmetries.

# H.2 Iteration and propagation schemes

## H.2.1 Electronic ground state

The HF equations (3.20a) and the Kohn-Sham equations (3.24) constitute a non-linear eigenvalue problem. This requires in any case an iterative solution. A conceptually straightforward strategy is to employ a standard solution scheme for the Schrödinger equation with given one-body Hamiltonian $\hat{h}$, e.g., a readily available eigenvalue solver for the equation (H.2). The mean-field Hamiltonian $\hat{h}$ depends on the set of occupied single particle states $\{\varphi_\alpha\}$. This is thus resolved iteratively. One starts with a reasonably guessed set $\{\varphi_\alpha^{(0)}\}$ and computes $\hat{h}^{(0)}$ from it. One then solves the Schrödinger equation for $\hat{h}^{(0)}$ which yields an improved set $\{\varphi_\alpha^{(1)}\}$. The steps are repeated until key quantities (energy, energy-variance) do not change between two iterative steps. This scheme, however, causes work overhead if the solution of the Schrödinger equation itself employs an iterative scheme. For then, it is better to merge these two iterations. A typical example for this simultaneous iteration scheme is the gradient step

$$\varphi_\alpha^{(n+1)} = \hat{\mathcal{O}}\left\{\varphi_\alpha^{(n)} - \hat{\mathcal{D}}\left(\hat{h}^{(n)} - (\varphi_\alpha^{(n)}|\hat{h}^{(n)}|\varphi_\alpha^{(n)})\right)\varphi_\alpha^{(n)}\right\} \tag{H.4}$$

where $\hat{\mathcal{O}}$ stands for ortho-normalization of the new set $\{\varphi_\alpha^{(n+1)}\}$ and $\hat{\mathcal{D}}$ is a convergence generating damping operator. The simplest choice is just a sufficiently small number $\hat{\mathcal{D}} = \eta$ which ensures convergence if it is smaller than twice the maximum representable energy. Much faster convergence can be achieved by appropriate tuning of $\hat{\mathcal{D}}$. For example, grid representations get their largest conceivable energies from the kinetic energy operator. Here it is a good idea to use $\hat{\mathcal{D}} = \eta/(\hat{T} + E_0)$ where $E_0$ is typically the depth of the potential. This step can proceed very fast with $\eta \approx 1$. For more details see [BLMR92]. Thus far we have discussed the solution of the stationary electronic problem. One also needs to optimize the ionic configuration such that a global minimum of the energy is reached. One typical scheme for that purpose is discussed in Section H.3.

## H.2.2 Dynamics

Time dependent problems require appropriate schemes for propagating the ions and electrons. The classical equations of motion for ionic propagation are usually solved by one of the standard propagation schemes for ordinary differential equations, for example the Verlet algorithm

[Ver67] (often called leap-frog). More elaborate methods are rarely needed because the electronic time scale sets the pace and this means that one can employ rather small time steps $\delta t$ for ionic propagation. The choice of propagation scheme for the electronic wavefunctions depends, of course, on the chosen numerical representation. For the example (H.3) of a grid in coordinate space (but not only for this example) and for a strictly local mean field, a very efficient stepping scheme is the time-splitting method [FFS82]

$$\varphi_\alpha(\mathbf{r},t+\delta t) = \exp\left(-\imath\frac{\delta t}{2}U(\mathbf{r},t+\delta t)\right)\exp\left(-\imath\delta t\hat{T}\right)\exp\left(-\imath\frac{\delta t}{2}U(\mathbf{r},t)\right)\varphi_\alpha(\mathbf{r},t) \tag{H.5}$$

The action of the local operator is trivial in coordinate space. The workload consists in the kinetic term. It is evaluated easily in Fourier space. In the case of finite differences, it can be separated into three one-dimensional propagators in the $x$-, $y$-, and $z$-direction. The nice thing with this time-splitting scheme is that the action of the local phase factor does not change the local density. Thus the density $\rho(\mathbf{r},t+\delta t)$ is already known after the evaluation of the kinetic propagator. This allows one to compute the $U(\mathbf{r},t+\delta t)$ for the last step without iteration. Another advantage of the step (H.5) is that it is unitary and this preserves ortho-normality of the set $\{\varphi_\alpha\}$. An often used alternative are the Crank-Nicholson or Peaceman-Rachford steps. These rely on an approximate separation of the propagation into a succession of three one-dimensional steps, for a general discussion see [PTVF92] and for applications in cluster dynamics e.g. [CRSU00].

## H.3 Details on simulated annealing

In the context of structure optimization, simulated annealing with Monte-Carlo techniques was discussed in Section 3.4.3. We exemplify here the method for the electrons treated at the level of LDA(-SIC). A random sampling of the full state including the electronic wavefunctions is tedious. The electronic minimum for a fixed ionic configuration is usually well defined and there exist efficient iteration techniques to reach it. In a sort of adiabatic picture, the random walk is applied only to the ions ($\mathbf{R}$) while the electrons follow immediately the actual ionic shape. The path is thus dictated by the energy $E(\mathbf{R},\phi_\mathbf{R})$ which is a function of $\mathbf{R}$ alone because the electronic state $\Phi_\mathbf{R}$ is now the ground state of the stationary KS equation at fixed $\mathbf{R}$. It is still rather cumbersome to optimize $\Phi_\mathbf{R}$ for each trial shot. The changes in one random step are very small. We can exploit the adiabatic theorem that the small energy change can be computed from the expectation value with the old $\Phi_\mathbf{R}$. It is only after a certain series of steps that one needs to update the electronic state from scratch. The computations are much simplified by introducing a single-ion energy

$$e_J\left(\mathbf{R},\phi_\mathbf{R}\right) = \sum_\alpha(\varphi_\alpha|\hat{V}_{\mathrm{PsP}}(\mathbf{r}_{\mathrm{el}},\mathbf{R}_J)\varphi_\alpha) + \sum_{I\neq J}\frac{e^2 Z_I Z_J}{|\mathbf{R}_I-\mathbf{R}_J|} \tag{H.6}$$

where the $\varphi_\alpha$ are the single-electron states of $\phi_\mathbf{R}$. The adiabatic theorem then associates $e'_J - e_J$ with a small change of ion $J$. The emerging Monte-Carlo scheme for simulated annealing

**Table H.1:** Outline of simulated annealing for optimizing the ionic configuration in connection with stationary KS states for the electrons.

- starting values $\Phi_{\mathbf{R}}$, $\mathbf{R}$, $\Delta$, and $T$
- loop1
  - store previous $E_{\text{prev}} \leftarrow E_{\text{total}}(\mathbf{R}, \phi_{\mathbf{R}})$
  - compute total ionic energy $E_{\text{loop}} = \sum e_J$
  - loop2
    - store $E_{\text{old}} \leftarrow E_{\text{loop}}$
    - loop3
      - random $J \in \{1, \ldots, N_{\text{ion}}\}$, $i \in \{x, y, z\}$
      - trial $\tilde{\mathbf{R}}'$: $R'_{J,i} = R_{J,i} + \Delta\,\texttt{random}(0,1)$
      - compute $e'_J = e_J\left(\mathbf{R}', \phi_{\mathbf{R}}\right)$
      - new trial energy $E' = E_{\text{loop}} + (e'_J - e_J)$
      - Metropolis:
        - $\texttt{random}(0,1) < \exp\left((E_{\text{loop}} - E')/T\right)$
        - $\Longrightarrow \quad \mathbf{R} \leftarrow \mathbf{R}'\,,\ e_J \leftarrow e'_J\,,\ E_{\text{loop}} = E'$
    - enddo3
    - reduce $\Delta \leftarrow \Delta(1 - \kappa_\Delta)$, $T \leftarrow T(1 - \kappa_T)$
    - update $E_{\text{loop}} = \sum e_J\left(\mathbf{R}', \phi_{\mathbf{R}}\right)$
    - terminate loop2 if $|E_{\text{loop}} - E_{\text{old}}| < \epsilon_{\text{ion}}$
  - enddo2
  - update $\phi = \phi_{\mathbf{R}}$ with KS
  - terminate if $|E_{\text{prev}} - E_{\text{total}}| < \epsilon_{\text{tot}}$
- enddo1

is summarized in Table H.1. The random steps of the ions are applied in the innermost loop. The core of this loop consists in the Metropolis step where $\exp\left((E_{\text{loop}} - E')/T\right)$ decides on acceptance or rejection of the trial configuration $\mathbf{R}'$. The step is regulated by the numerical parameters $\Delta$ (= size of trial steps) and $T$ (= temperature). Large $\Delta$ and $T$ allow one to explore a large configuration space but certainly miss the details of the minimum. The strategy is to start from large values and to reduce them gradually as the solution improves. This is regulated by the reduction parameters $\kappa_\Delta$ and $\kappa_T$ which are applied after each completed inner loop. The further numerical parameters are the numbers for each of the three loops and the termination criteria. The multitude of driver parameters shows that the method is very versatile and can accommodate many different situations. This also means that one needs experience to find appropriate driver parameters. For more details see, e.g., [PTVF92].

## H.4 The test-particle method for Vlasov-LDA and VUU

The Vlasov and VUU equations are phase space equations which involve the phase space density $f(\mathbf{r}, \mathbf{p}, t)$. They are thus to be solved in a six-dimension space (for phase space) as a function of time. This represents a gigantic task and a direct resolution in phase space is close to impossible. Furthermore, it is well known that a direct resolution of the Vlasov equation may be numerically problematic due to the tendency to filamentation [SL85]. A standard solution to this problem is to use the so called test-particle method. The phase space density $f(\mathbf{r},\mathbf{p},t)$ is represented by an ensemble of $N_{\rm tp}$ smooth particles

$$f(\mathbf{r},\mathbf{p},t) = \frac{N_{\rm el}}{N_{\rm tp}} \sum_{n=1}^{N_{\rm tp}} g_r\left(\mathbf{r}-\mathbf{r}_n(t)\right) g_p\left(\mathbf{p}-\mathbf{p}_n(t)\right) \quad . \tag{H.7a}$$

There are different choices for the smoothing functions. Most widely used and very efficient are normalized Gaussians

$$g_r(\mathbf{r}) = \left(2\sigma_r^2/\pi\right)^{3/2} \exp\left(-\frac{\mathbf{r}^2}{2\sigma_r^2}\right) \quad , \tag{H.7b}$$

$$g_p(\mathbf{p}) = \left(2\sigma_p^2/\pi\right)^{3/2} \exp\left(-\frac{\mathbf{p}^2}{2\sigma_p^2}\right) \quad , \tag{H.7c}$$

$$\sigma_r \sigma_p = 1 \quad . \tag{H.7d}$$

The relation (H.7d) is not compulsory for a numerical implementation. But it adds a crucial piece of physics background as the folding functions (H.7b) and (H.7b) are directly related to the Husimi picture and the mutual dependence (H.7d) of the widths is the necessary condition to implant the minimum quantum smoothing into the semi-classical Vlasov-LDA or VUU. The ansatz (H.7) maps the Vlasov or VUU equation into Hamiltonian equations for the coordinates $(\mathbf{r}_n, \mathbf{p}_n)$ of the test particles. The spatial folding (H.7b) modifies the local KS potential such that the test particles effectively experience a folded potential $g{\star}U_{\rm KS}$ [LSR95, DLRS97]. This folding seems to be an appropriate means to maintain the quantum smoothness in the classical treatment. There remains $\sigma_r$ as one free parameter. An appropriate choice delivers close agreement of results between Vlasov-LDA and full TDLDA, even when ionic dynamics is taken into account (Vlasov-MD, VUU-MD) [GRS02]. The test-particle method becomes particularly advantageous for evaluating the collision term Eq. (3.30b) in VUU. The nine-fold integration over momentum spaces is replaced by a double sum over pairs of test particles plus a stochastic evaluation of the scattering events, for details see [BD88]. This is still an awful lot of work, but just manageable. Computations can be accelerated by grouping the large number of test particles into smaller sub-ensembles and restricting collisions only to members of the same ensemble [GSR01].

The Vlasov equation carries no tag on the quantum statistics of the propagated particles. It is a classical propagation related to Boltzmann statistics. Numerical fluctuations due to the use of a finite number of test particles induce an artificial collision term which turns out to be a Boltzmann collision term. The related relaxation drives the phase-space distribution towards Boltzmann equilibrium which is very wrong in many-Fermion systems at high density and low temperature [RS95, LRS95]. The relaxation time increases with the number of test particles. Vlasov-LDA can thus be stabilized for a certain propagation time by using a sufficiently large $N_{\rm tp}$. Times of the order of $50 - 200$ fs can be reliably well propagated with an acceptable expense (typically $N_{\rm tp}/N_{\rm el} \sim 8000$). For longer analyzing times or to be on the safe side with Vlasov-LDA, one can introduce a stabilization term which ensures that the Pauli principle, here in the form $f \leq 2$, is never violated [DRS97]. Interestingly enough, the problem disappears if one uses VUU. The Ühling-Uhlenbeck collision term Eq. (3.30b) is tailored for Fermions. Its Pauli blocking provides sufficient stability for long time studies. Even at low temperatures, where the phase space factor suppresses almost all collisions, the few remaining collision events suffice to maintain Fermionic stability.

## H.5 On the solution of the TDTF equation

We discuss a few more details onf the time-dependent Thomas-Fermi equation (3.34). A particularly efficient and robust way to solve this equation is by means of the Madelung transform [Mad27]. Density and velocity-generator field are mapped into one complex field

$$\Phi(\mathbf{r},t) = \sqrt{\rho(\mathbf{r},t)}\exp\left(\imath\hbar\chi(\mathbf{r},t)/m\right) \tag{H.8a}$$

which allows one to summarize the two equations (3.34) into one Schrödinger-like equation

$$\begin{aligned} \imath\hbar\Phi &= -\frac{\hbar^2}{2m}\Delta\Phi + \left(U_{\mathrm{KS}}^{(\mathrm{LDA})} + \frac{\delta T_0}{\delta\rho} + U_{\mathrm{Mad}}\right)\Phi \quad , \\ U_{\mathrm{Mad}} &= \frac{\hbar^2}{2m}\frac{\Delta\sqrt{\rho}}{\sqrt{\rho}} \quad . \end{aligned} \tag{H.8b}$$

The Madelung potential causes trouble if $\rho$ becomes zero. This happens unfortunately in the simple Thomas-Fermi solution (3.33). We have learned in Vlasov-LDA that the Husimi folding allows one to maintain the minimum quantum smoothness. We apply here the same folding with (H.7b) to the Thomas-Fermi density. This provides a much better approach to the quantum mechanical density and it renders the Madelung potential to be well defined everywhere [Dom98]. This modification has a similar effect as the (properly chosen) Husimi folding in the Vlasov equation.

# Bibliography


[AARS96] Y Abe, S Ayik, P-G Reinhard, and E Suraud, Phys. Rep. **275** (1996), 49.

[AC77] R A Aziz and H H Chen, J. Chem. Phys. **67** (1977), 5719.

[ADDP98] P Alvarado, J Dorantes-Davila, and G M Pastor, Phys. Rev. B **58** (1998), no. 18, 12216.

[ADK86] M V Ammosov, N B Delone, and V P Krainov, Sov. Phys. JETP **64** (1986), 1191.

[AEM+95] M H Anderson, J R Ensher, M R Matthews, C E Wieman, and E A Cornell, Science **269** (1995), 198.

[AH65] I V Abarenkov and V Heine, Phil. Mag. **12** (1965), 529.

[AJ02] P H Acioli and J Jellinek, Phys. Rev. Lett. **89** (2002), 213402.

[AL67] N W Ashcroft and D C Langreth, Phys. Rev. **155** (1967), 682.

[All96] A P Allivisatos, Science **271** (1996), 933.

[AM76] N W Ashcroft and N D Mermin, *Solid State Physics*, Saunders College, Philadelphia, 1976.

[And+00] J. Andruszkow et al., Phys. Rev. Lett. **85** (2000), 3825.

[AP78] C H Aldrich and D Pines, J. Low Temp. Phys. **32** (1978), 689.

[ARS02a] K Andrae, P-G Reinhard, and E Suraud, Peprint, submitted to Eur. Phys. J. D, 2002.

[ARS02b] K Andrae, P-G Reinhard, and E Suraud, Peprint, J. Phys. B **35** (2002), 1.

[ARS03] K Andrae, P-G Reinhard, and E Suraud, Peprint, Preprint, submitted to Phys. Rev. Lett., 2003.

[AS98] D C Athanasopoulos and K E Schmidt, J. Phys. Chem. A **102** (1998), 1615.

[ASB+95] F Alasia, L Serra, R A Broglia, N V Giai, E Lipparini, and H E Roman, Phys. Rev. B **52** (1995), 8488.

[ASG+00] J Anton, K Schulze, D Geschke, W-D Sepp, and B Fricke, Phys. Lett. A **268** (2000), 85.

[AW93] G Audi and A H Wapstra, Nucl. Phys. A **565** (1993), 1.

[Bal75] R Balescu, *Equilibrium and Non-Equilibrium Statistical Mechanics*, Wiley, New York, 1975.

[BB74] R Balian and C Bloch, Ann. Phys. (NY) **85** (1974), 514.

[BB81] P F Bortignon and R A Broglia, Nucl. Phys. A **371** (1981), 405.

[BB94] G F Bertsch and R A Broglia, *Oscillations in Finite Quantum Systems*, Cambridge University Press, Cambridge, 1994.

[BB97] M Brack and R K Bhaduri, *Semiclassical Physics*, Addision-Wesley, Reading, 1997.

[BBFP02] M Barat, J C Brenot, J A Fayeton, and Y J Picard, J. Chem. Phys. **117** (2002), 1497.

[BBG96] P Blaise, S A Blundell, and C Guet, Phys. Rev. B **55** (1996), 15856.

[BBG+02] M Bianchetti, P F Buosante, F Ginelli, H E Roman, R A Broglia, and F Alasia, Phys. Rep. **357** (2002), 459.

[BC95] C Bréchignac and Ph Cahuzac, Comm. At. Mol. Phys. **31** (1995).

[BCC+90] C Brechignac, Ph Cahuzac, F Carlier, N de Frutos, and J Leygnier, J. Chem. Phys. **93** (1990), 7449.

[BCC+94] C Bréchignac, Ph Cahuzac, F Carlier, M de Frutos, R N Barnett, and Uzi Landman, Phys. Rev. Lett. **72** (1994), 1636.

[BCC+02] C Bréchignac, P Cahuzac, F Carlier, C Colliex, J Leroux, A Masson, B Yoon, and U Landman, Phys. Rev. Lett. **88** (2002), 196103–1.

[BCCL01] C Brechignac, Ph Cahuzac, B Concina, and J Leygnier, Euro. Phys. J. D **16** (2001), 91.

[BCdF+92] C Bréchignac, P Cahuzac, M de Frutos, N Kebaïli, J Leygnier, J P Roux, and A Sarfati, NATO ASI Series E: Applied Sciences, vol. I, p. 369, Kluwer, Dordrecht, 1992.

[BCdH94] I M L Billas, A Châtelain, and W A de Heer, Science **265** (1994), 1682.

[BCK+93] J Borggreen, P Chowdhury, N Kebaili, L Lundsberg-Nielsen, K Luetzenkirchen, M B Nielsen, J Pedersen, and H D Rasmussen, Phys. Rev. B **48** (1993).

[BD88] G F Bertsch and S Das Gupta, Phys. Rep. **160** (1988), 190.

[Bec84] D E Beck, Sol. St. Comm. **49** (1984), 381.

[Bec88] A D Becke, Phys. Rev. A **38** (1988), 3098.

[Bei95] A Beiser, *Concepts of Modern Physics*, McGraw-Hill, New York, 1995.

[Ber98] S Berry, Nature **393** (1998), 212.

[BG93] S A Blundell and C Guet, Z. f. Physik D **28** (1993), 81.

[BG95] S A Blundell and C Guet, Z. f. Physik D **33** (1995), 153.

[BGM94] A Bonasera, F Gulminelli, and J Molitoris, Phys. Rep. **243** (1994), 1.

[BGR00] U Becker, O Gessner, and A Rödel, J. El. Spectr. and Rel. Phen. **108** (2000), 189.

[BGS+00] D Babikov, E A Gislason, M Sizun, F Aguillon, and V Sidis, J. Chem. Phys. **112** (2000), 7032.

[BHK89] P F Bernath, K H Hinkle, and J J Keady, Science **244** (1989), 562.

[BHS82] G B Bachelet, D R Hamann, and M Schlüter, Phys. Rev. B **26** (1982), 4199.

[Bin01] C Binns, Surf. Sci. Rep. **44** (2001), 1.

[BJR00] F Balzer, S D Jett, and H-G Rubahn, Solid Films **372** (2000), 78.

[BK78] K H Bennemann and J B Ketterson (eds.), *The Physics of Liquid and Solid Helium, Part I & II.*, Wiley, New York, 1978.

[BKFK89] V Bonacic-Koutecky, P Fantucci, and J Koutecky, J. Chem. Phys. **91** (1989), 3794.

[BKPF+96] V Bonacic-Koutecky, J Pittner, C Fuchs, P Fantucci, M F Guest, and J Koutecky, J. Chem. Phys. **104** (1996), 1427.

[BLGC02] D Bruhwiler, C Leiggener, S Glaus, and G Calzaferri, J. Phys. Chem. **106** (2002), 3770.

[BLM79] O Bohigas, A M Lane, and J Martorell, Phys. Rep. **51** (1979), 267.

[BLMA92] J Borstel, U Lammers, A Mananes, and J A Alonso, Lecture Notes in Physics **404** (1992), 327.

[BLMR92] V Blum, G Lauritsch, J A Maruhn, and P-G Reinhard, J. Comp. Phys **100** (1992), 364.

[BLNR91] R N Barnett, U Landman, A Nitzan, and G Rajagopal, J. Chem. Phys. **94** (1991), 608.

[BLR91] R N Barnett, U Landman, and G Rajagopal, Phys. Rev. Lett. **67** (1991), 3058.

[BLW+99] B Bescos, B Lang, J Weiner, V Weiss, E Wiedemann, and G Gerber, Euro. Phys. J. D **9** (1999), 399.

[BM92] B H Bransden and M R C McDowell, *Charge Exchange and Theory of Ion-Atom Collisions*, Clarendon, Oxford, 1992.

[Boy50] S F Boys, Proc. Roy. Soc. London A **200** (1950), 542.

[BPBB01] B Baguenard, J C Pinar, C Bordas, and M Broyer, Phys. Rev. A **63** (2001), 023204.

[BPHH96] C Bordas, F Paulig, H Heln, and D L Huestis, Rev. Sci. Instr. **67** (1996), 2257.

[BR85] M A Bolorizadeh and M E Rudd, Phys. Rev. A **33** (1985), 882.

[BR94] K Boyer and C K Rhodes, J. Phys. B **27** (1994), L633.

[BR97] J Babst and P-G Reinhard, Z. f. Physik D **42** (1997), 209.

[BR01] F Balzer and H-G Rubahn, Nanotechnology **12** (2001), 105.

[Bra83] R H Brandsen, *Atomic Collision Theory*, Benjamin, Reading, 1983.

[Bra89] M Brack, Phys. Rev. B **39** (1989), 3533.

[Bra93] M Brack, Rev. Mod. Phys. **65** (1993), 677.

[Bra97] M Brack, The Scientific American **277** (1997), 50.

[BRS+84] D Bohle, A Richter, W Steffen, A E L Dieperink, N Lo Iudice, F Palumbo, and O Scholten, Phys. Lett. B **137** (1984), 27.

[BRS02] T Berkus, P-G Reinhard, and E Suraud, Int. J. Mol. Sci. **3** (2002), 69.

[BSC+96] S A Buzza, E M Snyder, D A Card, D E Folmer, and A W Castleman Jr, J. Chem. Phys. **105** (1996), 7425.

[BSK97] K Boucke, H Schmitz, and H-J Kull, Phys. Rev. A **56** (1997), 763.

[BSM+00] J Bansmann, V Senz, R-P Methling, R Röhlsberger, and K H Meiwes-Broer, Mat. Sci. Eng. C **19** (2000), 305.

[BT89] G F Bertsch and D Tomanek, Phys. Rev. B **40** (1989), 2749.

[CA80] D M Ceperley and B J Alder, Phys. Rev. Lett. **45** (1980), 566.

[Cal91] J Callaway, *Quantum Theory of the Solid State*, Academic, London, 1991.

[CB77] W J Chesnavich and M T Bowers, J. Chem. Phys. **66** (1977), 2306.

[CFH+00] H Cederquist, A Fardi, K Haghighat, A Langereis, H T Schmidt, S H Schwartz, J C Levin, I A Sellin, H Lebius, B Huber, M O Larsson, and P Hvelplund, Phys. Rev. A **61** (2000), 022712.

[CG82] K Chang and W Graham, J. Chem. Phys. **77** (1982), 4300.

[CG97] K Capelle and E K U Gross, Phys. Rev. Lett. **78** (1997), 1872.

[CGC00] J Cernihcaro, J R Goiciecha, and E Caux, Astrophys. J. **534** (2000), L199.

[CGIS00] J-P Connerade, L G Gerchikov, A N Ipatov, and S Senturk, J. Phys. B **33** (2000), 5109.

[CHH+00] E E B Campbell, K Hansen, K Hoffmann, G Korn, M Tchaplyguine, M Wittmann, and I V Hertel, Phys. Rev. Lett. **84** (2000), 2128.

[CHH02] C Champion, J Hanssen, and P A Hervieux, Phys. Rev. A **65** (2002), 922710.

[Cle85] K Clemenger, Phys. Rev. B **32** (1985), 1359.

[Cle91] K Clemenger, Phys. Rev. B **44** (1991), 12991.

[CP85] R Car and M Parinello, Phys. Rev. Lett. **55** (1985), 2471.

[CPCV95] F Catara, Ph, Chomaz, and N VanGiai, Z. f. Physik D **33** (1995), 219.

[CRS97] F Calvayrac, P-G Reinhard, and E Suraud, Ann. Phys. (NY) **255** (1997), 125.

[CRSU00] F Calvayrac, P-G Reinhard, E Suraud, and C A Ullrich, Phys. Rep. **337** (2000), 493.

[CSS02] P Claas, F Stienkemeier, and C P Schulz, Preprint, 2002.

[CTB89] M Cohen, A G G M Tielens, and J D Bregman, Astrophys. J. Lett. **344** (1989), L13.

[CTCS90] O Cheshnovsky, K J Taylor, J Conceicao, and R E Smalley, Phys. Rev. Lett. **64** (1990), 1785.

[CWFJ02] D A Card, E S Wisniewski, D E Folmer, and A W Castleman Jr, J. Chem. Phys. **116** (2002), 3554.

[DAR+01] Ph Dugourd, R Antoine, D Rayane, I Compagnon, and M Broyer, J. Chem. Phys. **114** (2001), 1970.

[Dat95] S Datta, *Electronic Transport in Mesoscopic Systems*, Cambridge University Press, Cambridge, 1995.

[DB88] S S Dietrich and B L Bermann, At. Data Nucl. Data Tab. **38** (1988), 199.

[DDK+02] N H Damrauer, C Dietl, G Krampert, S-H Lee, K-H Jung, and G Gerber, Euro. Phys. J. D **20** (2002), 71.

[DDM86] G Durand, J-P Daudey, and J-P Malrieu, J. de Phys. (Paris) **47** (1986), 1335.

[DDPS81] T G Dietz, M A Duncan, D E Powers, and R E Smalley, J. Chem. Phys. **74** (1981), 6511.

[DDR+96] T Ditmire, T Donnelly, A M Rubenchik, R W Falcone, and M D Perry, Phys. Rev. A **53** (1996), 3379.

[DDTMB01] T Diederich, T Döppner, J Tiggesbäumker, and K-H Meiwes-Broer, Phys. Rev. Lett. **86** (2001), 4807.

[Deh90] H Dehmelt, Rev. Mod. Phys. **62** (1990), 525.

[DG90] R M Dreizler and E K U Gross, *Density Functional Theory: An Approach to the Quantum Many-Body Problem*, Springer-Verlag, Berlin, 1990.

[DGGM+99] G Durand, J Giraud-Girard, D Maynau, F Spiegelmann, and F Calvo, J. Chem. Phys. **110** (1999), 7871.

[DGPS99] F Dalfovo, S Giorgini, L P Pitaevskii, and S Stringari, Rev. Mod. Phys. **71** (1999), 463.

[DGRS00] A Domps, E Giglio, P-G Reinhard, and E Suraud, J. Phys. B **33** (2000), L333.

[dH93] Walt A de Heer, Rev. Mod. Phys. **65** (1993), 611.

[DJB87] H L Davis, J Jellinek, and R S Berry, J. Chem. Phys. **86** (1987), 6456.

[DKSM01] D Dalacu, E Klemberg-Sapieha, and L Martinu, Surf. Sci. **472** (2001), 33.

[DLRS97] A Domps, P L'Eplattenier, P-G Reinhard, and E Suraud, Ann. Phys. (Leipzig) **6** (1997), 455.

[DLS+97] S Dobosz, M Lezius, M Schmidt, P Meynadier, M Perdrix, D Normand, J-P Rozet, and D Vernhet, Phys. Rev. A **56** (1997), R2526.

[DMA+95] K B Davis, M-O Mewes, M R Andrews, N J van Druten, D S Durfee, D M Kurn, and W Ketterle, Phys. Rev. Lett. **75** (1995), 3969.

[Dom98] A Domps, Ph.D. thesis, Université Paul Sabatier, Toulouse, 1998.

[DRHPT90] J Dupont-Roc, M Himbert, N Pavloff, and J Treiner, J. Low Temp. Phys. **81** (1990), 31.

[DRS97] A Domps, P-G Reinhard, and E Suraud, Ann. Phys. (NY) **260** (1997), 171.

[DRS98a] A Domps, P-G Reinhard, and E Suraud, Phys. Rev. Lett. **81** (1998), 5524.

[DRS98b] A Domps, P-G Reinhard, and E Suraud, Phys. Rev. Lett. **80** (1998), 5520.

[DRS00] A Domps, P-G Reinhard, and E Suraud, Ann. Phys. (NY) **280** (2000), 211.

[DS96] F Duplàe and F Spiegelmann, J. Chem. Phys. **105** (1996), 1492.

[DSN+02] B Dubertret, P Skourides, D J Norris, V Noireaux, A H Brivanlou, and A Libchaber, Science **298** (2002), 1759.

[DST+98] T Ditmire, E Springate, J W G Tisch, Y L Shao, M B Mason, N Hay, J P Marangos, and M H R Hutchinson, Phys. Rev. A **57** (1998), 369.

[DST00] D Durand, E Suraud, and B Tamain, *Nuclear Dynamics in the Nucleonic Regime*, Institute of Physics, London, 2000.

[DTMB02] T Diederich, J Tiggesbümker, and K H Meiwes-Broer, J. Chem. Phys. **116** (2002), 3263.

[DTS+97a] T Ditmire, J W G Tisch, E Springate, M B Mason, N Hay, J P Marangos, and M H R Hutchinson, Phys. Rev. Lett. **78** (1997), 2732.

[DTS+97b] T Ditmire, J W G Tisch, E Springate, M B Mason, N Hay, R A Smith, J Marangos, and M H R Hutchinson, Nature **386** (1997), 54.

[EFSP00] B J Eves, F Festy, K Svensson, and R E Palmer, Appl. Phys. Lett. **77** (2000), 4223.

[Eka84] W Ekardt, Phys. Rev. Lett. **52** (1984), 1925.

[Ell98] S Elliott, *The Physics and Chemistry of Solids*, Wiley, New York, 1998.

[Eng72] R Englman, *The Jahn-Teller Effect in Molecules and Crystals*, Wiley, London, 1972.

[Eng86] P C Engelking, J. Chem. Phys. **85** (1986), 3103.

[EP91] W Ekardt and Z Penzar, Phys. Rev. B **43** (1991), 1331.

[ESH$^+$02] C Ellert, M Schmidt, H Haberland, V Veyret, and V Bonacic-Koutecky, J. Chem. Phys. **117** (2002), 3711.

[ESR81] O Echt, K Sattler, and E Recknagel, Phys. Rev. Lett. **47** (1981), 1121.

[ESS$^+$95] C Ellert, M Schmidt, C Schmitt, T Reiners, and H Haberland, Phys. Rev. Lett. **75** (1995), 1731.

[EZK98] G E Engelmann, J C Ziegler, and D M Kolb, Surf. Sci. Lett. **401** (1998), L420.

[FA34] E Fermi and E Amaldi, Accad. Ital. Rome **6** (1934), 117.

[Fai87] F H M Faisal, *Theory of Multiphoton Processes*, Plenum Press, New York, 1987.

[FAL$^+$02] N Felidj, J Aubard, G Levi, J R Krenn, M Salerno, G Schider, B Lamprecht, A Leitner, and F R Aussenegg, Phys. Rev. B **65** (2002), 075419.

[FAS$^+$97] B Fricke, J Anton, K Schulze, W-D Sepp, and P Kuerpick, Nuc. Instr. Meth. B **124** (1997), 211.

[Fer28] E Fermi, Z. f. Physik **48** (1928), 73.

[FFS82] M D Feit, J A Fleck, and A Steiger, J. Comp. Phys. **47** (1982), 412.

[FIM$^+$00] P W Fry, I E Itskevich, D J Mowbray, M S Skolnick, J J Finley, J A Barker, E P O'Reilly, L R Wilson, I A Larkin, P A Maksym, M Hopkinson, M Al-Khafaji, J P R David, A G Cullis, G Hill, and J C Clark, Phys. Rev. Lett. **84** (2000), 733.

[For$^+$96] D Forney et al., J. Chem. Phys. **104** (1996), 4954.

[FP93] S Frauendorf and V Pashkevich, Z. f. Physik D **26** (1993), 98.

[FPM93] E S Fois, J I Penman, and P A Madden, J. Chem. Phys. **98** (1993), 6352.

[FR97] O Frank and J Rost, Chem. Phys. Lett. **271** (1997), 367.

[FSCR96] L Feret, E Suraud, F Calvayrac, and P-G Reinhard, J. Phys. B **29** (1996), 4477.

[FV82] J Friedrich and N Vögler, Nucl. Phys. A **373** (1982), 192.

[FW71] A L Fetter and J D Walecke, *Quantum Theory of Many-Particle Systems*, McGraw-Hill, New York, 1971.

[GAB$^+$94] F Garcias, J A Alonso, M Barranco, J M Lopez, A Mananes, and J Nemeth, Z. f. Physik D **31** (1994), 275.

[GBD$^+$95] T Goetz, M Buck, C Dressler, F Eisert, and F Traeger, Appl. Phys. A **60** (1995), 607.

[GCKS95] J Guan, M E. Casida, A M Koester, and D R Salahub, Phys. Rev. B **52** (1995), 2184.

[GDP96] E K U Gross, J F Dobson, and M Petersilka, Top. Curr. Chem. **181** (1996), 81.

[GFF$^+$02] F Gobet, B Farizon, M Farizon, M J Gaillard, J P Buchet, M Carr, P Scheier, and T D Märk, Phys. Rev. Lett. **89** (2002), 183403.

[GGA] `http://crab.rutgers.edu/ kieron`.

[GGKS01] R Gutierrez, F Grossmann, O Knospe, and R Schmidt, Phys. Rev. A **64** (2001), 013202.

[GH95] D H E Gross and P A Hervieux, Z. f. Physik D **35** (1995), 27.

[GIPS00] L G Gerchikov, A N Ipatov, R G Polozkov, and A V Solov'yov, Phys. Rev. A **62** (2000), 043201.

[GIS98] L G Gerchikov, A N Ipatov, and A V Solov'yov, J. Phys. B **31** (1998), 2331.

[GISG98] L G Gerchikov, A N Ipatov, A V Solov'yov, and W Greiner, J. Phys. B **31** (1998), 3065.

[GK90] E K U Gross and W Kohn, Adv. Quant. Chem. **21** (1990), 255.

[GL76] O Gunnarsson and B I Lundqvist, Phys. Rev. B **13** (1976), 4274.

[GLC$^+$01] M Gaudry, J Lermé, E Cottancin, M Pellarin, J-L Vialle, M Broyer, B Prével, M Treilleux, and P. Mélinon, Phys. Rev. B **64** (2001), 085407.

[Goe96] A Goerling, Phys. Rev. A **54** (1996), 3912.

[GPB91] M E Garcia, G M Pastor, and K H Benneman, Phys. Rev. Lett. **67** (1991), 1142.

[Gro01] D H E Gross, *Microcanonical Thermodynamics*, World Scientific, Singapore, 2001.

[GRS01] E Giglio, P-G Reinhard, and E Suraud, J. Phys. B **34** (2001), 1253.

[GRS02] E Giglio, P-G Reinhard, and E Suraud, Ann. Phys. (Leipzig) **11** (2002), 291.

[GRS03] E Giglio, P-G Reinhard, and E Suraud, Phys. Rev. A **67** (2003), 43202.

[GS95] B M Garraway and K-A Suominen, Rep. Prog. Phys. **58** (1995), 365.

[GSR01] E Giglio, E Suraud, and P-G Reinhard, Int. J. Phys. C **12** (2001), 1439.

[GST$^+$00] R E Grisenti, W Schllkopf, J P Toennies, G C Hegerfeldt, T Khler, and M Stoll, Phys. Rev. Lett. **85** (2000), 2284.

[GTH96] S Goedecker, M Teter, and J Hutter, Phys. Rev. B **54** (1996), 1703.

[GUG95] E K U Gross, C A Ullrich, and U J Gossmann, NATO ASI Series B, vol. 337, p. 149, Plenum Press, New York, 1995.

[GW64] M L Goldberger and K M Watson, *Collision Theory*, Wiley, New York, 1964.

[HAA$^+$00] D Hathiramani, K Aichele, W Arnold, K Huber, E Salzborn, and P Scheier, Phys. Rev. Lett. **85** (2000), 3604.

[Hab94] H Haberland (ed.), *Clusters of Atoms and Molecules 1- Theory, Experiment, and Clusters of Atoms*, vol. 52, Springer Series in Chemical Physics, Berlin, 1994.

[Har27] D R Hartree, Proc. Camb. Phil. Soc. **24** (1927), 89,111.

[HBL94] H Häkinnen, R N Barnett, and U Landman, Europhys. Lett. **28** (1994).

[HCN$^+$03] S Himadri, H S Chakraborty, R G Nazmitdinov, M E Madjet, and J M Rost, Preprint, 2003.

[Hei95] K Heinz, Rep. Prog. Phys. **58** (1995), 637.

[HG00] R Heinicke and J Grotemeyer, Appl. Phys. B **71** (2000), 419.

[HGGE97] P E Hodgson, E Galioli, and E Galioli-Erba, *Introductory Nuclear Physics*, Clarendon, Oxford, 1997.

[HIM93] H Haberland, Z Insepov, and M Moseler, Z. f. Physik D **26** (1993), 229.

[HK64] P Hohenberg and W Kohn, Phys. Rev. **136** (1964), 864.

[HKR$^+$02] U Hergenhahn, A Kolmakov, M Riedler, A R B de Castro, O Lofken, and T Moller, Chem. Phys. Lett. **351** (2002), 235.

[HM94] K Hansen and M Manninen, J. Chem. Phys. **101** (1994), 10481.

[HM96] H Häkinnen and M Manninen, Europhys. Lett. **34** (1996).

[Hoo91] W G Hoover, *Computational Statistical Mechanics*, Elsevier, Amsterdam, 1991.

[Hor91] M Horbatsch, Z. f. Physik D **21** (1991), S63.

[HPBK+98] M Hartmann, J Pittner, V Bonacic-Koutecky, A Heidenreich, and J Jortner, J. Chem. Phys. **108** (1998), 3096.

[HRU93] B D Hall, D Reinhard, and D Ugarte, Z. f. Physik D **26** (1993), 73.

[HS99] H Haberland and M Schmidt, Euro. Phys. J. D **6** (1999), 109.

[HvIY+93] H Haberland, B von Issendorff, Ji Yufeng, T Kolar, and G Thanner, Z. f. Physik D **26** (1993), 8.

[HW53] D L Hill and J A Wheeler, Phys. Rev. **89** (1953), 1102.

[HW00] H Haken and H C Wolf, *The Physics of Atoms and Quanta*, Springer, Berlin, 2000.

[HWB+01] M Hyslop, A Wurl, S A Brown, B D Hall, and R Monot, Euro. Phys. J. D **16** (2001), 233.

[HZ02] W Heerlein and G Zwicknagel, J. Comp. Phys. **180** (2002), 497.

[HZT02] W Heerlein, G Zwicknagel, and C Toeppfer, Phys. Rev. Lett. **89** (2002), 083202.

[IB00] K Ishikawa and T Blenski, Phys. Rev. A **62** (2000), 063204.

[IIT87] S Ichimaru, H Iyetomi, and S Tanaka, Phys. Rep. **149** (1987), 91.

[IM82] H Ibach and D L Mills, *Electron Energy Loss Spectroscopy and Surface Vibrations*, Academic, New York, 1982.

[Iud97] N. Lo Iudice, Prog. Part. Nucl. Phys. **28** (1997), 556.

[Jac62] J D Jackson, *Classical Electrodynamics*, Wiley, New York, 1962.

[JB93] N Ju and A Bulgac, Phys. Rev. B **48** (1993), 2721.

[JBB86] J Jellinek, T L Beck, and R S Berry, J. Chem. Phys. **84** (1986), 2783.

[JBD+00] D B Janes, M Batistuta, S Datta, M R Melloch, R P Andres, J Liu, N.-P Chen, T Lee, R Reifenberger, E H Chen, and J M Woodall, Superlattices and Microstructures **27** (2000), 555.

[JDK00] C J Joachain, M Dörr, and N Kylstra, Adv. Atom. Mol. Opt. Phys. **42** (2000), 225.

[JFA+01] T Jacob, B Fricke, J Anton, S Varga, T Bastug, S Fritzsche, and W Sepp, Euro. Phys. J. D **16** (2001), 257.

[JFS+02] T Jacob, S Fritzsche, W-D Sepp, B Fricke, and J Anton, Phys. Lett. A **300** (2002), 71.

[JG89] R O Jones and O Gunnarsson, Rev. Mod. Phys. **61** (1989), 689.

[JMG+98] M Bruchez Jr, M Moronne, P Gin, S Weiss, and A P Alivisato, Science **281** (1998), 2013.

[Kai01] A B Kaiser, Rep. Prog. Phys. **64** (2001), 1.

[KAM00] S Kümmel, J Akole, and M Manninen, Phys. Rev. Lett. **84** (2000), 3827.

[Kas28] L S Kassel, J. Phys. Chem. **32** (1928), 225.

[KB99] C Kohl and G F Bertsch, Phys. Rev. B **60** (1999), 4205.

[KB01] S Kümmel and M Brack, Phys. Rev. A **64** (2001), 022506.

[KBR98] S Kümmel, M Brack, and P-G Reinhard, Phys. Rev. B **58** (1998), 1774.

[KBR99] S Kümmel, M Brack, and P-G Reinhard, Euro. Phys. J. D **9** (1999), 149.

[KBRB00] S Kümmel, T Berkus, P-G Reinhard, and M Brack, Euro. Phys. J. D **11** (2000), 239.

[KCdH$^+$84] W D Knight, K Clemenger, W A de Heer, W A Saunders, M Y Chou, and M L Cohen, Phys. Rev. Lett. **52** (1984), 2141.

[KCdHS85] W D Knight, K Clemenger, W A de Heer, and W A Saunders, Phys. Rev. B **31** (1985), 2539.

[KCRS98] C Kohl, F Calvayrac, P-G Reinhard, and E Suraud, Surf. Sci. **405** (1998), 74.

[KEGC$^+$99] C Kohl, S M El-Gammal, F Calvayrac, E Suraud, and P-G Reinhard, Euro. Phys. J. D **5** (1999), 271.

[Kel64] L V Keldysh, Sov. Phys. JETP **20** (1964), 1307.

[KET98] M Klintenberg, S Edvardsson, and J O Thomas, J. Alloys and Compounds **275** (1998), 174.

[KFK91] V Bonačić Koutecký, P Fantucci, and J Koutecký, Chem. Rev. **91** (1991), 1035.

[KFR97] C Kohl, B Fischer, and P-G Reinhard, Phys. Rev. B **56** (1997), 11149.

[KGB95] W Klaus, M E Garcia, and K H Bennemann, Z. f. Physik D **35** (1995), 43.

[KHSW88] J Kübler, K-H Höck, J Sticht, and A R Williams, J. Phys. F **18** (1988), 469.

[KJSS00] O Knospe, J Jellinek, U Saalmann, and R Schmidt, Phys. Rev. A **61** (2000), 022715.

[KL99] M A Kornberg and P Lambropoulos, J. Phys. B **32** (1999), L603.

[KLI92] J B Krieger, Y Li, and G J Iafrate, Phys. Rev. A **45** (1992), 101.

[KLKD99] T Kirchner, H J Ludde, O J Kroneisen, and R M Dreizler, Nuc. Instr. Meth. B **154** (1999), 46.

[KLM95] M Koskinen, P O Lipas, and M Manninen, Nucl. Phys. A **591** (1995), 421.

[Klo71] C E Klots, J. Phys. Chem. **75** (1971), 1526.

[KMR95] C Kohl, B Montag, and P-G Reinhard, Z. f. Physik D **35** (1995), 57.

[KMR96] C Kohl, B Montag, and P-G Reinhard, Z. f. Physik D **38** (1996), 81.

[KN02] E Krotscheck and J Navarro, *Microscopic Approaches to Quantum Liquids in Confined Geometries*, World Scientific, Singapore, 2002.

[Koh97] C Kohl, Ph.D. thesis, F.-A. Universität Erlangen-Nürnberg, 1997.

[Kol95] M Kolbuszewski, J. Chem. Phys. **102** (1995), 3679.

[KR97] C Kohl and P-G Reinhard, Z. f. Physik D **39** (1997), 225.

[KR01] V P Krainov and A S Roshchupkin, J. Phys. B **34** (2001), L297.

[Kra96] K Krane, *Modern Physics*, Wiley, New York, 1996.

[Kro87] H W Kroto, Nature **329** (1987), 529.

[Kro00] E Krotscheck, J. Low Temp. Phys. **119** (2000), 103.

[KRS00] C Kohl, P G Reinhard, and E Suraud, Euro. Phys. J. D **11** (2000), 115.

[KS65] W Kohn and L J Sham, Phys. Rev. **140** (1965), 1133.

[KS67] P Kramer and M Saraceno, Lecture Notes in Physics **140** (1967).

[KS01] T Kunert and R Schmidt, Phys. Rev. Lett. **86** (2001), 5258.

[KSK+99] L Köller, M Schumacher, J Köhn, S Teuber, J Tiggesbäumker, and K-H Meiwes-Broer, Phys. Rev. Lett. **82** (1999), 3783.

[KV93] U Kreibig and M Vollmer, *Optical Properties of Metal Clusters*, vol. 25, Springer Series in Materials Science, 1993.

[KW02] S. Keppeler and R. Winkler, Phys. Rev. Lett. **88** (2002), 046401.

[KWSR97] J-H Klein-Wiele, P Simon, and H.-G Rubahn, Phys. Rev. Lett. **80** (1997), 45.

[KWSR99] J-H Klein-Wiele, P Simon, and H-G Rubahn, Optics Comm. **161** (1999), 42.

[KZ01] E Krotscheck and R Zillig, J. Chem. Phys. **115** (2001), 10161.

[LA00a] D Lang and Ph Avouris, Phys. Rev. Lett. **84** (2000), 358.

[LA00b] D Lang and Ph Avouris, Phys. Rev. B **62** (2000), 7325.

[Lam82] H Lamb, Proc. Londion Math. Soc. **13** (1882), 187.

[Lam+98] P Lambropoulos et al., Phys. Rep. **305** (1998), 205.

[Lan97] N D Lang, Phys. Rev. Lett. **79** (1997), 1357.

[LBA86] X P Li, J Q Broughton, and P B Allen, J. Chem. Phys. **85** (1986), 3444.

[LBP+93] J Lerme, Ch Bordas, M Pellarin, B Baguenard, J L Vialle, and M Broyer, Phys. Rev. B **48** (1993), 12110.

[LBW89] H Loewen, T Beier, and H Wagner, Europhys. Lett. **9** (1989), 791.

[LDG+01] E Lamour, S Dreuil, J-C Gauthier, O Gobert, P Meynadier, D Normand, M Perdrix, C Prigent, J M Ramillon, J-P Rozet, and D Vernhet, Proceedings of the SPIE - The International Society for Optical Engineering, vol. 4504, 2001, p. 97.

[LDNS98] M Lezius, S Dobosz, D Normand, and M Schmidt, Phys. Rev. Lett. **80** (1998), 261.

[Lie87] A Liebsch, Phys. Rev. B **36** (1987), 7378.

[LJ00] I Last and J Jortner, Phys. Rev. A **62** (2000), 013201.

[LLC+00] T Lee, J Liu, N-P Chen, R P Andres, D B Janes, and R Reifenberger, J. Nanopart. Res. **2** (2000), 345.

[LMP+00] J Lehmann, M Merschdorf, W Pfeiffer, A Thon, S Voll, and G Gerber, Phys. Rev. Lett. **85** (2000), 2921.

[LNH84] R A Larson, S K Ncoh, and D R Herschbach, Rev. Sci. Instr. **45** (1984), 1511.

[LNR+91] D L Lichtenberger, K W Nebesny, C D Ray, D R Huffman, and L D Lamb, Chem. Phys. Lett. **176** (1991), 203.

[LP88] E M Lifschitz and L P Pitajewski, *Physikalische Kinetik*, Lehrbuch der Theoretischen Physik, vol. X, Mir, Moscow, 1988.

[LPC+95] J Lerme, M Pellarin, E Cottancin, B Baguenard, J L Vialle, and M Broyer, Phys. Rev. B **52** (1995), 14163.

[LRMB91] G Lauritsch, P-G Reinhard, J Meyer, and M Brack, Phys. Lett. A **160** (1991), 179.

[LRS95] P L'Eplattenier, P-G Reinhard, and E Suraud, J. Phys. A **28** (1995), 787.

[LS89] E Lipparini and S Stringari, Phys. Rev. Lett. **63** (1989), 570.

[LS91] E Lipparini and S Stringari, Z. f. Physik D **18** (1991), 193.

[LSG02] A Lyalin, A Solov'yov, and W Greiner, Phys. Rev. A **65** (2002), 043202.

[LSR95] P L'Eplattenier, E Suraud, and P-G Reinhard, Ann. Phys. (NY) **244** (1995), 426.

[LSR02] C Legrand, E Suraud, and P-G Reinhard, J. Phys. B **35** (2002), 1115.

[LVC+02] M A Lebeault, J Viallon, J Chevaleyre, C Ellert, D Normand, M Schmidt, O Sublemontier, C Guet, and B Huber, Euro. Phys. J. D **20** (2002), 233.

[LVWW99] T Leisner, S Vajda, S Wolf, and L Wöste, J. Chem. Phys. **111** (1999), 1017.

[LZWR95] S Li, R Z Van Zee, W Wltmer, and K Raghavachani, Chem. Phys. Lett. **243** (1995), 275.

[Mad27] E Madelung, Z. f. Physik **40** (1927), 322.

[Mah93] G D Mahan, *Many Particle Physics*, Plenum, New York, 1993.

[Mar66] R A Marcus, J. Chem. Phys. **45** (1966), 2630.

[Mar93] T P Martin, Phys. Rep. **273** (1993), 199.

[MBB+91] T P Martin, S Bjornholm, J Borggreen, C Béchignac, P Cahuzac, K Hansen, and J Pedersen, Chem. Phys. Lett. **186** (1991), 53.

[MBB+03] F Megi, M Belkacem, M A Bouchene, E Suraud, and G Zwicknagel, J. Phys. B **36** (2003), 273.

[MBGL90] T P Martin, T Bergmann, H Göhlich, and T Lange, Chem. Phys. Lett. **172** (1990), 209.

[MBGL91] T P Martin, T Bergmann, H Göhlich, and T Lange, Z. f. Physik D **19** (1991), 25.

[MCR01] M E Madjet, H S Chakraborty, and Jan M Rost, J. Phys. B **34** (2001), L345.

[MCRS96] L Mornas, F Calvayrac, P-G. Reinhard, and E Suraud, Z. f. Physik D **38** (1996), 73.

[MEL+89] K M McHugh, J G Eaton, G H Lee, H W Sarkas, L H Kidder, J T Snodgrass, M R Manaa, and K H Bowen, J. Chem. Phys. **91** (1989), 3792.

[Mes70] A. Messiah, *Quantum Mechanics Vol. I*, Wiley, New York, 1970.

[MF91] A Maiti and L M Falicov, Phys. Rev. A **44** (1991), 4442.

[MGJ95] M Madjet, C Guet, and W R Johnson, Phys. Rev. A **51** (1995), 1327.

[MHM+95] B Montag, Th Hirschmann, J Meyer, P-G Reinhard, and M Brack, Phys. Rev. B **52** (1995), 4775.

[MI99] P Milani and S Iannotta, *Cluster Beam Synthesis of Nanostructured Materials*, Springer, Berlin, 1999.

[Mie08] G Mie, Ann. Phys. (Leipzig) **25** (1908), 377.

[MKPE98] M Mertig, R Kirsch, W Pompe, and H Engelhardt, Euro. Phys. J. D **9** (1998), 45.

[MLT+94] A McPherson, T S Luk, B D Thompson, A B Borisov, O B Shiryaev, X Chen, K Boyer, and C K Rhodes, Phys. Rev. Lett. **72** (1994), 1810.

[MLY01] M Moseler, U Landman, and C Yannouleas, Phys. Rev. Lett. **87** (2001), 053401.

[MM97] M Movre and W Meyer, J. Chem. Phys. **106** (1997), 7139.

[MNdSC+00] J R Marinelli, V Nesterenko, F F de Souza-Cruz, W Kleinig, and P-G Reinhard, Phys. Rev. Lett. **85** (2000), 3141.

[MPBS01] C Mayer, R Palkovits, G Bauer, and T Schalkhammer, J. Nanoparticle Res. **3** (2001), 361.

[MPC98] C Massobrio, A Pasquarello, and A Dal Corso, J. Chem. Phys. **109** (1998), 6626.

[MPT+00] M Merschdorf, W Pfeiffer, A Thon, S Voll, and G Gerber, App. Phys. A **71** (2000), 547.

[MR51] R A Marcus and O K Rice, J. Phys. Colloid. Chem. **55** (1951), 894.

[MR94] B Montag and P-G Reinhard, Phys. Lett. A **193** (1994), 380.

[MR95a] B Montag and P-G Reinhard, Z. f. Physik D **33** (1995), 265.

[MR95b] B Montag and P-G Reinhard, Phys. Rev. B **51** (1995), 14686.

[MRM94] B Montag, P-G Reinhard, and J Meyer, Z. f. Physik D **32** (1994), 125.

[MSK+01] R-P Methling, V Senz, E-D Klinkenberg, Th Diederich, J Tiggesbäumker, G Holzhöter, J Bansmann, and K H Meiwes-Broer, Euro. Phys. J. D **16** (2001), 173.

[MW95] W C Martin and W L Wiese, AIP, New York, 1995.

[MWL+01] A Mertig, R Wahl, M Lehmann, P Simon, and W Pompe, Euro. Phys. J. D **16** (2001), 317.

[MYB98] K Mishima, K Yamashita, and A Bandrauk, J. Phys. Chem. A **102** (1998), 3157.

[Mye77] W. D. Myers, *Droplet Model of Atomic Nuclei*, IFI/Plenum, New York and Washington and London, 1977.

[NBF+97] U Näher, S Björnholm, S Frauendorf, F Garcias, and C Guet, Phys. Rep. **285** (1997), 245.

[NEF00] N Nilius, N Ernst, and H-J Freund, Phys. Rev. Lett. **84** (2000), 3994.

[NHC+02] G E Ntamack, B A Huber, F Chandezon, M G K Njock, and C Guet, J. Phys. B **35** (2002), 2729.

[NHM90] H Nishioka, K Hansen, and B R Mottelson, Phys. Rev. B **42** (1990), 9377.

[Nil55] S G Nilsson, K. Dan. Vidensk. Selsk. Mat. Fys. Medd. **29** (1955), No. 16.

[Nit01] A Nitzan, Ann. Rev. Phys. Chem. **52** (2001), 681.

[NK84] M R Norman and D D Koelling, Phys. Rev. B **30** (1984), 5530.

[NKdSCI99] V O Nesterenko, W Kleinig, F F de Souza Cruz, and N Lo Iudice, Phys. Rev. Lett. **83** (1999), 57.

[NKP+02] R J Needs, P R C Kent, A R Porter, M D Towler, and G Rajagopal, Int. J. Quant. Chem. **86** (2002), 218.

[NKR02] V O Nesterenko, W Kleinig, and P-G Reinhard, Euro. Phys. J. D **19** (2002), 57.

[NMC95] V Natoli, R M Martin, and D Ceperley, Phys. Rev. Lett. **74** (1995), 1601.

[oAWA96] Commission of Atomic Weights and Isotopic Abundances, Pure and Appl. Chem. **68** (1996), 2339.

[ODDP99] M A Ojeda, J Dorantes-Davila, and G M Pastor, Phys. Rev. B **60** (1999).

[OPC98] T Oda, A Pasquarello, and R Car, Phys. Rev. Lett. **80** (1998), 3622.

[PAMB02] M Pletyukhov, Ch Amann, M Mehta, and M Brack, Phys. Rev. Lett. **89** (2002), 116601.

[Pau90] W Paul, Rev. Mod. Phys. **62** (1990), 531.

[Pau00a] H Pauly, *Atom, Molecule, and Cluster Beams I*, Springer, Berlin, 2000.

[Pau00b] H Pauly, *Atom, Molecule, and Cluster Beams II*, Springer, Berlin, 2000.

[PB99] G M Pastor and K H Bennemann, *Metal Clusters*, p. 211, Wiley, New York, 1999.

[PBB+91] J Pedersen, S Bjornholm, J Borggreen, K Hansen, T P Martin, and H D Rasmussen, Nature **353** (1991), 733.

[PBE96] J P Perdew, K Burke, and M Ernzerhof, Phys. Rev. Lett. **77** (1996), 3865.

[PE92] J M Pacheco and W Ekardt, Z. f. Physik D **24** (1992), 65.

[PEFS02] R E Palmer, B J Eves, F Festy, and K Svensson, Surf. Sci. **502-503** (2002), 224.

[Pet95] D Pettifor, *Bonding and Structure of Molecules and Solids*, Clarendon, Oxford, 1995.

[PFK+95] D Porezag, Th Frauenheim, Th Köhler, G Seifert, and R Kaschner, Phys. Rev. B **51** (1995), 12947.

[PG99] L Plagne and C Guet, Phys. Rev. A **59** (1999), 4461.

[PGM+00] M Perner, S Gresillon, J März, G von PLessen, J Feldmann, J Porstendorfer, K-J Berg, and G Berg, Phys. Rev. Lett. **85** (2000), 792.

[Pha92] L Phair, Nuc. Phys. A **548** (1992), 489.

[PHH+98] M F Politis, P A Hervieux, J Hanssen, M E Madjet, and F Martin, Phys. Rev. A **58** (1998), 367.

[PJ98] L Poth and A W Castleman Jr., J. Phys. Chem. A **102** (1998), 4075.

[PK59] J C Phillips and L Kleinmann, Phys. Rev. **116** (1959), 287.

[PK60] J C Phillips and L Kleinmann, Phys. Rev. **118** (1960), 1153.

[PK88] G Pacchioni and J Koutecky, J. Chem. Phys. **88** (1988), 1066.

[PKK97] M Protopapas, C H Keitel, and P I Knight, Rep. Prog. Phys. **60** (1997), 389.

[PKZB99] J P Perdew, S Kurth, A Zupan, and P Blaha, Phys. Rev. Lett. **82** (1999), 2544.

[PL65] P Pechukas and J C Light, J. Chem. Phys. **42** (1965), 3281.

[PN66] D Pines and P Nozières, *The Theory of Quantum Liquids*, W A Benjamin, New York, 1966.

[PNB02] A Proykova, D Nikolova, and R S Berry, Phys. Rev. B **65** (2002), 085411.

[Pop99] J A Pople, Rev. Mod. Phys. **71** (1999), 1267.

[PPW86] V R Pandharipande, S C Pieper, and R B Wiriniga, Phys. Rev. B **34** (1986), 4571.

[PRS00] A Pohl, P-G Reinhard, and E Suraud, Phys. Rev. Lett. **84** (2000), 5090.

[PRS01] A Pohl, P-G Reinhard, and E Suraud, J. Phys. B **34** (2001), 4969.

[Pru78] E Prugovecki, Ann. Phys. **110** (1978), 102.

[PSD+01] H Portales, L Saviot, E Duva, M Fujii, S Hayashil, N Del Fatti, and F Vallée, J. Chem. Phys. **115** (2001), 3444.

[PTCBF96] J Parker, K T Taylor, C Clark, and S Blodgett-Ford, J. Phys. B **29** (1996), L33.

[PTS90] J P Perdew, H Q Tran, and E D Smith, Phys. Rev. B **42** (1990), 11627.

[PTVF92] W H Press, S A Teukolsky, W T Vetterling, and B P Flannery, *Numerical Recipes*, Cambridge University Press, Cambridge, 1992.

[PW92] J P Perdew and Y Wang, Phys. Rev. B **45** (1992), 13244.

[PY89] R G Parr and W Yang, *Density-Functional Theory of Atoms and Molecules*, Oxford University Press, Oxford, 1989.

[PZ81] J P Perdew and A Zunger, Phys. Rev. B **23** (1981), 5048.

[RA91] U Röthlisberger and W Andreoni, J. Chem. Phys **94** (1991), 8129.

[RBA91] A Rubio, L C Balbas, and J A Alonso, Z. f. Physik D **19** (1991), 93.

[RBG90] P-G Reinhard, M Brack, and O Genzken, Phys. Rev. A **41** (1990), 5568.

[RCK84] E A Rohlfing, D M Cox, and A Kaldor, J. Chem. Phys. **81** (1984), 3322.

[RCK$^+$99] P-G Reinhard, F Calvayrac, C Kohl, S Kümmel, E Suraud, C A Ullrich, and M Brack, Euro. Phys. J. D **9** (1999), 111.

[Re95] T Reiners and etal, Phys. Rev. Lett. **74** (1995), 1558.

[Rei92] P-G Reinhard, Phys. Lett. A **169** (1992), 281.

[RESH97] T Reiners, C Ellert, M Schmidt, and H Haberland, Phys. Rev. Lett. **74** (1997), 1558.

[RF95] P-G Reinhard and H Flocard, Nucl. Phys. A **584** (1995), 467.

[RFB95] S M Reimann, S Frauendorf, and M Brack, Z. f. Physik D **34** (1995).

[RFV84] P-G Reinhard, J Friedrich, and N Voegler, Z. f. Physik A **316** (1984), 207.

[RG84] E Runge and E K U Gross, Phys. Rev. Lett. **52** (1984), 997.

[RG87] P-G Reinhard and K Goeke, Rep. Prog. Phys. **50** (1987), 1.

[RG92] P-G Reinhard and Y K Gambhir, Appl. Phys. Lett. **1** (1992), 598.

[RGB96a] P-G Reinhard, O Genzken, and M Brack, Ann. Phys. (Leipzig) **5** (1996), 576.

[RGB96b] P-G Reinhard, O Genzken, and M Brack, Ann. Phys. (Leipzig) **5** (1996), 1.

[RHUM93] D Reinhard, B D Hall, D Ugarte, and R Monot, Z. f. Physik D **26** (1993), 76.

[RLB01] M Rusek, H Lagadec, and T Blenski, Phys. Rev. A **63** (2001), 013203.

[RLWL$^+$99] Z Roller-Lutz, Y Wang, H O Lutz, T Bastug, T Mukoyama, and B Fricke, Phys. Lett. A **262** (1999), 66.

[RM02] S M Reimann and M Manninen, Rev. Mod. Phys. **74** (2002).

[RMM$^+$99] H Ruhl, A Macchi, P Mulser, F Cornolti, and S Hain, Phys. Rev. Lett. **82** (1999), 2095.

[RNS$^+$02] P-G Reinhard, V O Nesterenko, E Suraud, S El Gammal, and W Kleinig, Phys. Rev. A **66** (2002), 013206.

[Row70] D J Rowe, *Nuclear Collective Motion*, Methuen, London, 1970.

[RP99] V Russier and M P Pileni, Surf. Sci. **425** (1999), 313.

[RPSWB97] C Rose-Petruck, K J Schafer, K R Wilson, and C P J Barty, Phys. Rev. A **55** (1997), 1182.

[RR27] O K Rice and H G Ramsperger, J. Am. Chem. Soc. **49** (1927), 1672.

[RS80] P Ring and P Schuck, *The Nuclear Many-Body Problem*, Springer, Berlin, 1980.

[RS92] P-G Reinhard and E Suraud, Ann. Phys. (NY) **216** (1992), 98.

[RS93] A Rubio and Ll Serra, Phys. Rev. B **48** (1993), 18222.

[RS95] P-G Reinhard and E Suraud, Ann. Phys. (NY) **239** (1995), 193 and 216.

[RS98] P-G Reinhard and E Suraud, Euro. Phys. J. D **3** (1998), 175.

[RS99] P-G Reinhard and E Suraud, Resonance dynamics in metal clusters and nuclei, in "Cluster Physics", edited by W. Ekardt, Wiley, New York, 1999, p. 211.

[RS00] P-G Reinhard and E Suraud, Int. J. Mol. Sci. **1** (2000), 92.

[RS01a] P-G Reinhard and E Suraud, Appl. Phys. B **73** (2001), 401.

[RS01b] P-G Reinhard and E Suraud, Laser Phys. **11** (2001), 566.

[RS02] P-G Reinhard and E Suraud, Euro. Phys. J. D **21** (2002), 315.

[RSU98] P-G Reinhard, E Suraud, and C A Ullrich, Euro. Phys. J. D **1** (1998), 303.

[RTY86] P-G Reinhard, C Toepffer, and H L Yadav, Nucl. Phys. A **458** (1986), 301.

[RWGB92] P-G Reinhard, S Weisgerber, O Genzken, and M Brack, Lecture Notes in Physics **404** (1992), 254.

[RWGB94] P-G Reinhard, S Weisgerber, O Genzken, and M Brack, Z. f. Physik A **349** (1994), 219.

[Sue63] G Süssmann, *Einführung in die Quantenmechanik I*, Bibliographisches Institut, Mannheim, 1963.

[SAH+99] A Sanchez, S Abbet, U Heiz, W-D Schneider, H Häkkinen, R N Barnett, and U Landman, J. Phys. Chem. **103** (1999), 9573.

[SB93] H K Schweng and H M Boehm, Phys. Rev. B **48** (1993), 2037.

[SBBN93] L Serra, R A Broglia, M Barranco, and J Navarro, Phys. Rev. A **47** (1993).

[SBGL93] Ll Serra, G B Bachelet, N Van Giai, and E Lipparini, Phys. Rev. B **48** (1993).

[SBT+97] T Schaich, J Braun, J P Toennies, M Buck, and C Woll, Surf. Sci. **385** (1997), L958.

[SFW+02] C Sönnichsen, T Franzl, T Wilk, G von Plessen, and J Feldmann, Phys. Rev. Lett. **88** (2002), 077402.

[SGB+91] U Serra, F Garcias, M Barranco, J Navarro, and N VanGiai, Z. f. Physik D **20** (1991), 277.

[SH53] R T Sharp and G K Horton, Phys. Rev. **30** (1953), 317.

[SHD+01] M Schmidt, T Hippler, J Donges, W Kronmüller, B v Issendorff, H Haberland, and P Labastie, Phys. Rev. Lett. **87** (2001), 203402.

[SHTM99] M Sales, J M Hernandez, J Tejada, and J L Martinez, Phys. Rev. B **60** (1999), 14557.

[SIC95] T Seidemann, M Yu Ivanov, and P B Corkum, Phys. Rev. Lett. **75** (1995), 2819.

[SKBG00] G Seifert, M Kaempfe, K-J Berg, and H Graener, Appl. Phys. B **71** (2000), 795.

[SKH+01] M Schmidt, R Kusche, T Hippler, J Donges, W Kronmüller, B v Issendorff, and H Haberland, Phys. Rev. Lett. **86** (2001), 1191.

[SKK+97] M Schmidt, R Kusche, W Kronmüller, B von Issendorff, and H Haberland, Phys. Rev. Lett. **79** (1997), 99.

[SKLW99] L Schweikhard, S Krückeberg, K Lützenkirchen, and C Walther, Euro. Phys. J. D **9** (1999), 15.

[SKvIH98] R Schlipper, R Kusche, B von Issendorff, and H Haberland, Phys. Rev. Lett. **80** (1998), 1194.

[SKvIH01] R Schlipper, R Kusche, B von Issendorff, and H Haberland, Appl. Phys. A **72** (2001), 255.

[SL85] M Salimullah and Y G Lui, Phys. Rev. A **31** (1985), 4005.

[SL97] Ll Serra and E Lipparini, Z. f. Physik D **42** (1997), 227.

[Sla29] J C Slater, Phys. Rev. **34** (1929), 1293.

[Sla51] J C Slater, Phys. Rev. **81** (1951), 385.

[SLF+91] H R Siekmann, Ch Luder, J Faehrmann, H O Lutz, and K-H Meiwes-Broer, Z. f. Physik D **20** (1991), 417.

[SMR80] K Sattler, J Mühlback, and E Recknagel, Phys. Rev. Lett. **45** (1980), 821.

[SNBB02] A G Sunderland, C J Noble, V M Burke, and P G Burke, Comp. Phys. Comm. **145** (2002), 311.

[SNHG+92] D Steinmüller-Nethl, R A Höpfel, E Gornik, A Leitner, and F R Aussenegg, Phys. Rev. Lett. **68** (1992), 389.

[SP95] F Spiegelmann and R Poteau, Comm. At. Mol. Phys. **31** (1995), 395.

[SPF+97] A Sieck, D Porezag, Th Frauenheim, M R Pederson, and K Jackson, Phys. Rev. A **56** (1997), 4890.

[SPMR98] F Spiegelmann, R Poteau, B Montag, and P-G Reinhard, Phys. Lett. A **242** (1998), 163.

[SR97] L Serra and A Rubio, Phys. Rev. Lett. **78** (1997), 1428.

[SR00] E Suraud and P-G Reinhard, Phys. Rev. Lett. **85** (2000), 2296.

[SR02a] U Saalmann and M Rost, Phys. Rev. Lett. **89** (2002), 133401.

[SR02b] Ch Siedschlag and J M Rost, Phys. Rev. Lett. **89** (2002), 173401.

[SRS02] Ll Serra, P-G Reinhard, and E Suraud, Euro. Phys. J. D **18** (2002), 327.

[SS96] U Saalmann and R Schmidt, Z. f. Physik D **38** (1996), 153.

[SS98] U Saalmann and R Schmidt, Phys. Rev. Lett. **80** (1998), 3213.

[STLS68] K S Singwi, M P Tosi, R H Land, and A Sjölander, Phys. Rev. **176** (1968), 589.

[Sug98] S Sugano, *Microcluster Physics*, Springer, Berlin, 1998.

[SV01] F Stienkemeier and A F Vilesov, J. Chem. Phys. **115** (2001), 10119.

[SWB+02] Q Sun, Q Wang, T M Briere, V Kumar, Y Kawazoe, and P Jena, Phys. Rev. B **65** (2002), 235417.

[Sza85] L Szasz, *Pseudopotential Theory of Atoms and Molecules*, Wiley, New York, 1985.

[Sze55] P Szepfalusy, Acta Phys. Hung. **5** (1955), 325.

[Tak85] K Takahashi, Phys. Rev. Lett. **55** (1985), 645.

[Tak86] K Takahashi, J. Phys. Soc. Japan **55** (1986), 762.

[Tak89] K Takahashi, Prog. Theor. Phys. Supp. **98** (1989), 109.

[TDF$^+$01] S Teuber, T Döppner, T Fennel, J Tiggesbäumker, and K H Meiwes-Broer, Euro. Phys. J. D **16** (2001), 59.

[Tho26] L Thomas, Proc. Camb. Phil. Soc. **23** (1926), 195.

[Tho60] D J Thouless, Nucl. Phys. **21** (1960), 225.

[TKMBL93] J Tiggesbäumker, L Köller, K H Meiwes-Broer, and A Liebsch, Phys. Rev. A **48** (1993), R1749.

[TM91] N Troullier and J L Martins, Phys. Rev. B **43** (1991), 1993.

[TPL$^+$91] K-D Tsuei, E W Plummer, A Liebsch, E Pehlke, K Kempa, and P Bakshi, Surf. Sci. **247** (1991), 302.

[TS76] J D Talman and W F Shadwick, Phys. Rev. A **14** (1976), 36.

[TSG97] D Timpel, K Scheerschmidt, and S H Garofalini, J. Non-Cryst. Solids **221** (1997), 187.

[TTXB02] O C Thomas, W Theng, S Xu, and K H Bowen, Phys. Rev. Lett. **89** (2002), 213403.

[Tul90] J C Tully, J. Chem. Phys. **93** (1990), 1061.

[TVW01] J P Toennies, A F Vilesov, and K B Whaley, Physics Today **54** (2001), 31.

[Ull00] C A Ullrich, J. Mol. Struct. (THEOCHEM) **501-502** (2000), 315.

[URS97] C A Ullrich, P-G Reinhard, and E Suraud, J. Phys. B **30** (1997), 5043.

[URS00] C A Ullrich, P-G Reinhard, and E Suraud, Phys. Rev. A **65** (2000), 053202.

[UU32] E A Uehling and G E Uhlenbeck, Phys. Rev. **108** (1932), 1175.

[vBH72] U von Barth and L Hedin, Solid State Phys. **5** (1972), 1629.

[VBST02] T Vartanyan, J Bosbach, F Stietz, and F Träger, Appl. Phys. B **73** (2002), 391.

[VCF$^+$00] C Voisin, D Christofilos, N Del Fatti, F Vallée, B Prövel, E Cottancin, J Lermé, M Pellarin, and M Broyer, Phys. Rev. Lett. **85** (2000), 2200.

[Ver67] L Verlet, Phys. Rev. **159** (1967), 98.

[vFBvL$^+$02] M van Faassen, P L Boeij, R van Leeuwen, J A Berger, and J G Snijders, Phys. Rev. Lett. **88** (2002), 186401.

[VH73] R Vandenbosch and J R Huizenga, *Nuclear Fission*, Academic Press, New York, 1973.

[VHHS01] M Vogel, K Hansen, A Herlert, and L Schweikhard, Phys. Rev. Lett. **87** (2001), 013401.

[Vio90] V E Viola, J. Chem. Educ. **67** (1990), 723.

[vL01] R van Leeuwen, Int. J. Mod. Phys. B **15** (2001), 1969.

[Vla50] A A Vlasov, *Many Particle Theory and Its Applications to Plasma*, Gordon & Breach, New York, 1950.

[Vol84] D. Vollhardt, Rev. Mod. Phys. **56** (1984), 99.

[VUC97] G Vignale, C A Ullrich, and S Conti, Phys. Rev. Lett. **79** (1997), 4878.

[Wab$^+$02] H Wabnitz et al., Nature **420** (2002), 482.

[WavI02] G Wrigge, M A Hoffmann, and B von Issendorff, Phys. Rev. A **65** (2002), 063201.

[WBGT99] T Wenzel, J Bosbach, A Goldmann, and F Träger, Appl. Phys. B **69** (1999), 513.

[WBT93] Y Wang, G F Bertsch, and D Tomanek, Z. f. Physik D **25** (1993), 181.

[WCF+98] K Wohrer, M Chabot, R Foss, D Gardes, P A Hervieux, F Calvayrac, P-G Reinhard, and E Suraud, Nucl. Inst. Meth. B **146** (1998), 29.

[WDD+99] C Walther, G Dietrich, W Dostal, K Hansen, S Krckeberg, K Ltzenkirchen, and L Schweikhard, Phys. Rev. Lett. **83** (1999), 3816.

[Wei37] V Weisskopf, Phys. Rev. **52** (1937), 295.

[Wei78] M Weissbluth, *Atoms and Molecules*, Academic Press, San Diego, 1978.

[WGC+98] T A Winninghama, H P Gillisb, D A Choutovc, K P Martinc, J T Moorea, and K Douglasa, Surf. Sci. **406** (1998), 221.

[WH93] J. Buttet W. Harbich, S. Fedigro, Z. f. Physik D **26** (1993), 138.

[WIHS02] S P Webb, T Iordanov, and S Hammes-Schiffer, J. Chem. Phys. **117** (2002), 4106.

[Won70] C.W. Wong, Nucl. Phys. A **147** (1970), 563.

[WR91] S Weisgerber and P-G Reinhard, Phys. Lett. A **158** (1991), 407.

[WR92] S Weisgerber and P-G Reinhard, Z. f. Physik D **23** (1992), 275.

[WR93] S Weisgerber and P-G Reinhard, Ann. Phys. (Leipzig) **2** (1993), 666.

[YB92] C Yannouleas and R Broglia, Ann. Phys. (NY) **217** (1992), 105.

[YB96] K Yabana and G F Bertsch, Phys. Rev. B **54** (1996), 4484.

[YB97] K Yabana and G F Bertsch, Z. f. Physik D **42** (1997), 219.

[YB98] K Yabana and G F Bertsch, Phys. Rev. A **58** (1998), 2604.

[YBL98] C Yannouleas, E Bogachec, and U Landman, Phys. Rev. B **57** (1998), 4872.

[YH96] F Yang and J H Hamilton, *Modern Atomic and Nuclear Physics*, McGraw-Hill, New York, 1996.

[YH00] D-S Yang and P A Heckett, J. Elec. Spec. and Rel. Phen. **106** (2000), 153.

[YL95] C Yannouleas and U Landman, Phys. Rev. B **51** (1995).

[YOGI94] Y Yamaguchi, Y Osamure, J D Goddard, and H E Schaefer III, *A New Dimension to Quantum Chemistry*, Oxford University Press, Oxford, 1994.

[YPB90] C Yannouleas, J M Pacheco, and R A Broglia, Phys. Rev. B **41** (1990), 41.

[YPC+87] S H Yang, C L Pettiette, J Conceicao, O Cheshnovsky, and R E Smalley, Chem. Phys. Lett. **139** (1987), 233.

[YSR+02] G Yalovega, A V Soldatov, M Riedler, M R Pedersen, A Kolmakov, C Nowak, and T Moller, Chem. Phys. Lett. **356** (2002), 23.

[YVB93] C Yannouleas, E Vigezzi, and R A Broglia, Phys. Rev. B **47** (1993), 9849.

[YZB98] Y Hengtai Yu, T Zuo, and A D Bandrauk, J. Phys. B **31** (1998), 1533.

[ZB95] T Zuo and A D Bandrauk, Phys. Rev. A **52** (1995), R2511.

[Zew94] A H Zewail, *Femtochemistry, Vol. I & II*, World Scientific, Singapore, 1994.

[ZGB99] K Zickfeld, M E Garcia, and K H Bennemann, Phys. Rev. B **59** (1999), 13422.

[ZSS+94] F S Zhang, F Spiegelmann, E Suraud, V Fraysse, R Poteau, R Glowinski, and F Chatelin, Phys. Lett. A **193** (1994), 75.

[ZTR99] G Zwicknagel, C Toepffer, and P-G Reinhard, Phys. Rep. **309** (1999), 117.

[Zwe+00] J Zweiback et al., Phys. Rev. Lett. **85** (2000), 3640.

# Index

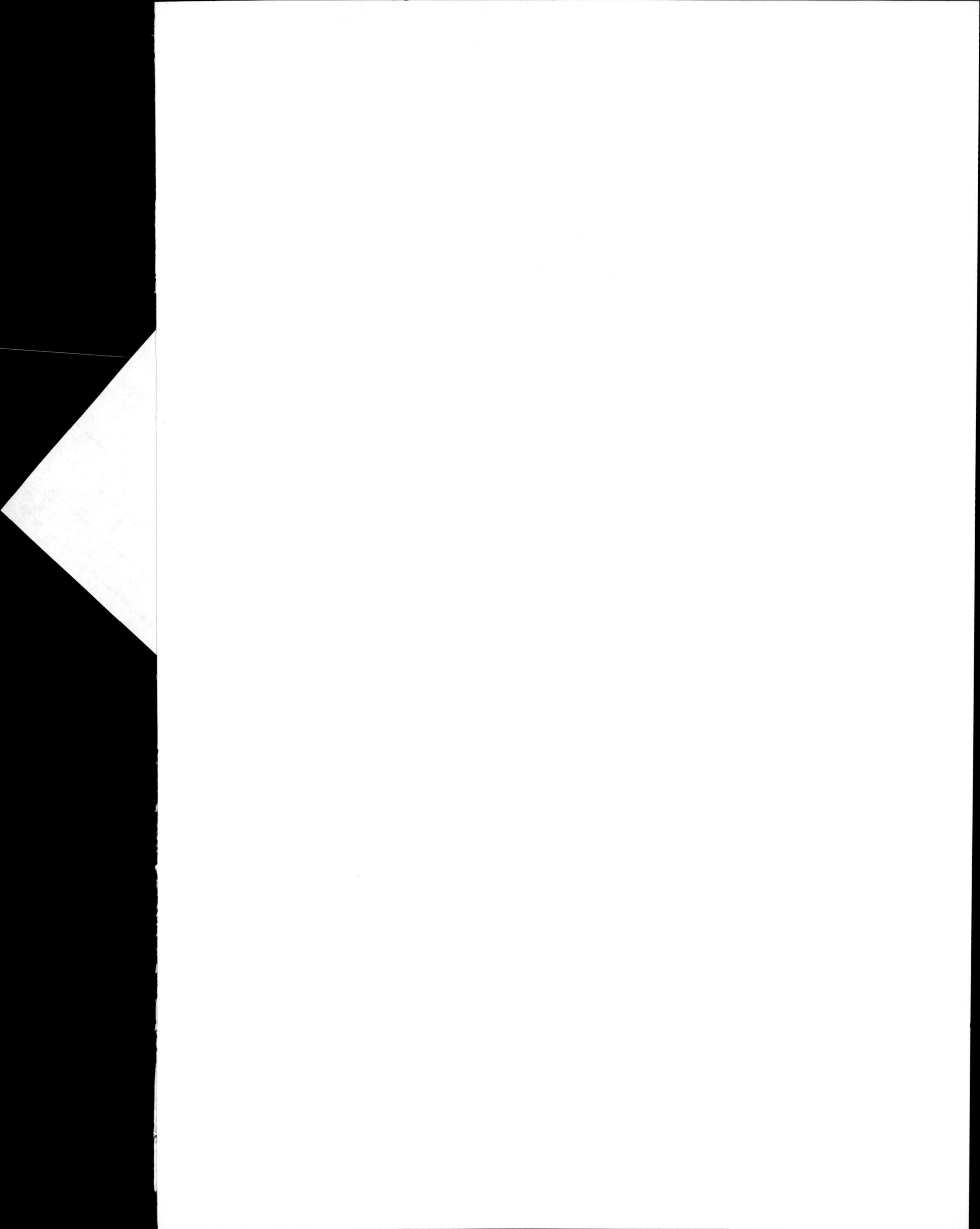

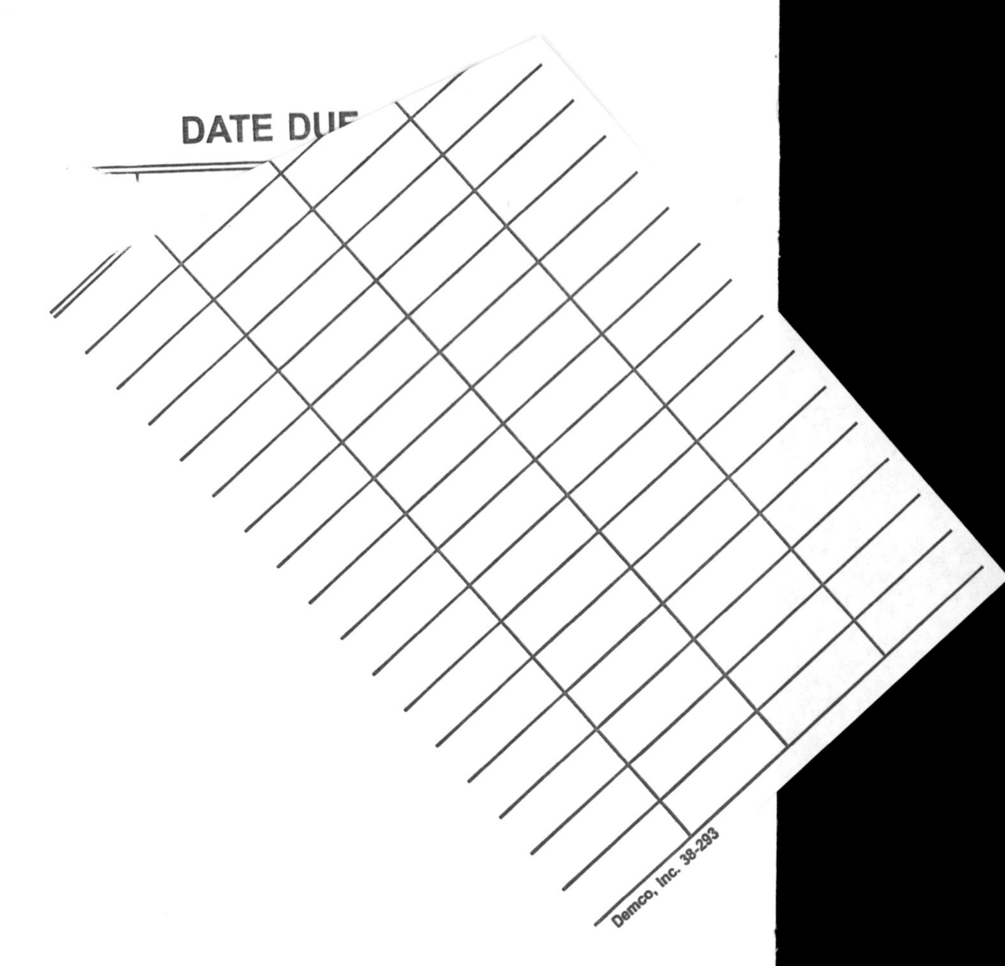
DATE DUE
Demco, Inc. 38-293